PROCEEDINGS OF SPIE

SPIE—The International Society for Optical Engineering

Laser Resonators III

☑ P9-DTV-988

Alexis V. Kudryashov
Alan H. Paxton
Chairs/Editors

26–28 January 2000
San Jose, California

Sponsored and Published by
SPIE—The International Society for Optical Engineering

Volume 3930

SPIE is an international technical society dedicated to advancing engineering and scientific applications of optical , photonic, imaging, electronic, and optoelectronic technologies.

The papers appearing in this book compose the proceedings of the technical conference cited on the cover and title page of this volume. They reflect the authors' opinions and are published as presented, in the interests of timely dissemination. Their inclusion in this publication does not necessarily constitute endorsement by the editors or by SPIE. Papers were selected by the conference program committee to be presented in oral or poster format, and were subject to review by volume editors or program committees.

Please use the following format to cite material from this book:

Author(s), "Title of paper," in *Laser Resonators III*, Alexis V. Kudryashov, Alan H. Paxton, Editors, Proceedings of SPIE Vol. 3930, page numbers (2000).

ISSN 0277-786X
ISBN 0-8194-3547-3

Published by
SPIE—The International Society for Optical Engineering
P.O. Box 10, Bellingham, Washington 98227-0010 USA
Telephone 360/676-3290 (Pacific Time) • Fax 360/647-1445

Printed in the United States of America.

Contents

Conference Committee

Conference Chairs

Alexis V. Kudryashov, Russian Academy of Sciences
Alan H. Paxton, Air Force Research Laboratory

Program Committee

Yurii A. Anan'ev, St. Petersburg State Technical University (Russia)
Hans J. Eichler, Technische Universität Berlin (Germany)
Pierre Galarneau, Institut National d'Optique (Canada)
Vladimir S. Ilchenko, Jet Propulsion Laboratory
Richard D. Jones, National Institute of Standards and Technology
James R. Leger, University of Minnesota/Twin Cities
Vittorio C. Magni, Politecnico di Milano (Italy)
Yaakov Shevy, University of Miami
Michel Tetu, Université Laval (Canada)
Anthony A. Tovar, Eastern Oregon University
Horst Weber, Technische Universität Berlin (Germany)
Koji Yasui, Mitsubishi Electric Corporation (Japan)
Luis E. Zapata, Lawrence Livermore National Laboratory

Session Chairs

Laser Resonators
Alexis V. Kudryashov, Russian Academy of Sciences
Alan H. Paxton, Air Force Research Laboratory

Active and Adaptive Laser Resonators
James R. Leger, University of Minnesota/Twin Cities
Luis E. Zapata, Lawrence Livermore National Laboratory

Resonators Including Nonstandard Optical Elements
Horst Weber, Technische Universität Berlin (Germany)
Koji Yasui, Mitsubishi Electric Corporation (Japan)

Characterization of Laser Beams
Richard D. Jones, National Institute of Standards and Technology
Alexis V. Kudryashov, Russian Academy of Sciences (Russia)

Solid State Laser Resonators
Anthony A. Tovar, Eastern Oregon University
Alan H. Paxton, Air Force Research Laboratory

Laser Resonator Design
Hans J. Eichler, Technische Universität Berlin (Germany)
Alan H. Paxton, Air Force Research Laboratory

Microresonators and Whispering-Gallery Modes I
Alexis V. Kudryashov, Russian Academy of Sciences
Pierre Galarneau, Institut National d'Optique (Canada)

Microresonators and Whispering-Gallery Modes II
Vladimir S. Ilchenko, Jet Propulsion Laboratory
Luis E. Zapata, Lawrence Livermore National Laboratory

Fiber Lasers and Resonators
Pierre Galarneau, Institut National d'Optique (Canada)
Alan H. Paxton, Air Force Research Laboratory

Introduction

This volume contains a selection of the papers presented at the third conference on laser resonators, held in San Jose during January 26-28, 2000. The design and investigation of the properties of various types of laser resonators continues to be one of the most important subjects in the contemporary research of laser physicists. Although the principles of traditional laser resonators were developed about 40 years ago, new aspects of this field are still actively investigated. The design of any new laser certainly includes work on the laser cavity. For many types of lasers, the cavity plays a large role in determining the beam quality and the frequency spectrum of the output laser radiation.

This year we had a two-and-a-half-day conference with 49 papers distributed among 9 sessions. As in the previous conferences, we had a very interesting session on solid state laser resonators with invited papers from European, Japanese, and U.S. scientists. Additionally, we organized a very successful session on laser beam characterization discussing the problems of measuring the M^2 factor as well as different aspects of the existing norms and rules for laser beam characterization. Of course, the questions of the laser cavity with nonstandard optical elements (adaptive mirrors, diffractive mirrors, variable reflectivity mirrors) were a focus of attention for the conference participants. Moreover, we had eight papers by the leading groups of scientists from the U.S., Japan, and Europe who have worked in the rapidly developing field of microsphere laser resonators. The papers indicated that this field is not purely an academic one, but also has the potential to provide interesting new solutions for some of the problems in applied physics and optical communications. The conference showed the huge potential in the development of microsphere resonators, and new results should be expected in the near future.

We would like to thank the members of our program committee, the session chairs, and all the participants for making the conference an outstanding forum for the exchange of ideas and knowledge.

Alexis V. Kudryashov
Alan H. Paxton

SESSION 1

Laser Resonators

Invited Paper

Diode-Pumped Solid-State Lasers: Resonator New Style for the 2000

A.Agnesi and G.C.Reali

INFM-Dipartimento di Elettronica, Universita' di Pavia

Via Ferrata 1, 27100 Pavia, Italia

Email: reali@ele.unipv.it; Telephone: +39 0382 505212

ABSTRACT

A selective review of the developments of laser resonators relevant to practical applications is given, with a main concern to recent advances connected with diode-pumped solid state laser revolution. An attempt to foresee some possible future developments is made.

Keywords: Laser resonators, solid-state lasers, diode-pumping, diffractive optics

1. INTRODUCTION

Since its first demonstration in 1960 by Maiman, the laser has found widespread applications in diverse areas including science, medicine, optical communications and remote sensing, and material processing. The basic understanding of the properties of lasers and their radiation requires knowledge of the physics of optical resonators. The laser beam characteristics as well as efficiency and alignment sensitivity are mainly determined by the resonator. Its design, together with the efficient energy coupling of the pump into the laser-active medium, rapidly becomes laser engineer's main concern to better the overall laser performances.

Infact, without the optical resonator the radiation emitted by the laser-active medium (of any type) could hardly be used. It is the feedback and spatial filtering action provided by the resonator that shapes the emission to be narrow-band, low-divergence and high-intensity, either one or all together these features being important in the applications.

Linear stable and unstable resonators represent the larger fraction of laser resonators used, in the past as well as in the present. In a limited number of laser design, special resonator concepts have also been used, including prism resonators, Fourier-transform resonators, variable-reflectivity-mirror resonators, hybrid (stable-unstable) resonators, ring resonators, resonators for annular media, phase-conjugation mirror resonators: some of these might play an important role in the future. Conventional (stable and unstable) resonators are widely discussed in monographs [1,2] and textbooks [3,4]. In this review, we will not deal with laser resonator designs employed for ultrashort-pulse generation, in the range from ps to fs time durations, a very specialized field primarily interesting for scientific applications. We concentrate on few selected, less conventional resonator designs, that in our opinion are fairly representative of the efforts in resonator developments over the years. In addition, dealing with DPSS lasers, we present only few realizations among the several proposed schemes, chosen to be representative of the innovative trends more than to be emblem of the top performance.

Most common lasers are designed with (geometrically) stable resonators, with an appropriately dimensioned intra-cavity pinhole that allows to select only the lowest order, gaussian-shaped mode. In this case, the beam cross-section is determined by the laser wavelength and resonator parameters, and, for the common practical choices of lengths and resonator geometry, it tends to be rather small compared with the commonly available section of the active medium, thus extracting only a small fraction of the stored energy. Achieving larger mode volumes for high energy/power operation is possible either operating the laser in multi-transverse mode or using longer cavities, both solutions presenting problems like beam quality deterioration and higher divergence or increased alignment sensitivity. In this respect, it must be emphasized that high power operation not only requires a careful cavity design to optimize energy extraction but also that the cavity design be made dynamically stable against thermal and mechanical perturbations, as described in [5,6].

A possible alternative to stable resonators is a second class of laser cavities, invented by Siegman (see [4]) and called unstable resonators (UR), characterized by having open geometrical optical paths of the beam as it bounces back and forth inside the cavity. In these resonators, the cross-sectional area of the beam is easily made comparable with that of the active medium, even if it still comprises only one transverse mode, the next higher order mode being usually suppressed by a

In *Laser Resonators III*, Alexis V. Kudryashov, Alan H. Paxton, Editors,
Proceedings of SPIE Vol. 3930 (2000) ● 0277-786X/00/$15.00

careful design of the cavity. Sometimes, stable resonator designs have also been devised that mimic UR behavior retaining their advantages while mitigating some disadvantages.

2. LESS CONVENTIONAL RESONATORS

Among the class of large mode lasers we pick up some less conventional resonators that in our opinion are relevant either because of their originality or for having peculiar performances. The first such design is one owned by Quantel [7] in which the active medium of a stable resonator with polarization output coupling is used also as an amplifier to increase the output energy (fig.1).

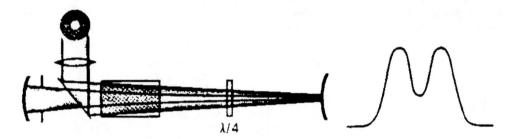

$\lambda/4$

fig.1: Quantel design

The basic resonator is a TEM$_{00}$ configuration formed by the concave rear mirror and partially reflective front surface of the quarter-wave plate. The aperture restricts the oscillator to the center of the rod and the resulting beam, polarized horizontally, is low energy. This beam exits the cavity through the quarter-wave plate and than re-enter again through it after reflection off the convex mirror. The expanded and diverging beam, orthogonally polarized with respect to the oscillator beam, is then amplified in the laser medium and coupled out by the polarizer. While the oscillator output is TEM$_{00}$, the beam exiting the polarizer is not due to the depleted gain at the rod center, which produces a dip in the output beam center. Fresnel rings are also present, and even in more severe fashion than in diffracton coupled URs since the beam is diverging as it re-enters the rod for amplification.

Another large mode resonator is the self-filtering UR (SFUR) [8]. The SFUR is a confocal negative branch resonator with an internal focus, making it a standing wave implementation of the self-imaging resonator proposed by Paxton and Salvi [9]. A diagram of the SFUR configuration is shown in fig.2.

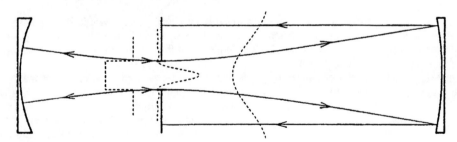

fig.2: Self-filtering Unstable Resonator

The origin of the self-imaging behavior is that the A and D terms in the half-round-trip matrices are zero,

$$\begin{bmatrix} A & B \\ C & D \end{bmatrix} = \begin{bmatrix} 0 & f_2 \\ -1/f_2 & 0 \end{bmatrix} \begin{bmatrix} 0 & f_1 \\ -1/f_1 & 0 \end{bmatrix} = \begin{bmatrix} m & 0 \\ 0 & 1/m \end{bmatrix}$$

(geometric magnification $m = -f_2/f_1$) so there is no complex curvature terms in the corresponding Fresnel integrals, and (to within a constant global phase factor) the round-trip kernel is real. In analytical terms, the focusing property of the mirror at each end of the resonator returns the Fourier transform of the field within the aperture to the plane of the aperture, and the aperture is imaged upon itself after a full round-trip through the resonator. The SFUR aperture at the focal plane is sized to clip the beam in both directions, and this two-way clipping affects the resonator performances in a unique way when the aperture radius a is equal to the radius of the first null of the Airy pattern of a uniformly illuminated aperture focused by the mirror with focal length f_1,

$$a = a_0 \equiv (0.61 f_1 \lambda)^{1/2}$$

If the illumination of the aperture is collimated and nearly uniform, as it is in the confocal arrangement considered here, the beam reflected back to the aperture by the shorter focal length mirror is approximately an Airy pattern, which is spatially filtered to allow only the field central spot up to its first zero to further propagate. In the second half of the round trip, the mode is magnified and recollimated, and then partially coupled out by some means just before hitting again the aperture. If $|m|$ is significantly greater than 1, the feedback portion of the fundamental mode is nearly uniform, and the propagation exactly replicates in the next round trip. The feedback is more peaked on axis for low $|m|$, and undergoes a sizeable clip on its way back through the aperture; nevertheless, the SFUR design is robust and able to handle various situations favourably down to the degeneracy limit of $|m|=1$, as well as reasonable mismatches of aperture sizing and positioning, and of cavity length. Numerically computed and experimentally detected beam profiles at the output are shown in fig.3.

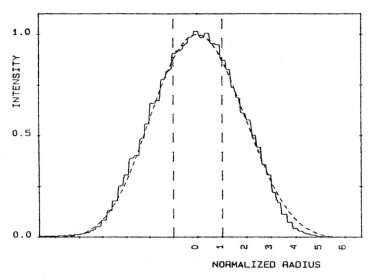

fig.3: SFUR oscillating mode: numerical and experimental

A further development was the diffraction filtered resonator DFR by Ananev and coworkers[10] and by Pax end Weston [11], who essentially noted that almost the same performances of the SFUR could be retained replacing the diaphragm with a flat mirror with the same dimension of the hole of the diaphragm, at the cost of a slightly less filtering action but with the advantage of making a shorter length resonator (cutting away the SFUR shorter length section).

As a final example of special resonator design we offer the variable-reflectivity-mirror (VRM) resonator, which was very successful in the past years. The theory of these resonators is amply treated in literature [2,3,4], and a practical implementation for a commercial laser oscillator with Gaussian VRM output coupler is described in [12]. The main advantage of VRM resonators is in the possibility of getting a very large and smooth fundamental mode for low to moderate magnifications, which means possibility of working with low-gain media. The disadvantage is that the peak mirror reflectivity is fixed and chosen to get a filled in output beam, without a dip in it, for the passive resonator; this makes the design stiff with respect to the possibility of increasing the pumping level, since high gain operation brings in a considerable level of gain saturation which changes the maximally flat condition.

The strong impulse in the development of new resonator configurations was pursued continously since the invention of the laser, with changing slope and objectives, beginning with stable resonators for low/medium power lasers, changing to unstable resonators for high power/energy systems, and turning the interest back again to low/medium power resonator design when diode-pumping appeared, causing a revolution in laser research and producing an irreversible trend in their use. Almost the entire history of laser development was repeated by this event and DPSS lasers rapidly became a mature technology. DPSS lasers started with low power oscillators (few mW), and the foreseen scaling to high powers and high energies pulses passed trough the idea of having one such DPSS oscillator with very stable operation as seeder, and a flashlamp pumped amplifier chain to rise it at the desired output power level.

One of the most sophisticated realization of this idea is Coherent Infinity [13]. This system is a real master-piece of laser technology and engineering, and is able to supply, at up to 100 Hz, 500 mJ pulses at 1064 nm and 250 mJ pulses at 532 nm, in clean low-divergence spatial mode and single longitudinal mode. The system is an hybrid MOPA: the master oscillator is a diode-pumped, Q-switched, single-frequency ring laser; the power amplifier is composed of an amplifier head, a relay-imaging system and a phase-conjugate mirror. A schematic of the system is shown in fig.4.

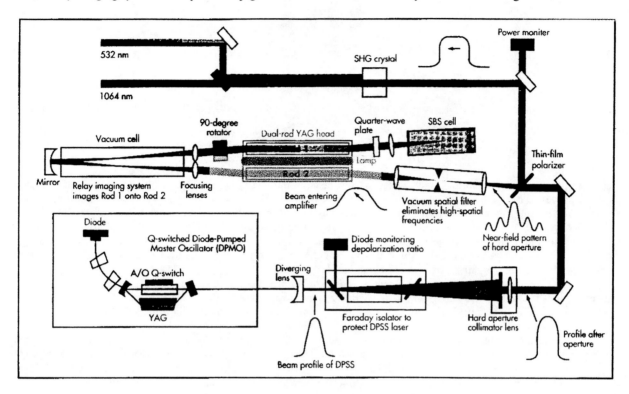

fig.4: Coherent Infinity laser system

3. DPSS LASERS: PUMPING SCHEMES, BEAM-SHAPING OPTICS, MODE-MATCHING AND ALL THAT

Since the end of the 1980s, DPSS lasers steadly became the most popular lasers in the 1-10 W average power range. This intermediate class of diode-pumped lasers is now widely used in most commercial applications such as industrial ones (marking, micro-machining, etc.), as well as in many scientific applications where DPSS lasers allow mode-locked lasers, parametric generation, and nonlinear optics experiment to be routinely done. The big advantages that DPSS lasers offer are: lower thermal dissipation, high efficiency, stability, long-lifetime, maintainance-free operation, compactness.

Advances in single-emitter high brightness multiwatt diode lasers as well as in linear bars and stacks of diode arrays have recently allowed significant improvements of diode-pumped sources in terms of pump efficiency, simplicity and reliability. Diode pumping is achieved either using longitudinal or transversal geometry (with several variants usually invented to bypass patents). Transversal, side-pumping is the direct descendant of the traditional optical pumping delivered using lamps, and it is generally less economic than end-pumping even if it is believed that it will be the necessary solution for high power systems. The preferred pump geometry in the 1-10 W range of pump powers is by far end-pumping, as simple and affordable solutions for thermal management of the optically excited laser rod (or, in few cases, slab) easily allows output power levels of ⁻10 W with 30-60% optical efficiency, depending on the laser material used and the emission wavelength as well as the operating regime (pulsed or cw). The significantly lower cost of diode arrays, combined with many available solutions for efficient beam reshaping and fiber coupling [14,15,16] (fig.5)

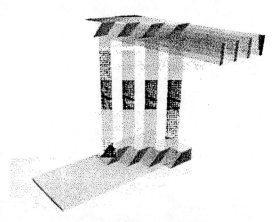

fig.5: An example of beam-shaping device

make end-pumping convenient for as much as 30 W power input into the laser crystal, considering the basic limits due to pump brightness, laser gain, crystal absorption and thermal fracture. The use of high-brightness, beam conditioned diode arrays, allowing narrow and extended pump volumes has been especially effective for pumping thermally sensitive materials such as Ho-Tm 2-µm lasers [17], as well as for exciting quasi-three levels transitions in Nd lasers near 946 nm [18], which are interesting for blue generation through second harmonic conversion. Traditional laser materials operating at their high-gain transitions, such as Nd:YAG, Nd:YLF and Nd:YVO$_4$, only require a sufficient pump beam homogeneity as well as readily available brightness from the fiber output.

The most standard implementation of the concept of end-pumping involves a fiber-coupled diode array, an optional optical imagery, and a stable laser resonator comprising a linear two-mirror cavity with the laser crystal at one end). Moderate power scaling is easily achieved with an only slightly more complicated geometry, the folded "Z" resonator, owing to double-longitudinal pumping from both crystal ends [19]. The non-spherical thermal aberration accompanying the pump profile produces appreciable losses on the laser mode, hence in multiwatt average power regimes the laser mode is generally chosen slightly smaller than the pump diameter, to favour the TEM$_{00}$ selection. Comparable performance and efficiency are achieved in many different resonator configurations [20,21,22], when the thermal lens and its non-spherical aberration are properly included in a simple ABCD study. Sometime thermal loads are increased by the particular operating regime chosen with a given active medium: either the emission of longer wavelengths (with smaller slope efficiency) or low-repetition rate Q-switching regimes with higher output couplings favours phonon-assisted relaxation processes producing extra heat to be dissipated [23]. Diffraction losses due to thermal aberrations are especially critical when exploiting optimum intracavity second harmonic generation (SHG), and require the resonator design be carefully optimized for their minimization; furthermore, noiseless intra-cavity SHG requires either single longitudinal mode or strong (hundreds) multi-longitudinal mode operation, this last condition implying a rather long cavity design.

End-pumping with more cost-effective pump geometries than fiber-coupling, such as a simple collimated linear array directly focused inside the active medium require a smarter resonator design making an efficient single mode extraction along the slow-axis direction of the diode bar. This has been achieved both employing cylindrical optics in linear cavities [24] and re-introducing the concept of unstable resonators in minilasers [25]. Although single transverse mode selection is not mandatory for many practical applications, high performance frequency conversion as well as stable pulsed mode operation (either mode-locking or Q-switching) critically depend on the laser beam quality. It is especially recommended that the resonator be unstable along the slow axis direction, where the single mode selection is more challenging. These concepts were successfully exploited in a passively Q-switched laser based on a simple end-pumped linear resonator, which was designed for either high-average power operation or high pulse peak power for efficient harmonic generation and parametric conversion [26].

In the last few years we have also been eyewitnesses of another rapid progress in the laser field: the new fiber laser technology, employing (mainly) end-pumped double-clad fibers, as shown in fig.6, by means of high brightness diode arrays (possibly with beam shaping).

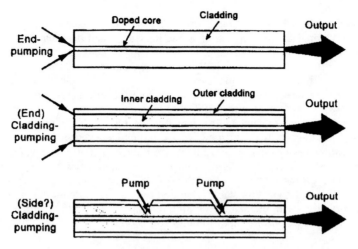

fig.6: Fiber laser pump concepts: end-pumping, double-clad end-pumping and double-clad side pumping

The Yb:glass fiber laser is already commercially available at few ten watts of output power [27,28,29], and owes its success to its extreme simplicity, high reliability and compactness. Upscaling to as much as 100 W has already been reported [30]. Single mode fiber lasers are presently limited to cw operation as the high peak power of Q-switching pulses comparable to those routinely generated for industrial applications in more traditional laser solutions would easily damage the small fiber core, even if this could rapidly change, if single mode requirement is relaxed, since first reports are now appearing on Q-switched fiber lasers (up to 2 mJ at 0.5 kHz, 100-ns width), employing large mode area fibers with externally-coupled acousto-optical devices [31]. Of course, it is expected that in the near future the modulator could be fully integrated into the fiber laser itself.

4. HIGH POWER DPSS LASERS (100 W AND BEYOND)

Material limitations arising at few tens watts output power become even more challenging when going up to the 100-W target and beyond. These power levels are of great interest in the industry, where CO_2 lasers still dominate. However, owing to peculiar characteristics of solid-state lasers such as improved beam quality and focusability, together with compactness, longer lifetime and easier operation, make the latter preferable in many circumstances. Although in principle the end-pumping geometry might be still applied for an initial up-scaling, through replication of Z-folded laser head modules[19] within the same resonator, it became soon apparent that different pump geometries such as side-pumping definitely have an edge over longitudinal excitation. Indeed, a more even distribution of the pump power into the laser rod (or slab) along the cavity axis makes possible to design a resonator for a given output power as well as helping in heat removal keeping a safety margin over the fracture limit due to thermal stresses (fig.7).

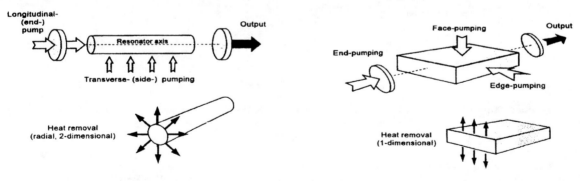

fig.7: Pump/laser geometries: rod and slab

Phase aberrations arising from the heavily pumped laser materials can be minimized to some extent using diffusion-bonded laser crystals, to mitigate the thermal stresses originating at the interfaces, and insuring more efficient heat removal [32]. Side-pumping optimized for uniform radial gain shaping, as well as careful reduction of depolarization stresses using quartz rotators in symmetric cavities, allowed as much as 208 W (diffraction-limited) [33]. Slab geometries are also increasingly attractive, allowing even more efficient thermal management. Side-pumped modules excited by collimated diode bars generating gain sheets in which the laser beam makes few passes to efficiently extract the stored energy, can be added in series (intra- or extra-cavity) for power scaling toward 100-W [34]. This geometry is particularly appealing for its ability to manage the increased thermal load in low-repetition rate (kHz), high energy (few mJ) Q-switching operation.

A different approach to the slab design, borrowed from CO_2 technology, was recently proposed. In this case the pumping was still longitudinal, but the power up-scaling was allowed by stacking few diode bars and imaging the fast-axis-collimated distribution inside the slab, thus yielding a thin gain sheet into a few tens millimetres long Nd:YAG slab. Efficient energy extraction in a diffraction limited beam was provided by the small gain aperture along the fast-axis direction in which the resonator was designed for stable operation (taking into account the thermal lensing), whereas in the orthogonal direction (with virtually absent thermal effects) the resonator was unstable with a magnification appropriately chosen to allow nearly optimum 40-W power output through edge-coupling [35]. These two pumping phylosophies are shown schematically in fig.8.

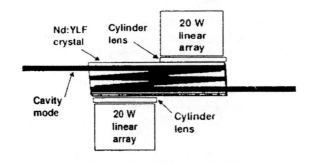

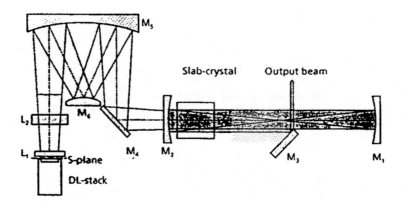

Fig.8: Slab side- end end-pumping

Owing to its reduced quantum defect and favourable thermal properties, Yb:YAG has attracted much interest as laser material for very high power applications. Because of its quasi-three levels nature, to date Yb:YAG has been mostly limited to pump schemes employing either high-brightness microchannel-cooled multistacks coupled by optical ducts [36], or to the thin-disk approach [37], whose basic advantage is to avoid transverse thermal gradients by removing heat from the face of the disk. These two geometries are shown in fig.9.

The former technique, much simpler in principle, has allowed significant power upscaling paralleling advances in both diode technology (compact and high brightness packages) and laser material processing (diffusion bonding and highly doped rods).

Though geometrically complex, the latter approach is less critically dependent on top-end technological achievements: it requires a sufficiently thin disk of highly doped Yb:YAG (from which the deposited heat is longitudinally removed, with small residual thermal lensing) to be pumped multipass (10-20 passes) from multiple-mirrored fiber-coupled diodes. The resonator is a simple linear stable cavity, and the whole design complexity is transferred to the pump module. Multi 100-W in a nearly diffraction limited beam has been recently achieved. Yb:YAG is also much attractive for high-energy Q-switching operation, owing to its millisecond lifetime and high saturation fluence.

Finally, it is expected that advances in the other extreme, an extremely elongated laser rod, that is fiber laser technology, will soon allow upscaling well beyond 100-W, notwithstanding the limitation imposed by the present technology.

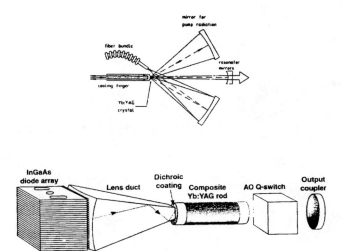

fig.9: Two high power laser implementations: disk and duct-rod

5. FUTURE FOR UNSTABLE RESONATORS AND DIFFRACTIVE-OPTICS?

What next? As we look into the years 2000 we can dream wide over future developments; infact, these developments could come to realization even before the dreams are over, given the extremely fast technological progress of the last years.

The first development we mention is the further push toward miniaturization: we have not delt with microchips [38] in this review, mainly addressed to resonators, because of their simple design in microchips, but these deserve a lot of attention because of their high performances and the variaty of applications they satisfy.

Another important research direction is associated with power-scaling of DPSS lasers, where there is little evidence that the development in high-power diode lasers can stop and put a limit to this climb up. In this perspective, thermal problems become "the" issue, and with them laser beam quality, thus resonator physics, is re-starting laser engineer's activity to look for compatible very high-power solutions.

Of course the first problem will be the optical quality and beam reshaping of the diode lasers themselves, which in industrial, medical and enviromental applications could directly provide high power, multi-kW coherent sources. Diode beam shaping have amply been demonstrated to reformat the diode output of the linear arrays into almost circular shape with a significant improvement in brightness distribution.

Another research direction could be the return to an extensive use of URs, since DPSS lasers will soon have gains very close or greater than flash-lamp pumped systems, repeating their hystory very closely. In particular hybrid resonator schemes could be very effective and cheap solutions for industrial applications where diode-beam shaping optics are likely to be too costly for a massive use, and where diodes are essentially used as narrow spectral band lamps.

A possible alternative, also explored in standard laser sources, either gas or solid-state, could be the use of intracavity laser mode reshapers by means of diffractive optical (DO) components. Recent results, by Friesem and co-workers[39] and also by other researchers, show how the cavity mode could be shaped and chosen so that it can fully exploit peculiar gain distributions inside the active medium (fig.10).

These DOs are both robust and easy to design for any form of transmission, and could be adapted to confine the field distribution in the region of the active medium where the gain is highest.

These are only few obvious directions for developments. Indeed, this is a field in which settled questions are few, and widely different approaches can co-exist, being plenty of room for further innovations.

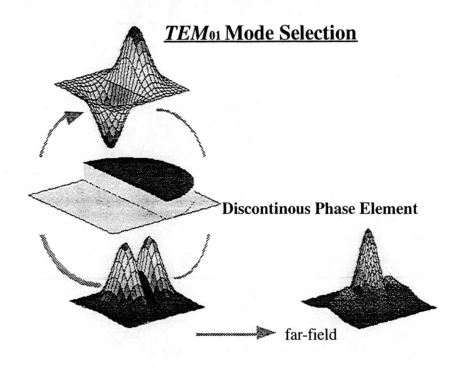

fig.10: Action of DO on laser mode

REFERENCES

1.   Yu. Ananev, *Laser Resonators and the Beam Divergence Problem*, Adam Hilger (1992)
2.   N. Hodgson and H. Weber, *Optical Resonators*, Springer Verlag (1997)
3.   W. Koechner, *Solid State Laser Engineering*, Springer Verlag (1995)
4.   A. E. Siegman, *Lasers*, University Science Books, Palo Alto (1986)
5.   H. P. Kortz, R. Ifflander, H. Weber, Appl. Optics, **20**, 4124 (1981)
6.   V. Magni, Appl. Optics, **25**, 2030 (1986)
7.   Quantel S.A., US Patent 4276519, High Yield Diffraction Limited Laser Oscillator, Jun 30, 1981
8.   P.G.Gobbi and G.C.Reali, Opt. Comm., **52**, 195, (1984)
9.   A.H.Paxton and T.C.Salvi, Opt.Comm., **26**, 305 (1978)
10.   Y. A. Ananev, S. G. Anikichev, and V. D. Solovev, Opt. Spectrosc. (USSR), **68**, 719 (1990)
11.   P. Pax and J. Weston, IEEE J. of Quantum Electr., **27**, 1242 (1991)
12.   Continuum Inc., US Patent 4918704, Q-switched solid state pulsed laser with injection seeding and a gaussian output coupling mirror, Apr 17, 1990
13.   Coherent Laser Group, Product Profiles: http://www.cohr.com/clg/products/pulsed.html
14.   W. A. Clarkson, D. C. Hanna, Opt. Lett. **21**, 375 (1996)

15. K. Du, M. Baumann, B. Ehlers, H. G. Treusch, P. Loosen, *OSA Trends in Optics and Photonics*, Vol. 10, Advanced Solid State Lasers, C. R. Pollock and W. R. Bosenberg, eds. (Optical Society of America, Washington, DC 1997), pp. 390-393

16. S. Yamaguchi, T. Kobayashi, Y. Saito, K. Chiba, Opt. Lett. **20**, 898 (1995)

17. C. Bollig, R. A. Hayward, W. A. Clarkson, D. C. Hanna, Opt. Lett. **23**, 1757 (1998)

18. W. A. Clarkson, R. Koch, D. C. Hanna, Opt. Lett. **21**, 737 (1996)

19. W. L. Nighan, N. Hodgson, E. Cheng, D. Dudley, in *Conference on Lasers and Electro-Optics*, OSA Technical Digest (Optical Society of America, Washington, DC 1999), paper CMA1

20. A. Agnesi, P. G. Gobbi, G. Reali, in *Conference on Lasers and Electro-Optics*, Vol. 11, 1997 OSA Technical Digest Series (Optical Society of America, Washington, DC) 1997, paper CFO2

21. M. D. Selker, T. J. Johnston, G. Frangineas, J. L. Nightingale, D. K. Negus, in *Conference on Lasers and Electro-Optics*, Vol. 9, 1996 OSA Technical Digest Series (Optical Society of America, Washington, DC) 1996, paper CPD21

22. W. L. Nighan, S. B. Hutchinson, D. Dudley, M. S. Keirstead, in *Conference on Lasers and Electro-Optics*, Vol. 9, 1996 OSA Technical Digest Series (Optical Society of America, Washington, DC) 1996, paper CWN2

23. M. Pollnau, W. A. Clarkson, D. C. Hanna, in *Conference on Lasers and Electro-Optics*, Vol. 6, 1998 OSA Technical Digest Series (Optical Society of America, Washington, DC) 1998, paper Ctu11

24. J. Zehetner, Opt. Commun. **117**, 273 (1995)

25. H. Liu, S.-H. Zhou, Y. C. Chen, Opt. Lett. **23**, 451 (1998)

26. A. Agnesi, S. Dell'Acqua, E. Piccinini, G. Reali, G. Piccinno, IEEE J. of Quantum Electron. **34**, 1480 (1998)

27. Polaroid, Product Profiles: http://www.polaroid.com/products/oem/laser

28. SDL Inc, Product Profiles: http://www.sdli.com/products/systems

29. IRE-POLUS group, Product Profiles: http://www.ire-polusgroup.com/irepolus.htm

30. V. Dominic, S. MacCormack, R. Waarts, S. Stevens, S. Bicknese, R. Dohle, E. Wolak, P. S. Yeh, E. Zucker, in *Conference on Lasers and Electro-Optics*, OSA Technical Digest (Optical Society of America, Washington, DC, 1999), paper CPD11

31. J. A. Alvarez-Chavez, H. L. Offerhaus, J. Nilsson, P. W. Turner, W. A. Clarkson, D. J. Richardson, Opt. Lett. **25**, 37 (1999)

32. R. Beach, R. Reichert, W. Benett, B. Freitas, S. Mitchell, S. Velsko, J. Davin, and R. Solarz, Opt. Lett., **18**, 1326 (1993)

33. Y. Hirano, Y. Koyata, S. Yamamoto, K. Kasahara, T. Tajime, Opt. Lett. **24**, 679 (1999)

34. Q-peak, Applied Photonics Systems: http://www.qpeak.com/Research/crd.htm

35. K. Du, N. Wu, J. Xu, J. Giesekus, P. Loosen, R. Poprawe, Opt. Lett. **23**, 370 (1998)

36. E. C. Honea, R. J. Beach, S. C. Mitchell, P. V. Avizonis, Opt. Lett. **24**, 154 (1999)

37. A. Giesen, in *Novel Lasers and Devices-Basic Aspects*, OSA Technical Digest (Optical Society of America, Washington, DC, 1999), paper LTuC1-1

38. J. J. Zayhowski, in *Handbook of Photonics*, M. C. Gupta ed., 326, CRC Press (1997)

39. R. Oron, Y. Danziger, N. Davidson, A. Friesem, E. Hasman, Appl. Phys. Lett., **74**, 1373 (1999)

Phase conjugation at SBS of pulse laser radiation with using the kinoform optics

F. A. Starikov, Yu. V. Dolgopolov, G. G. Kochemasov, S. M. Kulikov, V. K. Ladagin, S. A. Sukharev

Russian Federal Nuclear Center – Institute of Experimental Physics
607190 Sarov (Arzamas-16), Nizhny Novgorod Reg., Russia

ABSTRACT

Using the most complete in literature physical model of the non-steady-state stimulated Brillouin scattering (SBS), the numerical study is carried out of phase conjugation (PC) in the SBS-mirror that consists of an angular selector of Stokes radiation, an ordered raster of small lenses, a main focusing lens, and an SBS-cell. The ordered raster with controlled varying of its parameters allows to perform the effective angular filtering of non-conjugated Stokes component, to reduce the local light loads, and to avoid the competitive nonlinear effects. An optimal configuration of such SBS-mirror has been determined. It has the unique properties as compared with the current SBS-mirrors. It fixedly yields the PC quality that is near to an ideal (the PC coefficient is about of or more than 95%) at the selector transmittance 50-70% and any level of SBS saturation, i.e. any reflection coefficient. In the SBS-mirrors of different types the high PC quality in the focused beams takes place at the high reflection coefficients only that is difficult to realize as a rule. The first experimental data obtained at a Nd laser facility show the validity of the simulation results. The developed conception of SBS-mirror with unique properties can be applied for the improvement of wide class of industrial lasers.

Key words: phase conjugation, stimulated Brillouin scattering.

1. INTRODUCTION

The phase conjugation (PC) based on stimulated Brillouin scattering (SBS) has been widely used in double-pass laser amplifiers. After the first pass, the amplified laser beam, distorted due to various optical inhomogeneities in the amplifier, is incident on the PC-mirror. The purpose of PC-mirror is to create a reflected Stokes beam with the wave front being a conjugate to the incident one and, thus, to compensate the optical inhomogeneities of amplifier during the reverse pass. The PC fidelity and reflectivity of PC-mirror have a crucial influence on the performance of such system.[1-4]

The main component of PC-mirror is an SBS-cell. It was also a usual way to use a random phase plate located ahead of SBS-cell. The aberrated laser beam after passing through the random phase plate acquires a small-scale speckled structure. This promotes the reduction of "snake-like" distortions of the Stokes wave in the SBS-cell and increases the PC fidelity. Then the use of SBS-cell in the form of a rather long light waveguide allows obtaining a practically perfect PC fidelity of the statistically uniform speckled beam, but unfortunately at substantial laser intensities the lightguide damage resistance is a challenge. In order to avoid this and to solve the problem of exceeding an experimental SBS threshold, the speckled laser beam is tightly focused by a lens into the cell with a gaseous or liquid SBS-medium. However, the focused speckled laser beam becomes a statistically nonuniform one, and its bell-shaped mean envelope being more pronounced in the Stokes beam. This leads to a decrease in the PC fidelity. For example, in the SBS steady-state linear mode, when the Stokes beam weakly affects the laser beam, the PC coefficient does not exceed 30%.

On the other side, there are a few experimental investigations of double-pass amplifiers where the aberrated laser beam experience ordered spatial distortions after or during the first pass. First, the ordered distortions may be introduced into the laser beam through the use of an ordered phase plate in the PC-mirror scheme instead of the random one. In this case the phase plate is a kinoform raster of identical Fresnel lenses.[5-8] Second, the ordered distortions can appear in the laser beam during its first pass because of a specific construction of the amplifier, for example, if the amplifier is made in the form of an ordered fibre bundle.[9]

We have recently carried out a numerical investigation[10] of PC at the steady-state SBS of an aberrated laser beam in the case of PC-mirror composed of an angular selector of Stokes radiation, an ordered raster of small lenses, a main focusing lens and an SBS-cell. As a result of the investigation, a new effect of extremely low noising of Stokes beam has been found. Its es-

In *Laser Resonators III*, Alexis V. Kudryashov, Alan H. Paxton, Editors,
Proceedings of SPIE Vol. 3930 (2000) ● 0277-786X/00/$15.00

sence is in the fact that the angular selection of Stokes radiation leads to the nearly perfect PC when the SBS-mirror arrangement is optimal. For this purpose the input window of SBS-cell should be placed in a certain intermediate region between the focal plane of rasher's lenses and the focal plane of main lens. It has been shown also that the usage of the Gaussian random phase plate, which was traditional in the PC research, instead of ordered one results in the substantially poorer PC fidelity and unprincipled role of angular selection.

Calculation[10] have been performed in an approximation of steady-state linear (unsaturated) SBS when the Stokes field does not affect the laser pump one. But in the experiments the SBS saturation shows itself in any event. Moreover, the operation in the saturated regime of SBS is one of the aims of the experiment as it leads to an increase of the SBS-mirror reflection coefficient. Further, for modeling the experiments on PC at SBS of pulsed laser beams it is needed to additionally allow for the non-steady-state transient processes related to the finite time of hyper-sound relaxation in the SBS-medium. The usage of simplified models of SBS often results in the problems of interpretation of the experimental data or in the failure of quantitative comparison with the experiment. Below the most complete in literature physical and numerical model of non-steady-state transient SBS is described that takes into account all main physical effects and is more close to experimental situation. Carrying on our recent paper,[10] the model is applied to the study of the quality of PC at SBS of laser beam passed through an ordered raster of small lenses in the SBS-mirror. Main attention is paid to the determination of the SBS-mirror arrangement giving the PC quality that is near to an ideal in the wide range of laser pump power.

2. TRANSIENT SBS MODEL

2.1. Equations of non-steady-state SBS with saturation

The SBS is described by the equations for the pressure variation p' and electromagnetic field $\mathbf{E}$:

$$\frac{\partial^2 p'}{\partial t^2} - \nabla^2 \left(v_s^2 p' + \Gamma \frac{\partial p'}{\partial t} - \frac{Y}{8\pi\rho\beta_s} \mathbf{E}^2 \right) = 0, \tag{1}$$

$$\frac{n^2}{c^2} \frac{\partial^2 \mathbf{E}}{\partial t^2} - \nabla^2 \mathbf{E} = -\frac{1}{c^2} Y\beta_s \frac{\partial^2 (p'\mathbf{E})}{\partial t^2}, \tag{2}$$

where $Y=\rho(\partial\varepsilon/\partial\rho)_s$ is the parameter of nonlinear coupling, $\beta_s=1/(\rho v_s^2)$ is the compressibility of the medium, $v_s^2=(\partial p/\partial\rho)_s$, v_s is the adiabatic hyper-sound speed, $\Gamma=(4\eta/3+\eta')/\rho^0$ is the damping constant, η and η' are the shear and bulk viscosity, n is the refractive index.

The light scattered by SBS (Stokes component) has a frequency shift $\Delta\omega=-(2nv/c)\sin(\varphi/2)$ with respect to the incident laser light, where φ is scattering angle. The SBS-active medium extended along the direction of laser pump propagation ensures the almost backward reflection of Stokes component. The light field in a scalar approximation can be presented as the sum of two counter-propagating waves running along the z axis:

$$E = \frac{1}{2} \left(A_L e^{i(\omega_L t - k_L z)} + A_S e^{i(\omega_S t + k_S z)} + c.c. \right), \tag{3}$$

where indexes «L» and «S» refer to incident pump laser and reflected Stokes waves, respectively. The resonant hyper-sound wave is written in the form:

$$p' = \left(p e^{i(\Omega t - qz)} + c.c. \right), \tag{4}$$

where $\Omega=\omega_L-\omega_S$ and $q\cong 2k_L$ are the frequency and wave number of hyper-sound, p is the hypersonic complex amplitude.

The amplitudes A_L, A_S and p are regarded to be slowly varying in comparison with the exponents in (3), (4). The frequencies and wave numbers of laser and Stokes waves are considered to coincide. Substituting (3), (4) into (1), (2), we obtain the system of equations:

$$\frac{n_0}{c} \frac{\partial A_L}{\partial t} + \frac{\partial A_L}{\partial z} - \frac{1}{2ik} \Delta_\perp A_L + \frac{ik}{2} \left(\frac{n^2}{n_0^2} - 1 \right) A_L = -\frac{i}{2} pA_S, \tag{5}$$

$$\frac{n_0}{c} \frac{\partial A_S}{\partial t} - \frac{\partial A_S}{\partial z} - \frac{1}{2ik} \Delta_\perp A_S + \frac{ik}{2} \left(\frac{n^2}{n_0^2} - 1 \right) A_L = -\frac{i}{2} p^* A_L, \tag{6}$$

$$\frac{\partial^2 p}{\partial t^2} + 2 \left(\frac{1}{\tau} + i\Omega \right) \frac{\partial p}{\partial t} + \frac{2i\Omega}{\tau} p = \frac{2g\Omega}{\tau} A_L A_S^* + 2i\Omega S, \tag{7}$$

where $g=\omega_0\tau\Omega/[4\beta_S n_0^2 c^2]$ is the SBS gain coefficient, $\tau=2/(q^2\Gamma)^2$ is the hyper-sound relaxation time (phonon lifetime), n_0 is the mean refractive index defined by the relationship $k=n_0\omega_0/c$. The field amplitudes are normalized so that the radiation density fluxes are equal to $J_L=|A_L|^2$, $J_S=|A_S|^2$. If the time of varying A_L (the coherence time or the pulse duration in the case of coherent laser radiation) is considerably more than the hypersonic period $T=2\pi/\Omega$ then the equation (7) can be simplified:

$$\frac{\partial p}{\partial t} + \frac{p}{\tau} = -i\frac{g}{\tau}A_L A_S^* + S .\tag{8}$$

In the equations (5) and (6) the refraction due to non-uniformity of refractive index n not related to density variations in hypersonic wave is taken into account together with the amplification owing to the parametric build-up of hypersonic wave. This refraction can be linear as well as nonlinear when n depends on the laser and Stokes intensities. In the present work the refraction is neglected. The fluctuation Langevin force S is added to (7), (8) phenomenologically. It is delta-correlated in time:

$$<S(\mathbf{r}_1, z_1, t_1)S^*(\mathbf{r}_2, z_2, t_2)>=B_p(\mathbf{r}_1\text{-}\mathbf{r}_2, z_1\text{-}z_2)\delta(t_1\text{-}t_2).\tag{9}$$

where B_p is the spatial correlation function of the hyper-sound source. Its form assumes the statistical homogeneity of the SBS medium.

2.2. Modeling of the hyper-sound noise source

For solving the system of equations (5)-(8) it is needed to know the mean and correlation characteristics of hyper-sound noise, which should agree with the observable quantities. The interpretation and characterization of the initial conditions p_0 for the equations (7) and (8) and its first temporal derivative $\dot{p}_0$ for the equation (7) as well as the source power S are the main problems. These parameters must ensure a stationary mean level of the hyper-sound noise in the SBS-medium in the absence of the radiation. As a result of the study we obtain the following conditions:

$$<p_0(\mathbf{r}_1, z_1)p_0^*(\mathbf{r}_2, z_2)>=4\kappa\, F_p(\mathbf{r}_1\text{-}\mathbf{r}_2)\,\delta(z_1\text{-}z_2),\tag{10}$$

$$< \dot{p}_0(\mathbf{r}_1, z_1)\,\dot{p}_0^*(\mathbf{r}_2, z_2)>=\Omega^2<p_0(\mathbf{r}_1, z_1)p_0^*(\mathbf{r}_2, z_2)>,\tag{11}$$

$$<p_0(\mathbf{r}_1, z_1)\,\dot{p}_0^*(\mathbf{r}_2, z_2)>=0,\tag{12}$$

$$<S(\mathbf{r}_1, z_1, t_1)S^*(\mathbf{r}_2, z_2, t_2)>=(2/\tau)<p_0(\mathbf{r}_1, z_1)p_0^*(\mathbf{r}_2, z_2)>\delta(t_1\text{-}t_2),\tag{13}$$

where κ is the spontaneous scattering coefficient of the laser radiation measured in units of $1/(\text{cm}\cdot\text{sr})$, $F_p(\mathbf{r}_1\text{-}\mathbf{r}_2)$ is dimensionless transverse correlation function, which gives a unit axial brightness of Stokes radiation scattered spontaneously from laser beam. The coefficient κ is taken from the experimental data or estimated from the relationship for gaseous medium $\kappa=4\pi^2(n_0\text{-}1)^2/(\lambda_0^4 N_0)$, where N_0 is the molecular concentration.

Numerical integration of the system of equations (5)-(8) is conducted by a method of finite differences along the characteristics. A technique of the splitting on physical processes and transverse co-ordinates lies in the basis of integration method. For solving the parabolic equations (5), (6) a special difference scheme has been developed which has fourth or sixth order of accuracy and considerably exceeds the traditional spectral method in precision.

The adaptation of relationships (10) and (13) to a concrete numerical scheme gives the following results:

$$<p_0(\mathbf{r}_1, z_1)p_0^*(\mathbf{r}_2, z_2)>=4\kappa\, F_p(\mathbf{r}_1\text{-}\mathbf{r}_2)\,\delta_{z_1, z_2}/\Delta z,\tag{14}$$

$$<S(\mathbf{r}_1, z_1, t_1)S^*(\mathbf{r}_2, z_2, t_2)>=<p_0(\mathbf{r}_1, z_1)p_0^*(\mathbf{r}_2, z_2)>\beta\delta_{t_1, t_2},\tag{15}$$

where $\beta=(2/\tau\Delta t)[1\text{-}\alpha+\alpha^2/2\text{-}\alpha^3/8]/[1\text{-}\alpha/2]^2$, $\alpha=\Delta t/\tau$; Δz and Δt are the steps along the longitudinal co-ordinate and time, respectively.

A procedure of determining the distribution of initial random hyper-sound amplitude $p_0(\mathbf{r}, z)$ amounts to finding the random function f_p from its correlation function $F_p(\mathbf{r}')$.[11] Let's suppose that the statistics of hyper-sound noise amplitude is Gaussian. The complex function f_p is presented as $f_p(\mathbf{r}, z)=\varphi_r^{(n)}(\mathbf{r})+i\varphi_i^{(n)}(\mathbf{r})$, where $\varphi_r^{(n)}$, $\varphi_i^{(n)}$ are the statistically independent real functions. Their correlation function has the form:

$$<\varphi_r^{(n)}(\mathbf{r})\varphi_r^{(n)}(\mathbf{r})>=<\varphi_i^{(n)}(\mathbf{r})\varphi_i^{(n)}(\mathbf{r})>=F_p(\mathbf{r}')/2, \quad <\varphi_r^{(n)}(\mathbf{r})\varphi_i^{(n)}(\mathbf{r})>=0.$$

Index n denotes here the dimensionality, $n=2$ and $n=3$ correspond to 2D and 3D medium, respectively.

Let's set the delta-correlated field $\{\psi_\mu\}$ on the transverse numerical greed at $n=2$ so that $<\psi_{\mu_1}\psi_{\mu_2}>=\delta_{\mu_1\mu_2}$. At $n=3$ this field has the form $\{\psi_{\mu,\nu}\}$, and $<\psi_{\mu_1,\nu_1}\psi_{\mu_2,\nu_2}>=\delta_{\mu_1\mu_2}\delta_{\nu_1\nu_2}$. Each function $\varphi_r^{(n)}$, $\varphi_i^{(n)}$ is written as:

$$\varphi^{(2)}(x_\mu) = \sum_{\mu_1=-\infty}^{\mu_1=\infty} C_{\mu_1}^{(2)} \psi_{\mu+\mu_1}\,, \qquad \varphi^{(3)}(x_\mu, y_\nu) = \sum_{\mu_1=-\infty}^{\mu_1=\infty} \sum_{\nu_1=-\infty}^{\nu_1=\infty} C_{\mu_1\nu_1}^{(3)} \psi_{\mu+\mu_1,\nu+\nu_1}\,, \qquad (16)$$

where the expansion coefficients have the form:

$$C_\mu^{(2)} = (\sqrt{\Delta x / \pi}) \int_0^{\pi/\Delta x} \sqrt{\Phi^{(2)}(K_x)} \cos(\mu\Delta x K_x) dK_x$$

$$C_{\mu\nu}^{(3)} = \sqrt{\Delta x \cdot \Delta y / \pi^2} \int_0^{\pi/\Delta x} \int_0^{\pi/\Delta y} \sqrt{\Phi^{(3)}(K_x, K_y)} \cos(\mu\Delta x K_x) \cos(\nu\Delta y K_y) dK_x dK_y\,,$$

where, in turn, Δx, Δy are the steps on transverse greed, $\Phi^{(n)}(K)$ is the spatial spectrum of correlation function $<\varphi^{(n)}(\mathbf{r}_1)\varphi^{(n)}(\mathbf{r}_2)>$. Given the Gaussian function $F_p(\mathbf{r}')=[\pi\theta_S^2]^{(n-1)/2}\exp[-(r'/l_p)^2]$, where $l_p=2/k\theta_S$ is the transverse correlation length of hyper-sound source fluctuations, we find at $l_p \geq 2\Delta x, 2\Delta y$:

$$C_\mu^{(2)} = \left(2\pi\theta_s^2 / \pi\right)^{1/4} (\sqrt{2}\Delta x / l_p)\exp\left[-2\mu^2\Delta x^2 / l_p^2\right],$$

$$C_{\nu\mu}^{(3)} = \left(2\pi\theta_s^2 / \pi\right)^{1/2} (\sqrt{2}\Delta x\Delta y / l_p)\exp\left[-2(\mu^2\Delta x^2 + \nu^2\Delta y^2)/l_p^2\right].$$

It should be noted that for calculating $\varphi_r^{(n)}$ and $\varphi_i^{(n)}$ one can sum in (16) a small number of terms since the expansion coefficients quickly fall off with the index increase. This reduces the calculation time considerably.

Generally speaking, $l_p \sim v_s\tau \sim 0.1\text{-}10$ µm, but to set a very small value of l_p is often impossible and, moreover, to no purpose since the high-angular components of Stokes noise go away from the gain channel and give a very small contribution to the Stokes signal. However Δx must be rather small so that one can model correctly the initial stage of forming the Stokes wave.

2.3. Test calculations of the laser radiation scattering

We carry out the test analysis of SBS in two cases where it is possible to obtain the analytical dependencies and estimates. In the first case we consider the spontaneous scattering of the laser radiation by the hyper-sound noise. In the second case we describe the linear SBS when there is no depletion of the laser intensity. The calculation of SBS equations is conducted into the SBS-cell. Its input and output (for laser beam) windows is situated at $z=0$ and $z=L$, respectively. In the SBS-mirror configuration there are no the phase elements.

Under spontaneous scattering of the laser radiation by the fluctuations of hyper-sound amplitude, when $g=0$, there is no parametric build-up of the hypersonic oscillations. The calculation parameters are as follow. The Gaussian laser beam at the input of SBS-cell has the amplitude distribution:

$$A_L = A_{L0}\exp(-\mathbf{r}^2/a^2). \qquad (17)$$

This distribution does not vary temporally. The wavelength $\lambda=0.53$ µm corresponds to the second harmonics of a Nd laser. The medium parameters $\kappa=10^{-5}$ cm^{-1}sr^{-1}, $\tau=5$ ns, the SBS-cell length $L=80$ cm. The parameters of calculation $l_p=3\Delta x$, where $\Delta x=\Delta y=7.5\times10^{-3}$ cm. The diffraction is not taken into account and the medium is three-dimensional. In Fig.1 the dynamics of the power of spontaneous Stokes radiation $P_S(z,t)=\int J_S(\mathbf{r},z,t)d\mathbf{r}$ at the input of the SBS-cell at $z=0$ is shown. This value is normalized to the mean power $<P_{S0}>=[\kappa L J_{L0}\pi\theta_S^2]\pi a^2/2$. In Fig.1 the plot of the power averaged over the temporal interval 10τ is given as well. It is seen that the power fluctuates around the mean analytical level that tells about the adequate modeling of the hyper-sound noise source.

Next we compare the results of numerical calculations with the analytical solution obtained at linear SBS. This approximation takes place when the Stokes intensity is substantially lesser than the laser intensity and there is no noticeable depletion of the latter. They suppose that the amplification factor of Stokes intensity $G=gJ_Lz$ at linear SBS is not more than 30. The condition $G\approx30$ means the beginning of the influence of Stokes field on laser one and the obtaining of the energy reflection coefficient from SBS-mirror on a level of a few percent.

With no diffraction a solution of the equation (6) at $E_L=$const is found analytically. Under a determined description of the hyper-sound source, when $p_0=\tau S$, at the SBS-cell input it has the form of the integral:

$$A_S(z,t) = \frac{ip_0^* A_L}{2}\left\{\int_0^{\min(ct,z)} \exp(-y_1)\left[I_0\left(2\sqrt{\alpha y_1 y}\right) + \frac{y}{c\tau}\frac{I_1\left(2\sqrt{\alpha y_1 y}\right)}{\sqrt{\alpha y_1 y}}\right]dy + \frac{z}{c\tau}\int_z^{\max(ct,z)} \exp(-y_1)\frac{I_1\left(2\sqrt{\alpha y_1 z}\right)}{\sqrt{\alpha y_1 z}}dy\right\},$$

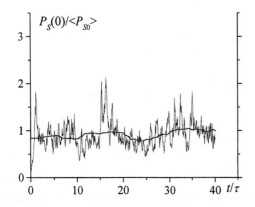

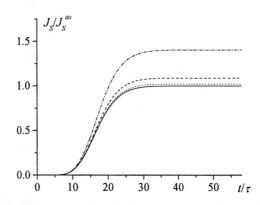

Fig.1. The power dynamics of spontaneous Stokes radiation at the input of SBS-cell. Solid bold curve corresponds to averaging over the interval 10τ.

Fig.2. The dynamics of density flux of Stokes radiation at the input of SBS-cell for gJ_L=0.368 cm^{-1} obtained analytically (—) and numerically at $gJ_L\Delta z$=0.184 ($\cdots$), 0.368 (- - -), 0.736 (-·-·-).

where $\alpha=gJ_L/2$, $y_1=(ct-y)/c\tau$, I_0 is the modified Bessel function of zeroth order. In Fig.2 the dynamics of Stokes intensity is shown at τ=5 ns, g=23 cm/GW, J_L=0.016 GW/cm^2, L=80 cm (i.e. G=29.4) and for the various longitudinal integration steps Δz. As the calculations shows, the relative numerical over-estimation of J_S in the steady-state limit is approximately equal to $(gJ_L\Delta z)^2G/50$.

3. PC OF GAUSSIAN BEAM AT SATURATED SBS

As it has been noted above, the condition $G{\approx}30$ corresponds to the experimental threshold of SBS. In all actual experiments the role of SBS saturation is noticeable in a varying degree. The SBS saturation is considered [1-4] to favour the PC fidelity improvement. The amplification factor of Stokes intensity at the focused Gaussian speckled laser beam in the steady-state regime can be expressed as

$$G^{(n)} = g \int_{z_1}^{z_2} J_L(\mathbf{r}=0;z)dz = gJ_{L0}z_0\left(\frac{a}{w_0}\right)^{n-1} \cdot \begin{cases} (\eta_2-\eta_1), & n=3; \\ \ln\left[\dfrac{\mathrm{tg}(\pi/4+\eta_2/2)}{\mathrm{tg}(\pi/4+\eta_1/2)}\right], & n=2, \end{cases}$$

where η_1 and η_2 are the input and output co-ordinates of SBS-cell, η=arctg(z/z_0), $z_d=ka^2/2N_L$ is the diffraction length, $z_0=z_d/(1+z_d^2/z_f^2)$ is the laser beam waist length, z_f is the focal length of focusing lens, N_L is the excess of laser divergence over the diffraction limit. Under the SBS threshold condition $G{\approx}30$ for 3D medium we have J_L^{th}=30/($gz_d\Delta\eta$), where $\Delta\eta=\eta_2-\eta_1$. In view of the relationship $P_L^{th}=J_L^{th}\pi a^2/2$ one obtains P_L^{th}=30λN_L/(2$g\Delta\eta$). At the location of some beam waists into the SBS-cell when $\Delta\eta{\approx}\pi$ we have $P_L^{th}{\approx}5\lambda N_L/g$.

Let's consider an ideal Gaussian laser beam with the E_L distribution at the input of SBS-cell (17) at a=0.047 cm. The input laser amplitude does not vary in time. We take the following parameters of SBS-medium: n_0=1.05, N_0=10^{21} cm^{-3}, v_s=145 m/s, κ=10^{-5} cm^{-1}sr^{-1}, τ=5 ns, g=23 cm/GW, the SBS-cell length L=80 cm. In the calculations below we the aforesaid laser beam is focused into SBS-cell with the help of a lens with focal length z_f=50 cm. In this case z_d=130.9 cm, z_0=16.7 cm, w_0=0.0168 cm, $\eta_1{\approx}$-1.2, $\eta_2{\approx}$1.14, $\Delta\eta$=2.34, N_L=1, therefore the threshold power equals $P_L^{th}{\approx}6.4\lambda/g$=1.5×10^{-5} GW and the threshold density flux is equal to J_L^{th}=4.3×10^{-3} GW/cm^2.

In the calculations the diffraction and random hyper-sound source are taken into account. The calculations are conducted at l_p=3Δx, Δx=7.5×10^{-3} cm. As we consider the coherent laser radiation, it is reasonable to introduce the PC coefficient $h(z,t)=|\iint A_L(\mathbf{r},z,t)A_S(\mathbf{r},z,t)d\mathbf{r}|^2/[P_S(z,t)P_L(z,t)]$, where $P_L(z,t)=\iint|A_L(\mathbf{r},z,t)|^2d\mathbf{r}$ and $P_S(z,t)=\iint|A_S(\mathbf{r},z,t)|^2d\mathbf{r}$ are the powers of laser and Stokes radiation, respectively. At an ideal PC we have h=1. The asymptotic values of PC coefficient and Stokes power at the SBS-cell input as well as laser power at the output are monitored in the calculations. In Fig.3 are shown the asymptotic PC coefficient at the SBS-cell input and the reflection coefficient (the ratio of asymptotic Stokes power at the SBS-cell input to the input laser power) as functions of laser power. The PC coefficient value obtained under the near-threshold SBS conditions agrees with the theory[2] and calculations[10,12] of the linear steady-state SBS. The PC quality with h<90% can not be considered as satisfactory. As it is seen from Fig.3, the PC fidelity that is near to the ideal with h>90% occurs under the level of the laser radiation reflection from SBS-cell that is more than 80%.

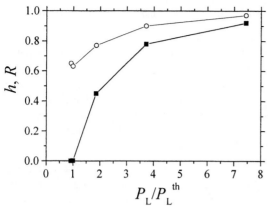

Fig.3. Reflection coefficient R (■) and asymptotic PC coefficient h (O) against the laser power.

Fig.4 demonstrates the laser and Stokes density flux distributions in the longitudinal section of SBS-cell at $J_{L0}=3.2\times10^{-2}$ GW/cm² at three temporal moments. It is seen how the Stokes radiation begins to form from the noise (Fig.4(d)), experiences the linear SBS stage (Fig.4(e)) and is "attracted" to the input window of SBS-cell under the deep SBS saturation (Fig.4(f)). It is interesting that the SBS saturation changes considerably the spatial structure of laser intensity in the SBS-cell (Fig.4(c)).

The transverse distributions of the laser and Stokes density fluxes at those three temporal moments are given in Fig.5. They show demonstrably the temporal rising of the PC quality. With the validation purpose, the calculations have been made in the case of $J_{L0}=3.2\times10^{-2}$ GW/cm² for three steps Δz so that $gJ_{L0}\Delta z=0.49$, 0.98, and 1.96. The calculations give that the noted variation of the step Δz does not practically affect the Stokes power and the PC quality. The energy balance takes place (i.e. the incident laser energy is equal to the sum of the laser energy passed through the SBS-cell and the output Stokes energy) with the accuracy of 0.15, 0.58, and 2.4%, respectively.

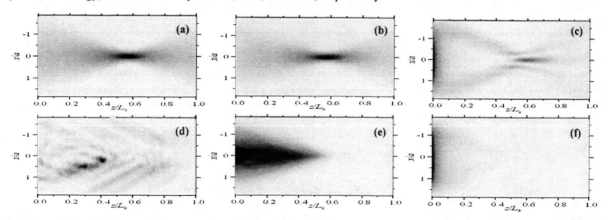

Fig.4. Longitudinal distributions of laser J_L (a-c) and Stokes J_S (d-f) density flux in the SBS-cell at $t/\tau=0.01$ (a, d), 1 (b, e) and 6 (c, f) obtained for $J_{L0}=0.032$ GW/cm².

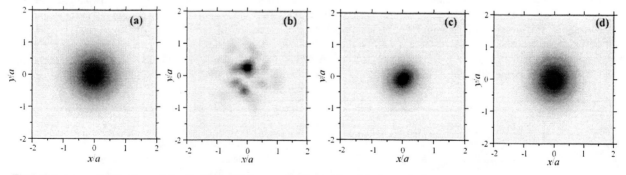

Fig.5. Transverse distributions of density flux of the laser radiation J_L (a) and Stokes radiation J_S at the SBS-cell input at $t/\tau=0.01$ (b), 1 (c) and 6 (d) obtained for $J_{L0}=0.032$ GW/cm².

The calculations show that the Stokes power, the hyper-sound amplitude and the PC coefficient at the SBS-cell input as well as the laser power at the SBS-cell output reach the asymptotic steady-state values over a temporal interval $\sim(5-20)\tau$. It depends on the SBS saturation depth. But it should be noted that the steady state in the strict sense of the word does not occur. At the long times of the calculation one can observe the seldom but deep fluctuations of the parameters of the radiation and hyper-sound and the PC quality even under the deep SBS saturation. We have carried out the additional calculations with a reduced transverse step. The results show that the aforesaid fluctuations are not the artifact.

4. SBS-MIRROR WITH THE RASTER OF SMALL LENSES

We have shown above that one can reach the high PC fidelity in the case of SBS-mirror, including the focusing lens and SBS-cell, only at the great reflection coefficients of the laser radiation from the SBS-mirror. But often it is not feasible since the considerable light loads result in the growth of parasitic competitive non-linear processes in the SBS-active medium (the thermal self-defocusing, the electrostriction self-focusing, the optical breakdown), which exert the negative influence on the SBS process to the point of its destruction. Therefore it is actual to obtain the near-to-ideal PC quality for the relatively small laser powers. With this purpose we will consider below more complicated SBS-mirror with an ordered phase plate in the form of a raster of small lenses and an angular selector of the Stokes radiation. As it recently shown under conditions of the linear steady-state SBS,[10] the raster is more preferable in comparison with the random phase plate.

The principal schematic arrangement of SBS-mirror with the raster of small lenses is shown in Fig.6. We suppose that the laser beam after its first pass in the amplifier has a half-width a and angular divergence θ_L. Its undesirable excess over the diffraction limit results from the optical nonuniformities in the actual amplifiers. According to Fig.6, the laser radiation is coupled into the angular selector consisting of two lenses and a pinhole situated between them. The pinhole plane is the place where the angular selection of radiation can occur. The angular spectrum of the radiation is cut in the selector by a value of the transmission angle θ_{sel}. The condition $\theta_{sel} > \theta_L$ is satisfied. After passing through the selector, the laser beam goes through the ordered raster of identical small lenses with a focal length f and a size d. Optical properties of the multiple Fresnel lens [7,8] do not differ practically from the conventional one. The distance between the phase plate and the main focusing lens is L_0. The main lens with focal length F focuses the aberrated and orderly disturbed laser radiation field into the SBS-cell of length L. The distance between the main lens and the input window of the SBS-cell is L_1.

The laser intensity after passing through the raster has the specific distributions in the zones I and II, where the laser beam is broken up into the ordered arrays of beamlets. One of them, zone I, is the focal region of main focusing lens (see Fig.6). An ordered spotted picture is formed in zone I. Each peak of this picture has a structure of angular distribution of the laser intensity incident on the raster. The angular interval between the peaks is equal to $\theta_d = \lambda/d$. The smooth envelop of the peaks depends on the amplitude transmission coefficient of the raster's cell and does not depend on the incident laser field under its large-scale nonuniformity. We take $d=0.1$ cm, $f=16$ cm. At $\lambda=1.315$ μm it gives 3×3 main peaks of the equal intensity in the zone I.[10] The other specific distribution of the laser intensity is in the zone II, i.e. the focal region of the raster's small lenses. Unlike the zone I, the envelope of small focal peaks in the zone II is proportional to the laser intensity incident on the raster.

The steady-state linear SBS calculations[10] have shown that the distribution of Stokes intensity in the pinhole plane consists of a discrete set of peaks or diffraction orders that is consistent with the experiments.[5-9] The angular separation between them equals θ_d like that in the laser pattern in the zone I. A perfect PC requires only the central Stokes order of diffraction conjugated to the laser beam. Angular selection in the pinhole plane removes the non-zero orders of diffraction, thereby increasing the PC fidelity. A novel encouraging effect of extremely low noising of the Stokes field was found.[10] Its essence is in the fact that the angular selection of central Stokes mode allows achieving a nearly perfect PC. For this purpose the input window of the SBS-cell should be placed in a certain region between the zone I and zone II, where the quasi-lightguide mosaic zones with the periodical distribution of the laser intensity are formed. If the arrangement of PC-mirror is optimal, the PC coefficient is more than 90-95% whereas the PC coefficient before the selection does not exceed the value of 30%. To increase the selector transmission coefficient, the SBS-cell should be moved towards the main focusing lens so that the zone II would be put within the SBS-cell. However, this increase of selection coefficient is accompanied by a noticeable decrease of PC fidelity after the selection.

To explore an effect of SBS saturation on the performance of the aforesaid SBS-mirror, we carry out the calculations in the 2D approximation that reduces the calculation time considerably. As it has been shown earlier,[10] the 2D approximation describes all the principal features of working the SBS-mirror considered. In this case the PC and selection coefficients for the 3D medium are close to the squared corresponding values obtained in the 2D approximation. The temporal profile of the laser power in calculations has a rectangular shape and the duration 12τ. We use the distribution (17) at $a=0.4$ cm. The calculations are conducted at $F=100$ cm, $\theta_{sel}=\theta_d/2$, $L_0=1$ cm and two distances L_1. At $L_1=1$ cm, when the both zones I and II are within the SBS-cell, the SBS-cell length is $L=200$ cm. At $L_1=68$ cm, when the only zone I is

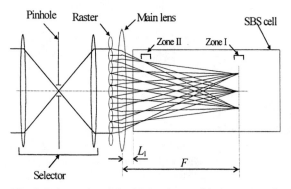

Fig.6. Schematic of the SBS-mirror with the raster of small lenses.

within the SBS-cell, the SBS-cell is shorter, L=132 cm. In the calculations the axial laser intensity J_{L0}=$|A_{L0}|^2$ varies. A few calculations is made for each J_{L0} value with the different realizations of the hyper-sound noise. The typical instantaneous distributions of the laser and Stokes density flux in the longitudinal section of the SBS-cell are shown in Fig.7 for L_1=1 cm и 68 cm. The quasi-lightguide zones, which are formed in the place of intersection of the neighboring laser beamlets existing in the zone I, are distinctly seen. At L_1=68 cm the region of the quasi-lightguide zones is situated in the vicinity of the input window of SBS-cell (see Fig.7(b)), ensuring an optimal SBS-mirror arrangement in terms of achieving the highest PC fidelity under the linear steady-state SBS conditions.[10] The dependence of the reflection coefficient R (i.e. the ratio of the Stokes energy to the incident laser energy), the PC coefficient after the selection h_{sel} and the selection coefficient k_{hsel} (i.e. the Stokes energy fraction passed through the pinhole) on the laser intensity are shown in Fig.8. The steady-state asymptotic values of h_{sel} and k_{sel}, which are maximal in time as a rule, are presented. For the PC coefficient at the SBS-cell input (unselected) the relationship h=$h_{sel}k_{sel}$ is fulfilled with a good accuracy.

As it is seen from Fig.8, the results of steady-state linear SBS calculations[10] agree with the new data obtained under the near-threshold SBS conditions when the reflection coefficient is equal to a few percent. The R fluctuations are rather small under the both arrangements of SBS-mirror. At L_1=1 cm the SBS threshold is less than that is at L_1=68 cm due to the longer SBS-cell length and the additional presence of bright zone II within the SBS-cell. At L_1=1 cm near the SBS threshold the PC quality is low but the selection losses are less than at L_1=68 cm. With increasing the reflection coefficient, the both h_{sel} and k_{sel} rise, nevertheless the PC quality is far from an ideal. The value h_{sel} is less than 90% at R=60% despite of a considerable rise of k_{sel}. The fluctuations of h_{sel} and k_{sel} are relatively small under any SBS saturation level.

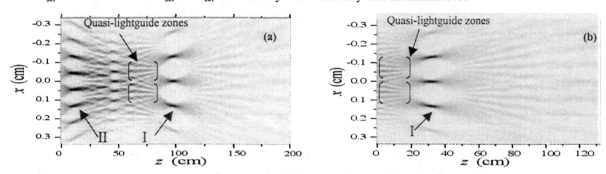

Fig.7. Longitudinal distributions of laser density flux J_L in the SBS-cell obtained for (a) L=200 cm, L_1=1 cm and (b) L=132 cm, L_1=68 cm.

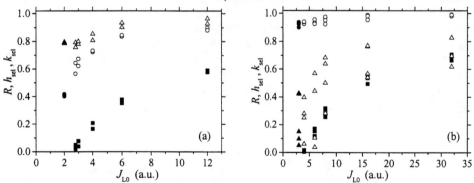

Fig.8. Numerical dependence of the reflection coefficient R (■), the steady-state PC coefficient after selection h_{sel} (O) and the selection coefficient k_{sel} (Δ) on the laser intensity at L_1=1 cm (a) and 68 cm (b). Symbols ● and ▲ correspond to the PC coefficient and the selection coefficient obtained within the framework of steady-state linear SBS model.[10]

At L_1=68 cm, h_{sel} is more than 90% near the SBS threshold that has been the most substantial result of the steady-state linear SBS calculations.[10] It is extremely important that the effect of SBS saturation does not deteriorate this result, moreover, with increasing the laser power the PC quality rises yet more approaching to the ideal. The fluctuations of PC coefficient are very small. But there is a considerable spread in the k_{sel} value that is explained as follows. The PC coefficient is near to the ideal in any quasi-lightguide zone.[10] As it is seen from Fig.7, each quasi-lightguide zone is formed by two neighboring laser beamlets existing in zone I. In each laser beamlet in the zone I the Stokes radiation is amplified practically independently, therefore the Stokes fields in the quasi-lightguide zones are not phased. In other words, there is no complete diffraction mixing of the Stokes radiation in the transverse section of the beam. Contra, at L_1=1 cm the output Stokes radiation, passed from the far field to the near one in the medium with considerable gain, is transversely coherentized that leads to the small spread in values of k_{sel}.

In the experiments one registers, as a rule, not instantaneous but temporally integrated data. For modeling the actual experimental situations it is needed to take into account a concrete temporal profile of the laser power and a non-steady-state effect. Calculations of the following series repeat the previous ones except for that the Gaussian profile of laser power is taken. The profile width is about 4τ, and the non-steady-state effects being showing themselves. The calculations with the same laser power but the different realizations of the hyper-sound noise are analogous to a series of flashes of a pulsed laser in the experiment. The typical dynamics of the laser power at the input and output of SBS-cell and the Stokes power at the input of SBS-cell is shown in Fig.9. Fig.10 illustrates the dynamics of the PC coefficient at the input of SBS-cell h and after the angular selection h_{sel} as well as the selection coefficient k_{sel}.

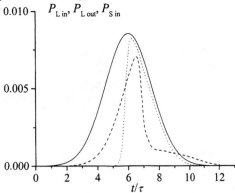

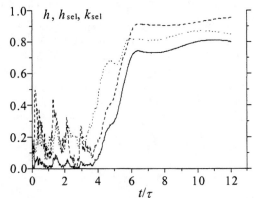

Fig.9. The dynamics of laser power at the input (—) and output (- - -) and Stokes power at the input (····) of the SBS-cell at L_1=1 cm, J_{L0}=24 a.u.

Fig.10. The dynamics of the PC coefficient at the SBS-cell input h (—) and after the angular selection h_{sel} (- - -) and the selection coefficient k_{sel} (····) at L_1=1 cm, J_{L0}=24 a.u.

Fig.11 gives the numerical dependence of the reflection coefficient from the SBS–mirror R, the selection coefficient k_{sel} and the PC coefficient after selection h_{sel} on the maximal axial laser intensity J_{L0} for the both arrangements of the SBS-mirror. Unlike Fig.8, here are given the h_{sel} and k_{sel} values, which have been averaged over the pulse (i.e. the values time-integrated with a weight of the output Stokes power and divided by the Stokes energy). It is interesting to compare Fig.11 with Fig.8. It is seen that the spread in the averaged-over-the-pulse values of k_{sel} increases at L_1=1 cm as well as at L_1=68 cm. The cause of this fact consists in the fluctuations of instantaneous value of k_{sel} within an initial stage of the Stokes beam amplification (although the maximal asymptotic values of k_{sel} in the different realizations are much more close each other). As for h_{sel}, at L_1=1 cm it decreases and the spread in its value rises. But at L_1=68 cm the time-averaged PC quality is near to the ideal at any level of SBS saturation. Thus the SBS-mirror configuration at L_1=68 cm is preferable in terms of the PC fidelity. Its disadvantage is in the significant spread in the k_{sel} value and in a great probability to obtain very small k_{sel}. In this case there is the danger of parasitic generation of amplified spontaneous emission in the laser amplifier during the second pass.

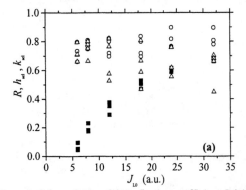

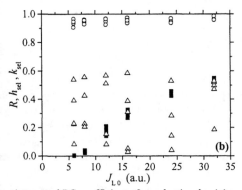

Fig.11. Numerical dependence of the reflection coefficient R (■), the time-integrated PC coefficient after selection h_{sel} (○) and the time-integrated selection coefficient k_{sel} (△) on the laser intensity at L_1=1 cm (a) and 68 cm (b).

We have begun the detailed experimental verification of the numerical results. The schematic of experimental setup is shown in Fig.12. The Nd laser consists of master oscillator and two amplifiers. At the output of the second amplifier a practically diffraction-limited laser beam is formed with λ=1.06 μm, energy 1.5 J and pulse duration 25 ns. The laser radiation is sent to a KD*P crystal where its frequency is doubled. Energy of the second harmonic radiation is ≤300 mJ, the pulse duration is 25 ns, the divergence is $3 \cdot 10^{-4}$ rad. On going to the second harmonic, the laser setup is isolated from the back-scattered SBS component. SBS is excited in a cell filled with a mixture of SF_6 and Xe. The SF_6 partial pressure is 1.5 atm and the total mixture pressure is 28 atm. The SBS gain coefficient in such a medium is g=0.023 cm/MW, the phonon lifetime is $\tau \approx 5$ ns. Gas filling into the SBS cell is realized through a special system containing a filter for purifying gases from dispersion micro-impurities and aerosol particles with a diameter that is not less than 0.1 μm. In the experiments we use an eight-level kinoform raster of Fresnel lenses with the diffraction efficiency of 95%. An image of the raster fragment is shown in Fig.13.

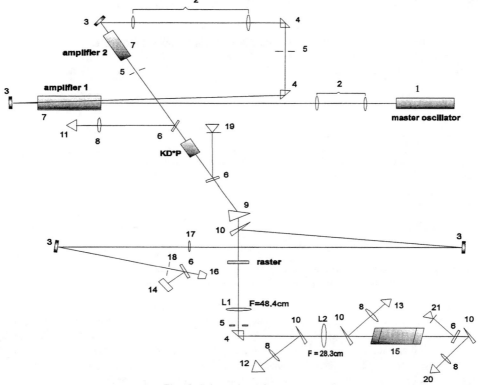

Fig.12. Schematic of the experimental setup.
(1) master oscillator; (2) matching telescopes; (3) dielectric mirrors; (4) turning prisms; (5) pinhole; (6) glass plates; (7) amplifiers; (8) focusing lenses; (9) dispersion prism; (10) glass wedges; (11-13, 20) calorimeters; (14) photographic camera; (15) SBS cell; (16) pyro-calorimeter; (17) measuring lens; (18) measuring diaphragm; (19, 21) coaxial photodetectors (CPDs); (L1 and L2) lenses focusing the pump radiation into the cell.

Depending on mutual location of the raster, SBS-cell, and focusing lenses, various versions of focusing the pump radiation into the cell are possible. The bright zones are located as follows: the zone II and the zone I. This location of zones is achieved by the following way: after the raster a two-lens-system consisting of lenses L1 and L2 is placed. The lens foci plane forming the raster is placed at focus of lens L1. Their image was transferred to the focal plane of lens L2. The zone I is

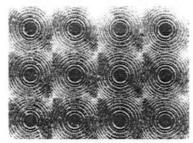

Fig.13. Photographic image of a fragment of the kinoform raster.

formed at an equivalent focus of the two-lens system 30.4 cm and located at a distance of 17.8 cm from the zone II. The beam diameter on the raster is 0.8 cm. By changing distance L_1 between lens L2 and the input window of the SBS cell, it is possible to realize various regimes of exciting SBS.

Figs 14 and 15 compare the experimental and numerical dependencies of the time-integrated selection coefficient k_{sel} and the reflection coefficient from the SBS–mirror R on the laser energy E_L for two arrangements of the SBS-mirror. At the first arrangement the both zones I and II are placed within the SBS-cell as in Fig.7(a), whereas at the second one only the zone I is there as in Fig.7(b). It is seen that simulation results agree with experimental data very well in regard to the behavior and the spread of k_{sel} and R. As for the time-integrated PC coefficient after selection h_{sel} (Fig.14,c), it is difficult to experimentally measure it exactly.

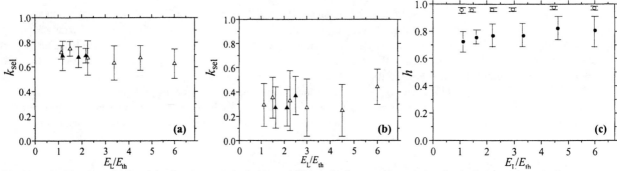

Fig.14. (a, b) The dependence of the time-integrated selection coefficient k_{sel} in experiment (▲) and calculations (Δ) on the laser energy at the arrangement of the zone II+zone I (a) and the zone I (b) within the SBS-cell. (c) Numerical dependence of the time-integrated PC coefficient after selection h_{sel} on the laser energy at the arrangement of the zone II+zone I (●) and the zone I (O) within the SBS-cell.

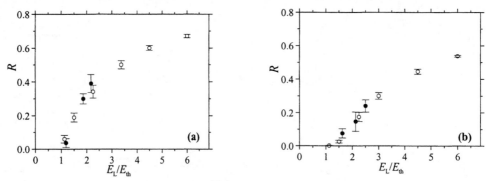

Fig.15. The dependence of the reflection coefficient R in experiment (●) and calculations (O) on the laser energy at the arrangement of the zone II+zone I (a) and the zone I (b) within the SBS-cell. $E_{th}=0.08$ J.

For the purpose of increasing and stabilizing k_{sel} at keeping the high PC quality we propose to modify the SBS-mirror configuration. As it follows from Fig.14, we must fulfil two conditions. First, the input window of the SBS-cell should be set in the field of quasi-waveguide zones for achieving the high PC quality. Second, the zones I and II should be within the SBS-cell for reaching more high and stable values of k_{sel}. Both the conditions can be satisfied at placing the zone II in the SBS-cell behind the zone I. In order to realize this, we move the raster (with the selector) away the main lens so that L_0 exceeds the distance $(F+f)$. In this case the zone II is outside of the SBS-cell but its image, which is $F[1+F/(L_0-F-f)]$ distant from the main lens, can fall within the SBS-cell behind the zone I. In the calculations we put L_0=241 cm, L_1=1 cm, therefore the zone I and the image of zone II are at the distance of 100 и 180 cm, respectively, from the main length, and at L=200 cm they being within the SBS-cell. The distribution of the laser density flux in the longitudinal section of the SBS-cell for the zone I+zone II arrangement within the SBS-cell are shown in Fig.16. As the calculations show (see Fig.17), such the SBS-mirror arrangement allows to ensure the excellent PC quality and to increase the selection level and stability.

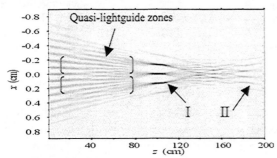

Fig.16. The longitudinal distributions of laser radiation density flux J_L in the SBS-cell at t/τ=1 obtained for L=200 cm, L_0=241 cm, L_1=0.

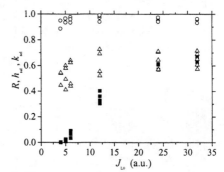

Fig.17. Numerical dependence of the reflection coefficient R (■), the time-integrated PC coefficient after selection h_{sel} (O) and the time-integrated selection coefficient k_{sel} (Δ) on the laser intensity for L=200 cm, L_0=241 cm, L_1=0.

5. CONCLUSIONS

In conclusion, we have obtained the following results on the phase conjugation (PS) at SBS of the focused laser beam.

The most complete in literature physical model and computer code of the non-steady-state SBS have been developed. The model is based on solving the parabolic equations for laser and Stokes fields and the equation for hyper-sound amplitude. The model embodies the 2D or 3D medium, the transient processes, the random hyper-sound source, the diffraction, the refraction, the absorption and self-action of the radiation, the effect of Stokes field on laser one (SBS with saturation). The test calculations show a good accuracy of the numerical method.

The 3D calculations have been carried out of the PC of a Gaussian beam of the long duration in the simplest SBS-mirror including the focusing lens and the SBS-cell. The dependencies of the reflection coefficient from the SBS-mirror and the PC fidelity on the laser intensity have been obtained. The improvement of PC quality takes place with rising the reflection coefficient. It has been found that the PC quality with the PC coefficient that is more than 90%, i.e. that is near to the ideal, can be obtained at the reflection coefficient that is more than 80%. This is usually intractable in the experiments due to growing the competitive nonlinear processes in the SBS-medium under the high laser powers.The PC has been studied of the Gaussian focusing laser beam with the pulse duration about 4τ, where τ is the hyper-sound relaxation time, in the SBS-mirror that includes additionally the ordered raster of small lenses and the angular selector of Stokes radiation. The raster allows to perform the effective angular filtration of non-conjugated Stokes component, to reduce the local light loads in the SBS-medium and to avoid the influence of undesirable nonlinear effects. Unlike the random phase plate, in the case of using the raster there are more possibilities to controllably change its characteristics. This allows to expand the spectrum of conditions of focusing the laser beam into the SBS-cell and to determine the most optimal ones. As a result of the calculations, an optimal arrangement of the SBS-mirror has been found with the unique properties. It gives the PC coefficient that is more than 90-95% at the selection coefficient 50-70% and any level of the SBS saturation, i.e. at any reflection coefficient. The first experimental data obtained at the Nd laser facility have shown the validity of the simulation results. The developed conception of SBS-mirror with the unique properties after the complete experimental verification can be applied for the improvement of wide class of industrial lasers.

ACKNOWLEDGEMENT

The present work was financially supported by ISTC within the framework of the project #591-98.

REFERENCES

1. *Optical Phase Conjugation*, R. Fisher, ed., Academic Press, N.Y., 1983.
2. B. Ya. Zel'dovich, N. F. Pilipetskii and V. V. Shkunov, *Principles of Phase Conjugation*, Springer-Verlag, Berlin, 1985.
3. V. I. Bespalov and G. A. Pasmanik, *Nonlinear Optics and Adaptive Laser System*, Nauka, Moscow, 1985.
4. V. V. Ragul'skii, *Wave Front Reversal at Stimulated Scattering of Light*, Nauka, Moscow, 1990.
5. S. T. Bobrov, K. V. Gratsianov, A. F. Kornev et al., "*Investigation* of possibility of phase conjugation fidelity improvement of SBS-mirrors at smooth aberrations", *Optics Spectrosc.* **62**, pp. 402-406, 1987.
6. K. V. Gratsianov, A. F. Kornev, V. V. Lyubimov , V. G. Pankov, "Phasing of laser beams at phase conjugation by SBS", *Optics Spectrosc.* **68,** pp. 617-619, 1990..
7. Yu. V. Dolgopolov, G. A. Kirillov, G. G. Kochemasov et al, "Iodine laser with SBS-mirror" *Proc. SPIE*, **1980**, pp. 23-30, 1992.
8. Yu. V. Dolgopolov, A. M. Dudov, L. I. Zykov, G. A. Kirillov, G. G. Kochemasov, S. M. Kulikov, V. M. Murugov, S. A. Sukharev and A. F. Shkapa, "High power iodine laser with high quality phase conjugation" *Izv. AN USSR ser. fiz.* **58**, pp. 35-40 (1994).
9. H. J. Eichler, B. Liu, M. Duelk, Z. Lu and J. Chen, "Phase conjugation behind an ordered multimode fibre bundle", *Optics Comm.* **123**, pp. 412-422, 1996.
10. G.G.Kochemasov and F.A.Starikov, "Novel features of phase conjugation at SBS of beams passed through an ordered phase plate ", *Optics Comm.* **170**, pp.161-174, 1999.
11. V. P. Kandidov, V. I. Ledenev, "Numerical modeling of random field of dielectric permittivity of turbulent atmosphere", Vestnik MGU, ser. fiz. **23**, 3 (1982); V. K. Ladagin, F. A. Starikov, and V. A. Volkov, "Amplified spontaneous emission at gain saturation: two investigation approaches", in *Advances in Laser Interaction with Matter and Inertial Fusion*, eds. G.Velarde, J.M.Martinez-Val, E.Minguez, J.M.Perlado (World Scientific, London, 1997), pp.546-549.
12. G. G. Kochemasov, V. D. Nikolaev, "On reproduction of spatial distribution of the pumping beam amplitude and phase in the course of SBS", *Kvantovaya Elektronika* **4**, 115 (1977).

Automatic CO_2 Laser of New Generation

Yuri Bulkin*, Anatoli Adamenkov*, Evgeni Kudryashov*, Boris Vyskubenko*,
Yuri Kolobyanin*, Alexey Kuznetzov**, Victor Masychev***, Pavel Kapustin***
*Russian Federal Nuclear Center – VNIIEF, Sarov, Nizhni Novgorod region, Russia
**Lebedev Institute, Moscow, Russia
***NPO "Istok", Fryazino, Moscow region, Russia

ABSTRACT

A novel automatic output stabilized tunable CO_2 laser has been developed providing a new quality level of investigations in the areas of atmosphere monitoring, active media diagnostics, spectroscopy, etc. The optimally designed and arranged laser is composed of sites and systems with reliable operational resources. Mean time till failure of the laser is 1000 h. About 0.3 cm^{-1} resolution of its high-Q and highly selective resonator allows extraction of most lines in sequence and hot bands without using a hot cell in the laser cavity. The laser is also distinguished among many other lasers by generating a high power (up to 50 W) at traditional lines and a great amount (near 160 totally) of sequence and hot lines (up to 10 W) in the 9-12 μm range which can be qualitatively changed in favor of sequence or hot lines by replacing some of the sealed-off active element or the whole set of them. Software and hardware developed for handling the laser operation offer the possibility to extract a needed set of lines in any succession and with a required run duration.

Keywords: Tunable laser, resonator, traditional and non-traditional lines.

1. INTRODUCTION

The work aim was to develop an automatic CO_2 laser tunable over lines in traditional, sequence, and hot bands and to create a hardware and software for handling its operation. The development was based on realization of the optimal optical structure of the laser cavity (without «hot» cell) and application of a whole series of experimental methods and approaches in order to meet the predetermined requirements.

Different possibilities of technical solutions assumed in the laser construction, compositions of laser media, and methods for retuning between generation lines have been corroborated on the models of the developed in the work CO_2 laser. The results obtained have been published /1-6/.

As was suggested, this activity should eventuate in an automatic tunable output-stabilized CO_2 laser capable to provide for a new quality level of investigations in many scientific areas: atmospheric monitoring, diagnostics of active media, spectroscopy, etc.

The first problem to be solved in this work was right radiation source selection to suit tunable CO_2 laser generating lines in five bands. We preferred a sealed-off electric-discharge tube made of molybdic glass with inner diameter 15 mm, discharge gap length 1 000 ÷ 1 500 mm, GaAs output windows, and filled with a mixture having optimal composition. This tube is a modification of GL-501-, GL-502-type active elements mass-produced in Russia. Our choice was justified by the following factors:

- these industrial sealed-off active elements offer high technical and exploitation characteristics and have reliable hermeticity, pressure and gas-composition stability, including the case of activating physical-chemical process by gas-discharge plasma; their average operational resource exceeds 10 000 hours with breakdown-free continuous operation being about 1 000 hours so that the given elements are the most reliable among all existing ones including waveguide, open-flow, etc.;

- they provide for high-Q and high selectivity resonator realizations, which could hardly be satisfied simultaneously with other active elements.

The second problem involved was the resonator construction optimal for tunable CO_2 laser. Need for sequence and hot lines steady generation puts severe requirements upon the resonator. It must be high-Q - since gain values typically range between 0.1 and 0.3 m^{-1} for sequence lines of electric-discharge lasers and between 0.05 and 0.1 m^{-1} for hot - and highly selective with resolution better than 0.4 cm^{-1} - since lines in the second sequence and basic bands are alternated. VNIIEF study of polarization features and other properties of laser reflection gratings showed that such resonators must be built on the following scope of experimental methods and approaches:

- application of high quality optics (mirrors and laser reflection gratings) with precisely abided by required parameters;

- special selection of laser reflection gratings according to their reflectance to first and zero orders for S- and P-polarized radiation;

- matching between polarization properties of laser reflection gratings and output windows of active elements;

24

In *Laser Resonators III*, Alexis V. Kudryashov, Alan H. Paxton, Editors,
Proceedings of SPIE Vol. 3930 (2000) ● 0277-786X/00/$15.00

The third problem brought about with further radiation spectrum enrichment was inverse parameters improvement in active media that is to be reached through multifactor optimization for laser values of interest: mixture composition, discharge parameters, extent of active element cooling, etc.. Nowadays, the most perfect technique to study CO_2 lasers active media is the "laser spectrograph" method /1-6/ developed at the VNIIEF. Experimental study based on this method gave the conditions optimal to generate lines in various bands of interest. It is possible to fill the sealed-off active element with a mixture the most appropriate for generating the maximum lines amount in all the five bands.

The fourth problem of the given work was automation of the facility that has necessitated the development of software and hardware means to handle laser operation.

2. MAIN LASER UNITS DESIGN AND FABRICATION

Three modifications of laser head frame were developed and fabricated around Invar rods to provide for high level of passive frequency stabilization.

Created special precession unit with reduction ratio N = 2160 can adjust any grating angular position within the 0°- 20° range with 3" accuracy corresponding to a step of the used step driver.

Three modifications of sealed-off gas-discharge tubes of the GL-501-type with active lengths L= 100, 125, 150 cm accordingly, GaAs Brewster-angle windows, and nontransmitting spherical mirrors with radius R = 3 and 10 m accordingly were chosen for the study. The third modification with active length 150 cm and nontransmitting copper mirror R = 10 m was required for broadening the spectrum generated in automatic operation regime at lines with large rotational numbers J in both traditional and non-traditional bands. The spherical mirrors' substrates were made of Invar and covered with Cu and silicium oxide protective film.

These are the parameters of the used reflection gratings:

substrate dimension..∅ 40 x 10 mm;
substrate material.. 32NKD (Invar);
dimension of grooved area........…............................. 25×25 (mm);
covering.. gold (Au);
grooves per mm...150;
working spectrum order...first;
blaze angle.. ~ 52⁰;
reflection index for S-polarized radiation at 10.6 µm
to 1-st order…...~ 90-91 %;
to zero order.....…...~ 5-6 %.

Laser radiation receiver was also constructed by ourselves. It operates in the lock-system to stabilize frequency and power in the developed automatic tunable CO_2 laser. The receiver consists of a diffuser, pyroelectric detector, electronics unit, and optical modulator placed prior the diffuser. All the elements of the receiver, including the modulator, are accommodated in the rack of the automatic tunable CO_2 laser.

3. SMALL-DIMENSION POWER SUPPLY AND NON-STANDARD ELECTRONICS PERFORMANCES

Construction of the CO_2 laser includes also a compact power supply to energize the following elements:
- laser active element of the GL-50-type with a controllable high-voltage DC current;
- control units of the step driver and lock system with stabilized DC voltages of various levels.

The major performances of this originally developed element are:
- source voltage - single-phase, 50 Hz , 220 V ± 15%.
- continuous operation - 8 hours and more.
- cooling type - forced air.
    Feeding of the gas-discharge tube of the laser:
- maximum output direct current - 40 mA (real resistance of discharge plasma - about 300 kΩ).
- instability of discharge current – at most 1%.
- output current control – from 25% to 100%.
- maximum output voltage - -20 kV.
- maximum output power - 500 W.
    Feeding of the step driver control and lock systems.
- stabilized output direct voltage - ±15 V (maximum load current - 200 mA).
- stabilized output direct voltage - 5 V (maximum load current - 1 A).

- stabilized output direct voltage - 230 V (maximum load current 10 mA).
- non-stabilized output direct voltage - 5.2 V (maximum load current - 2.5 A).
- tolerant deviation of stabilized voltage - ±5%.
- overall output voltage - 60 W.

The developed nonstandard electronics is to support laser operation in automatic regime. This apparatus incorporates three main units:
- receiver;
- lock;
- step driver control.

As was mentioned, the receiver unit is to detect laser radiation and it was implemented around a pyroelectric detector of the MG-30-type. To stabilize frequency and power of the CO_2 laser, we used the system of peak-holding control with support modulation. Preliminary tests of the lock system operating in the structure of the tunable CO_2 laser with the GL-501-type active element showed the following performances. Output power deviations did not exceed a percentage for a minute of laser operation while the frequency stability was about $10^{-4} cm^{-1}$. The lock system can be operated manually with the front panel on the power supply or remotely using master computer. Step-driver control unit adjusts angular position of the diffraction corner reflector with total turnabout zone of 40 000 steps; rotation angle per driver step is 3 arc seconds. The control pulses repetition rate exceeds 200 Hz. The unit is operated manually with the front panel on the power supply or remotely using master computer.

4. AUTOMATIC LASER OPERATION

The main objective of software support for the automatic tunable CO_2 laser is to get the laser through its operation cycle, which is to lock laser line of interest, adjust precisely laser cavity (by angle), and measure the output power at the selected line. A program package incorporating the root program LASER, data bank, and object library has been developed toward this end. The data bank collects the information on, approximately, 500 lines in six generation bands of the CO_2 laser (00^01-10^00, 00^01-02^00, 01^11-11^10, 01^11-03^10, 00^02-10^01, 00^02-02^01). The object library provides for procedures to operate the step driver, cater for the analog-to-digital converter, and other application programs. The step driver rotates the grating in the laser cavity providing thereby tuning in a chosen line. The analog-to-digital converter serves to put obtained experimental data (in particular, laser power) into personal computer memory.

Being started, the program LASER provides the user's menu that looks as shown in fig.1. This menu allows user to enter settings for laser retuning through generation lines, in particular, the retuning method. There are two possible methods: "by-line" and "by-step". In "by-line" method, retuning starts on the line in which the laser is tuned at the moment (reference line) towards the indicated, working, line. The lines must be entered in corresponding menu item along with grating rotation velocity. In this case the program interacts with the data bank for identifying the lines and recognizing the value and the sign of the necessary angular moving. In "by-step" method, the step driver simply rotates the grating by necessary steps in a specified direction and with a specified velocity.

All the settings being entered, user can proceed immediately with retuning procedure (functions of other menu items will be considered below). The program analyses and processes initial data and initiates the step driver. Progress and end of retuning process displays on information panel. User can interrupt the process in any moment or it stops itself in case of emergency when the step driver reaches the available limits. When the step driver stops and information is provided, the program is supposed to return back to its original menu, and the foregoing process can be again repeated.

Initial menu contains also three possibilities: "initial line quest", "precise tuning", and "power measurement" which directly connect with the laser operation cycle. Any combinations of these possibilities are available. After completing the appropriate procedures, the program returns to original menu.

The essence of initial line quest is as follows. A number of steps N the driver makes from the center of the line 1P(18) to run of one of the «end-position» detectors was determined experimentally. Seeking for the initial reference line, the step driver moves from some initial position to that of the «end-position» detectors run. Then it changes move direction to the opposite and returns by N steps backward. Here starts, automatically, the procedure of precise angular tuning of the laser cavity. During this procedure the step driver first moves back by several steps and then advances consecutively, by-step, angular position of the grating. The lock system runs with every step. The program fixes information on relative changes in output power in every step. This process completes after a certain amount of steps or in the case of generation vanishing. Hereinafter, the acquired data are being processed, and the step-by-step driver will rotate the grating to the position corresponding to the highest output power at a chosen line.

The described procedure of precise angular tuning can be activated independently (in autonomic regime) by setting the appropriate menu item with subsequent run of the program. In general, it is used for weak-lines laser operation and for measuring generation power.

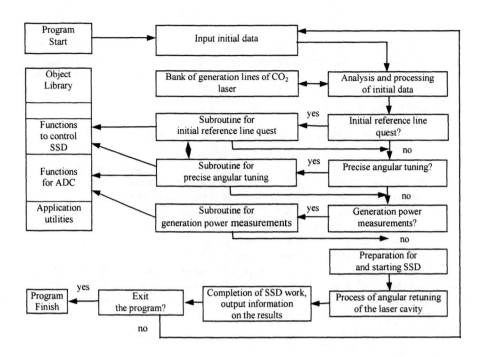

Program Start	Input initial data		
Object Library	Bank of generation lines of CO_2 laser	Analysis and processing of initial data	
Functions to control SSD	Subroutine for initial reference line quest	Initial reference line quest?	yes
Functions for ADC	Subroutine for precise angular tuning	Precise angular tuning?	yes
Application utilities	Subroutine for generation power measurements	Generation power measurements?	yes

(flowchart)

Program Start → Input initial data → Analysis and processing of initial data

Object Library

Bank of generation lines of CO_2 laser

Functions to control SSD

Subroutine for initial reference line quest ← yes ← Initial reference line quest?

no

Functions for ADC

Subroutine for precise angular tuning ← yes ← Precise angular tuning?

no

Application utilities

Subroutine for generation power measurements ← yes ← Generation power measurements?

no

Preparation for and starting SSD

Process of angular retuning of the laser cavity

Completion of SSD work, output information on the results

Exit the program? — yes → Program Finish

no

```
                    < initial  data >
    method of tuning                    by lines
    reference line                      1PS(33)
    working line                        1P(16)
    velocity of tuning                  100
    amount of steps                     3,500
    direction of tuning                 forward
    initial line quest                  no
    precise tuning                      yes
    power measurements                  yes
    program exit                        no

    select item - ↑↓,  set the choice - Enter,  start - Esc

    ─────────────── information ───────────────
    from line 1PS(33) to line1P(16)  -  1,743 steps backward
    covered steps: -1,743              successfully completed

                    press any key
```

Fig. 1   Block-scheme of the program laser and view of a screen at program laser srarting

As was previous case, user also activates power measurement with corresponding settings in the same name menu item and subsequent run of the program. The power is measured by a pyroelectric detector of the MG-30-type incorporated in the receiver of the lock system. Gained by matching amplifier detector's signals are put into computer memory via a 16-bit ADC of the LA-20-type. Upon statistical processing, the data are to be compared with calibration curve measured beforehand by a sample power meter of the IMO-4-type. Proceeding on this comparison, computer gives the absolute value of laser power. Block-diagram of the program LASER is given in fig.1.

5. STUDY OF SPECTRAL AND GENERATION CHARACTERISTICS OF THE LASER IN AUTOMATIC OPERATION

First of all it should be mentioned that generation and spectral characteristics performed in this paper were obtained in automatic operational regime only in accordance with the final objectives of the work. The work involved three sealed-off and one quasi-sealed-off active elements which spectral characteristics have been studied at experimental test-beds in automatic operation regime. The first sealed-off active element was filled with a mixture of the $1CO_2+1.6N_2+0.8Xe+5.4He$ composition under the total pressure 15 Torr and had a discharge gap about 100 cm long and a nontransmitting mirror with 3 m radius. The resonator length in this laser modification was about 1 370 cm. As a dispersing element and output mirror we used a reflection grating which characteristics were given in section 2.

A diaphragm placed between the grating and output mirror (Brewster) would provide for single-mode generation. The distance between the diaphragm and grating was 90 mm. Polarization plane of the Brewster window was adjusted perpendicular to the grating grooves. In automatic operational regime with this active element and resonator in use and for the discharge parameters I = 20 mA, U = 9 kV, we detected generation of 122 lines:

P_4 - P_{56} in 00^01 - 10^00 band (26 lines)
R_2 - R_{50} in 00^01 - 10^00 band (25 lines)
P_4 - P_{52} in 00^01 - 02^00 band (25 lines)
R_4 - R_{48} in 00^01 - 02^00 band (23 lines)
P_{25} - P_{43} in 00^02 - 10^01 band (10 lines)
P_{15}-P_{17}, P_{19}-P_{22}, P_{24}-P_{29} in 01^11 - 11^10 band (13 lines)

Single-mode generation power was a value from 0.1 W to ~4 W depending on the line and band index.

The second sealed-off element with an identical mixture under the total pressure 15 Torr and nontransmitting mirror with a 3 m radius had the 125 cm active length. The resonator was 222 cm long. The diaphragm was placed at 70 cm of the grating. Generation power measurements were taken at each of lines. Block-diagram of measurements is given in fig.2 The results obtained are shown in figs.3-4. Each drawing has power scale in its upper right corner. In automatic operation regime with this active element in use, we had generation of 127 lines:

P_2 - P_{56} in 00^01 - 10^01 band (28 lines)
R_4 - R_{52} in 00^01 - 10^01 band (25 lines)
P_4 - P_{52} in 00^01 - 02^00 band (25 lines)
R_2 - R_{50} in 00^01 - 02^00 band (25 lines)
P_{29} - P_{43} in 00^02 - 10^01 band (8 lines)
P_{12}, P_{14-17},P_{19-22},P_{24-35} in 01^11 -11^10 band (21 lines)

A great amount of hot and small of sequence lines is a specific feature of the given active element having 11.4 % CO_2. This is caused by a rather low vibrational temperature of asymmetric level in CO_2 molecules of the mixture. Single-mode generation power ranged here from 0.4 W to 12 W depending on the line and band index.

The generation and spectral characteristics were also obtained for the third sealed-off active element (see figs. 5-6) filled with a mixture of the $5.4\%CO_2+15.4\%N_2+5.8\%Xe+73.4\%He$ composition under the total pressure 15 Torr and with nontransmitting mirror having a 10 m radius. The resonator was about 250 cm long. The diaphragm was placed at 70 cm of the grating. A great amount of sequence lines (34) and, contrary, small of hot ones (3) is a specific feature of the given active element, which is expectable, in principle, for a mixture with a small CO_2 and high He concentrations and, consequently, with a rather high vibrational temperature of asymmetric level. Total of lines here – 129:

P_4 - P_{50}, P_{56} in 00^01 - 10^00 band (25 lines)
R_2 - R_{46} in 00^01 - 10^00 band (23 lines)
P_4 - P_{46} in 00^01 - 02^00 band (22 lines)
R_2 - R_{44} in 00^01 - 02^00 band (22 lines)
P_{15} - P_{43} in 00^02 - 10^01 band (15 lines)
R_5 - R_{29} in 00^02 - 10^01 band (13 lines)
P_{11} - P_{21} in 00^02 - 02^01 band (6 lines)
P_{19} in 01^11 - 11^10 band (1 line)

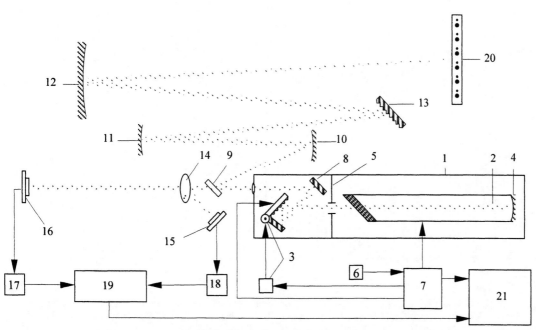

Fig.2. Block-diagram of output power measurements

1 - tunable CO_2 laser; 2 - active element; 3 - diffraction corner reflector on piezocorrector basis; 4 - nontransmitting mirror; 5 - diaphragm; 6 - pyrodetector (MG-30); 7 - power supply, lock, and step driver control units; 8 - rotatable mirror; 9 - NaCl plate covered with dielectric material; 10, 11, 12 - spherical mirrors; 13 - grating; 14 - chopper; 15,16 - pyrodetectors; 17, 18 - amplifiers; 19 - analog-to-digital converter; 20 - thermoscreen; 21 - personal computer.

P_4, P_6 in $01^11 - 03^10$ band (2 lines).

Single-mode generation power was a value from 0.2 W to 15 W depending on the line and band index.

Following the results obtained with these three tubes one can draw an evident conclusion that it is possible to generate one or another set of spectral lines by changing active mixture composition and resonator selectivity. If the resonator is equipped with a reflection grating, it have a resolution that depends on the principal mode sizes on the grating plane. For the third active sealed-off element case with the resonator length 250 cm and nontransmitting mirror of 10 m radius, resolution value is approximately 1.8 times higher than was the case with the second sealed-off element having the 222-cm resonator length and 3-m radius of nontransmitting mirror. Figs. 7, 8 shows the dependencies of the threshold gain factor k_p (m^{-1}) on the detuning $\Delta \nu$ from the autocollimation frequency value obtained for four different wavelengths. It is seen from the comparison between these two pictures that at the same detuning values the case with the resonator length 250 cm and 10-m radius offers the threshold values essentially higher (i.e. higher selectivity) than those shown in fig.8. Therefore, fig.5 demonstrates sequential lines in P-branch of 00^01-10^00 band right up to J=15, and fig.6 also demonstrates them in R-branch of 00^01-10^00 band, which is impossible for the case with the resonator length 222 cm and nontransmitting mirror radius 3 m (see figs. 3, 4). By comparison between figs 3 and 5 we can come to another conclusion that the power values for one and the same sequential lines (P_{29}, P_{31}, P_{33}, for instance) are much higher for a mixture of the $1CO_2+3N_2+1Xe+13.5He$ composition than those for the $1CO_2+1.6N_2+0.8Xe+5.4He$ – mixture. Instead, "hot" lines and traditional lines with high J (P_{52}, P_{54}) are almost fully absent in fig.5. This means that the power depends here on the mixture composition and spectral range. This is why we have tested then the quasi-sealed-off tube that allowed us to use different mixtures in the same active element for lasing needed spectral range. Thus, a mixture of the $1CO_2+1Xe+3N_2+13.5He$ composition, more optimal to generate lines in the second sequence band, was used for 8 < J < 46. Meanwhile, the $1CO_2+1.6N_2+0.8Xe+5.4He$ mixture should be used to generate far lines with small and large J in traditional bands. This tube had the discharge gap 125 cm and

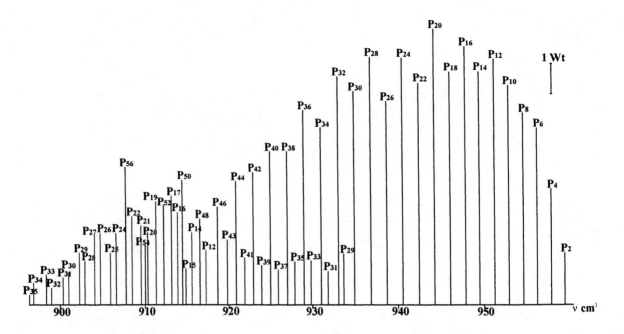

Fig.3 Single-mode power at various lines for 125 cm-discharge-gap sealed-off active element. Nontransmitting mirror radius 300cm. Mixture composition $CO_2:N_2:He:Xe=0.114:0.182:0.614:0.09$. Big lines - P-branch in 00^01-10^00 band, other - lines in P-branch of 01^11-11^10 and 00^02-10^01 bands.

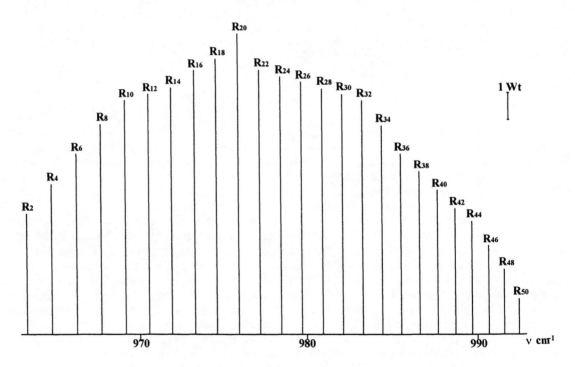

Fig.4. Single-mode power at various lines for 125 cm-discharge-gap sealed-off active element. Radius of nontransmitting mirror 300cm. Mixture composition $CO_2:N_2:He:Xe= 0.114:0.182:0.614:09$. R-branch of 00^01-10^00 band.

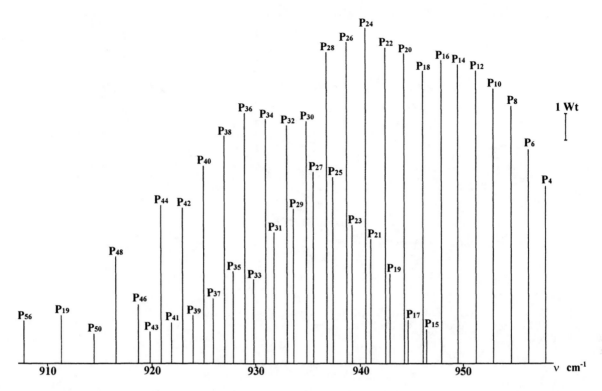

Fig.5 Single-mode power at various lines for the sealed-off active element with discharge gap L=150 cm. Radius of the nontransmitting mirror 1000 cm. Mixture composition $CO_2:N_2:He:Xe=0.054:0.1544:0.7336:0.058$ Big lines - P-branch in $00^01\text{-}10^00$ band; other - P-branch in $01^11\text{-}11^10$ and $00^02\text{-}10^01$ bands.

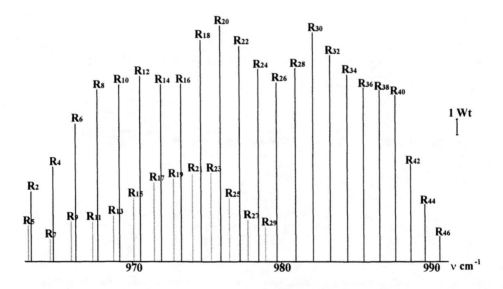

Fig.6. Power values at different lines in single-mode operation obtained for the sealed-off active element with discharge gap L=150 cm, radius of the nontransmitting mirror R=1000 cm, and a mixture of the $CO_2:N_2:He:Xe=0.054:0.1544:0.7336:0.058$ composition. Lines are marked as follows: big- R-branch lines in the $00^01\text{-}10^00$ band; small- R-branch lines in the $00^02\text{-}10^01$ band.

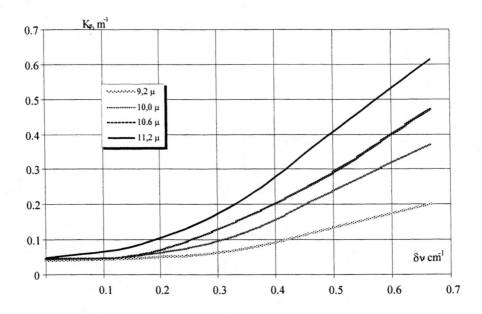

Fig.7   Calculative threshold gain factor k_p vs. detuning Δv relative to autocollimation frequency (resonator length – 250 cm, nontransmitting mirror radius – 10 m)

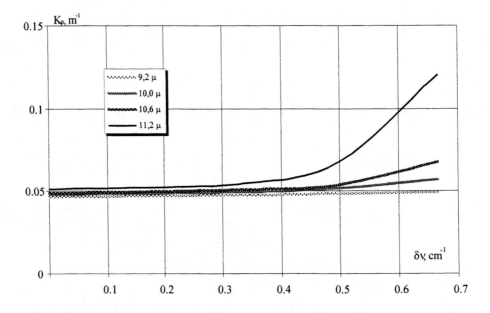

Fig.8   Calculative threshold gain factor k_p vs. detuning Δv relative to autocollimation frequency (resonator length – 222cm, nontransmitting mirror radius – 3 m)

Band $00^02 - 10^01$

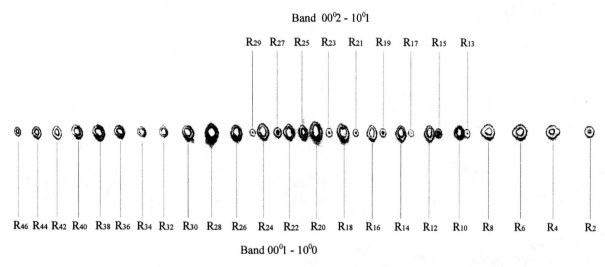

Fig.9   Spectrogram of R-branch lines in $00^01 - 10^00$ and $00^02 - 10^01$ bands

Band $01^11 - 11^10$

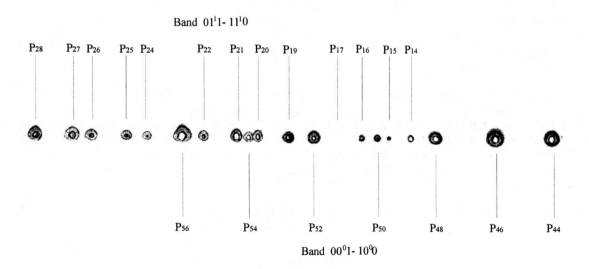

Fig.10 Spectrogram of P-branch lines of bands $00^01 - 10^00$ (j ⩾ 44) and $01^11 - 11^10$

nontransmitting mirror R = 10 m. The resonator length was about 222 cm. The diaphragm was placed at 70 cm of the grating. The gas-discharge tube was cooled to -20^0C with spirit solution. Fig. 9 shows the R-branch spectrum in $00^01 - 10^00$ and $00^02 - 10^01$ bands obtained on a thermoscreen placed in the focal plane of a panoramic specrograph for consecutive passing of the laser from the R_2 line towards R_{46}. Lines in first and second bands are generated consecutively in time starting from R_8. Fig. 10 shows spectrogram of P-branch lines in $00^01 - 10^00$ band ($J \geq 44$) and in "hot" $01^11 - 11^10$ band. In automatic operational regime with this active element we measured generation of 141 lines in four bands, as follows:

$P_4 - P_{56}$ in $00^01 - 10^00$ band (27 lines)
$R_2 - R_{50}$ in $00^01 - 10^00$ band (25 lines)
$P_4 - P_{52}$ in $00^01 - 02^00$ band (25 lines)
$R_2 - R_{50}$ in $00^01 - 02^00$ band (25 lines)
$P_{15} - P_{43}$ in $00^02 - 10^01$ band (15 lines)
$R_{13} - R_{31}$ in $00^02 - 10^01$ band (10 lines)
$P_{14} - P_{17}$, $P_{19} - P_{22}$, $P_{24} - P_{29}$ in $01^11 - 11^10$ band (14 lines).

Generation power ranged from 0.1 W up to 6 W depending on the line and band index.

6. CONCLUSION

Based on the results of our experimental study we have developed an automatic CO_2 laser that can be tuned readily in many diverse of lines in traditional, sequence, and hot bands. The laser cavity design was provided for high selectivity and Q so that its resolution was about 0.3 cm^{-1}. This ensures generation of sequence and hot lines without a hot cell. Total of lased lines was about 150.

These are the major advantages of the laser:
- sufficiently high output power (up to 20 W),
- great number of generation lines (~150) in both traditional and non-traditional (hot and sequence) bands in the 9-12 μm range,
- possibility to vary the generation spectrum by changing active sealed-off elements or mixture composition in an active sealed-off element,
- fully automatic tuning in any required generation line,
- possibility to scan over a predetermined series of generation lines in automatic operation.

Main characteristics of the tunable CO_2 laser:

- total of emission lines   at least 150 in five bands (00^01-10^00, 00^01-02^00, 00^02-10^01, 00^02-02^01, 01^11-11^10)
- generation regime   single-mode
- maximum output power per line   0.1 - 20 W
- frequency instability per line   at most 10^{-4} cm^{-1}
- power instability per line   at most 1%

A whole number of results connected with the work offers unquestionable promise for independent future evolution in many applications. The power supply and nonstandard electronic apparatus, carried out on the modern electronic basis so as to meet the highest technical requirements and fabricated as an universal united module, are capable of making an extensive contribution to researches and developments for many other electric-discharge gaseous lasers (CO-, N_2O-, NH_3- lasers, etc.). The methods provided for apparatus support and the program package for data collection, handling and control after laser parameters released in the course of the development can also be independently and widely used in different scientific areas associated with tunable CW lasers of various types.

ACKNOLEGEMENT
The work was being done under the ISTC Project #216 "Automatic Tunable CO_2 Laser"

REFERENCES
1)   Yu N Bulkin, Yu V Kolobyanin, Yu V Savin, V A Tarasov "Diagnostics of Active Media in Homogeneous and Mixture GDL by Probing CO_2 Lasers Tunable over the Lines in Conventional and Non-conventional Bands" *Quantum Electronics (Mosc.),* **18** 9 (1991)

2)   A A Adamenkov, Yu N Bulkin, V V Buzoverya, Yu V Kilobyanin, E A Kudryashov, V A Tarasov "Dagnostics of the Active Medium of a Waveguide CO_2 Laser by the "Laser Spectrograph" Method" Quantum Electronics **25**(1) 23-26 (1995)

3)   A A Adamenkov, Yu N Bulkin, Yu V Kolobyanin, E A Kudryashov, V A Tarasov, Yu V Savin "Diagnostics of the Active Medium of a Homogeneous CO_2 Gasdynamic Laser by the "Laser Spectrograph" Method". Quantum Electronics **25** (2) 108-110 (1995)

4)   A A Adamenkov, Yu N Bulkin, Yu V Kolobyanin, E A Kudryashov *Optica atmosperi (Moscow)* **8** 4 (1995)10.

5)   Yu N Bulkin, A A Adamenkov, E A Kudryashov, B A Vyskubenko, Yu V Kolobyanin. "Investigation of a Gasdynamic Laser by the "Laser Spectrograph" Method", *Proceedings of the International Conference on Lasers'96*, Editors: V.J. Corcoran&T.A. Goldman, 182-187, STS Press, Portland, 1997

6)   A A Adamenkov, V V Buzoverya, Yu N Bulkin, Yu V Kolobyanin, E A Kudryashov "Diagnostics of the Active Medium of an Electric-Discharge CO_2 Module by the "Laser Spectrograph" Method. *Proceedings of XI International Symposium on Gas Flow and Chemical Lasers and High-Power Laser Conference.* Editor: Howard J.Baker. V. 3092, 265-268, SPIE, Edinburg, UK, 1996

SESSION 2

Active and Adaptive Laser Resonators

Quasi Q-switch regime of CO$_2$ laser generation

Alexis V. Kudryashov, Vadim V. Samarkin, Alexander M. Zabelin*

Russian Academy of Sciences, Institute on Laser and Information Technologies (IPLIT RAN),
Adaptive Optics for Industry and Medicine Group
Dm. Ulyanov 4, bld. 2, apt. 13, Moscow, 117333, Russia,[1]
*Technolaser Ltd., Svyatoozerskaya 1, Shatura, 140700, Russia.

ABSTRACT

The results of the use of two types of adaptive mirrors to generate quasi Q-switch pulses of the CO$_2$ laser radiation are presented. The excess of the output power in peak above the average level in CW regime was more then 2.5 times. The power of the lasers that were used in the experiments was in the range of 1 – 3 kW CW. Also, the design of the different types of the active correctors used in the experiments is discussed.

Keywords: active corrector, intracavity laser beam control, Q-switch pulses.

1. INTRODUCTION

Adaptive optical elements are known to be used for the correction of a various types of wavefront distortions of the light beam penetrated through some inhomogeneous media. But starting from late 70th several types of experiments showing the efficiency of the use of adaptive systems to improve the laser beam quality were carried out[1]. Almost all these experiments were made on the CO$_2$ high power lasers (up to Mwatt) and the obtained results were really promising. Later, in 80th and 90th adaptive technique was applied to control for the radiation of various industrial lasers – YAG, excimer, copper-vapor[6]. But usually authors were trying to influence on the output mode structure of the laser beam by means of the intracavity active mirror. For CO$_2$ laser there is also the necessity and the possibility to use active intracavity mirror to obtain quasi Q-switch regime of generation. The idea of application of active corrector is to change the parameters of laser resonator, the stability factor as well as lazing characteristics. Usually Q-switch pulses in CO$_2$ lasers are obtained by putting mechanical shutter inside laser cavity. To avoid undesirable beam transformations the beam is to be focused inside laser resonator and shutter is to be put exactly in the focal plane[2]. To overcome these problems we suggest to use active deformable mirrors as the modulating element – the bimorph one or/and "cylindrical".

2. ACTIVE MIRRORS FOR LASER BEAM MODULATION.

2.1. Bimorph active corrector.

The traditional semipassive bimorph mirror consists of a glass, copper or quartz substrate firmly glued to a plate actuator disk made from piezoelectric ceramic (lead zirconium titan, PZT) (see fig. 1). Applying the electrical signal to the electrodes of the piezoceramic plate causes, for example, tension of the piezodisc. Glued substrate prevents this tension, and this results in the deformation of the reflective surface. To reproduce different types of aberrations with the help of such corrector usually the outer electrode is divided in several controlling electrodes, that have the shape of a part of a sector. The size as well as the number of such electrodes depends upon the number and the type of the aberrations to be corrected. In our work we usually used the geometry of the electrodes given on fig. 3. The behavior of the bimorph corrector (deformation of the surface when the voltage is applied to the particular electrode) is well described by the following equation[3]:

---

[1] Further author information –
Tel./Fax.: +7 095 1377136, E-MAIL: kud@laser.ru

38

In *Laser Resonators III*, Alexis V. Kudryashov, Alan H. Paxton, Editors,
Proceedings of SPIE Vol. 3930 (2000) ● 0277-786X/00/$15.00

$$D'\nabla^2\nabla^2 W + (\rho_1 h_1 + \rho_2 h_2)\frac{d^2 W}{dt^2} = \frac{d_{31}\nabla^2\widetilde{E}(x,y)E_1(2\Delta_1 h_1 - h_1^2)}{2(1-\nu)}$$

$$D' = \frac{E_2}{1-\nu^2}\left(\frac{\Delta_1^3}{3} + \frac{\Delta_2^3}{3} - \Delta_1^2 h_1 + \Delta_1 h_1^2 - \frac{h_1^3}{3}\right) + \frac{E_1}{1-\nu^2}\left(\Delta_1^2 h_1 - \Delta_1 h_1^2 + \frac{h_1^3}{3}\right)$$

$$\Delta_2 = \frac{E_2 h_2^2 + E_1(h^2 - h_1^2)}{2(E_2 h_2 + E_1 h_1)}; \quad \Delta_1 = h - \Delta_2$$

Here, h_1, h_2 – the thickness of a piezodisk, and substrate, E_1, E_2 – Young's modulus of a piezodisk and substrate, h – total thickness of the mirror, ν - the Poisson ratios, d_{31} – transverse piezo modulus, $\widetilde{E}(x,y)$ - the strength of the electric field applied uniformly to the given electrode. This equation was used to optimize radii r_1 and r_2 (fig. 3) for the best correction of the low order aberrations such as coma, astigmatism, spherical aberration.

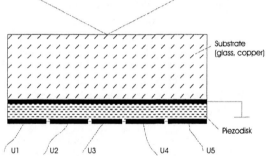

Fig. 1. Scheme of a semipassive bimorph corrector.

For high average power CO_2 lasers there is the problem of constructing controllable cooled mirrors, production of which is rather complicated. A corrector of this kind should satisfy a number of technical requirements: it should have the necessary optical strength, its service life should be long (~ 1000 hours), and it should be easy to construct and use. The mirror surface should be continuously deformable and the amplitude of displacement of the corrector surface should be $\sim\lambda/2$ (λ=10.6 μ - is the wavelength of the corrected radiation).

In our Group we have developed water cooled mirrors based on semi-passive bimorph piezoelement. They consisted of a copper (or molybdenum) plate 2.5 mm thick and 100 mm in diameter. One side of the plate was polished and used as a mirror, whereas two piezoelectric ceramic disks 0.3 mm thick and 50 and 46 mm in diameter were glued on the other side (fig. 2). First piezodisk was used to control for the curvature of the mirror surface and 17 electrodes were evaporated on the outer side of the second piezodisk to compensate for different aberrations of the wavefront. The cooling system of the corrector was of the waffle type. A copper plate consisted of two solded disks in which channels of 0.5 mm deep were formed for the circulation of the cooling liquid. The size of the contact areas between the plates was 3x3 mm. Fig. 3 shows the sample of such a corrector.

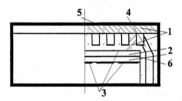

Fig. 2. Copper water cooled bimorph mirror. 1 - copper substrate; 2 - piezoceramic disks; 3 - control electrodes; 4 - canals for cooling water; 5 - reflecting surface; 6 - common electrode.

The static and dynamic characteristics of the mirror were determined by an interference method. The sensitivity of correctors was estimated from the displacement of the interference fringes at the center of the pattern when the voltage of 100 V was applied to all electrodes. It amounted to 1,7 μ.

The frequency of first resonance of our correctors was in the range of 3 - 4 kHz. An active mirror was tested under an optical load of CO_2 laser radiation with an average power density of 2.5 kW/cm^2 [4]. The corrector surface profile was determined using a shearing interferometer. This optical load produced practically no deformation of the mirror surface, indicating that the cooling system was effective. The rate of flow of cooling water was 400 mliter/min. There were no distortions of corrector response function under the action of this load. A similar test of an uncooled adaptive mirror resulted in considerable thermal deformation of the surface.

Fig. 3. A photo of a sample of a water cooled bimorph corrector

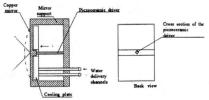

Fig. 4. Deformable "cylindrical" mirror.

These bimorph correctors were successfully used as an intracvity mirrors to control for the radiation of a CO_2.

2.1. "Cylindrical" active corrector.

Another type of the active mirror to be installed as one of the mirrors of the laser resonator was suggested to change the wavefront of the laser beam only in one direction by means of single actuator. The design of such mirror is given on fig. 4. The mirror itself is the rectangular copper plate. One side of this plate is polished to the optical quality. Piezoceramic column (actuator) is installed between the rib of rigidity (shown on fig. 4) and mirror support. The cooling system made of copper plate surrounded by the tube with cooling water is connected to deformable mirror through special thermo conductive pasta. Of course the deformation of the surface via applied voltage to piezoceramic column is not the spherical one, but has the shape close to cone in one direction. This is the main shortcoming of such a corrector. The amplitude of the deformation of the surface is 40 μ, for the aperture of the mirror 55 mm (long side of the rectangular). Thickness of the reflecting plate of such corrector – 4 mm. First resonance frequency – 15 kHz. Such a corrector was designed to be used in the scheme of the stable-unstable resonator to produce q-switch laser pulses.

3. RECEIVING THE Q-SWITCH REGIME OF CO_2 LASER GENERATION.

3.1. The use of the bimorph corrector.

When one needs to change some output parameters of the laser beam (for example get different output mode configuration or even change the regime of laser generation) you need to reconstruct all laser cavity or alter the block of power supply. Such operations are rather expensive and sometimes takes a long period of time. That is why the use of mirrors with controllable flexible surface in laser is of large interest. As for CO_2 lasers there is another interesting and promising field for active correctors – to obtain the Q-switch regime of generation of CW lasers. The problem here is that it is impossible to suggest some media that will change its transparency parameter on the wavelength of the CO_2 laser radiation (10.6 μ) and thus obtain Q-switch regime. So, the possible way of getting this regime is to use intracvity active corrector that will change the parameters of laser resonator (g_1g_2 parameter). In this case the time of change of the cavity should be about one order less then the time of the excited state of CO_2 gas (5 – 1.5 ms).

40

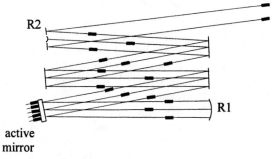

R2

active
mirror

R1

Fig. 5. Scheme of CO_2-laser resonator.

To understand the possibility of obtaining such a Q-switch regime of generation of industrial CW CO_2 we used Russian laser TL-5 developed and manufactured in Scientific Research Center for Technological Lasers Russian Academy of Sciences[8]. This laser is characterized by high beam quality (divergence of less then 1 mrad at 55 mm output beam diameter) and output power 5kW. Fig. 5 shows the scheme of the laser resonator. The unstable confocal resonator with magnification M=2 consists of two spherical mirrors with radii of curvature R_1 =-13 m and R_2 = 26 m and 4 flat folding mirrors between them. Resonator length was L=6.5 m. One of the flat mirrors was replaced by the cooled flexible mirror based on bimorph piezoelement. The output beam of this laser had the shape of a ring with outside aperture of 55 mm and inside aperture - 25 mm. Well known parameters of this laser cavity g_1 and g_2, taking into a consideration deformable mirror could be written in the following way:

$$g_1 = 1 - \frac{L}{R_1}, \quad g_2 = 1 - L\left(\frac{2}{R} + \frac{1}{R_2}\right)$$

where R is the radius of curvature of flexible mirror under control voltage, L=6.5 m - the length of laser cavity. For specified resonator parameter (with 1/R=0) we have g_1 =1.5 and g_2 =0.75.

When a static voltage +300 V was applied to all electrodes of the flexible mirror radius of curvature of its surface R changed from plus to minus 200 m. The value g_2 was changing correspondingly in the range of 0.69-0.82. In the dynamic regime of mirror action the cosine voltage with frequency in the range of 1 - 10 kHz was applied to mirror electrodes. At the mirror resonance frequency - 3.8 kHz the deformation of mirror increased up to R= +50 m and g_2 value was changed from 0.49 to 1.0. The magnification M of unstable non-confocal resonator could be given as :

$$M = \frac{g_1 g_2 + \sqrt{g_1 g_2 (g_1 g_2 - 1)}}{g_1 g_2 - \sqrt{g_1 g_2 (g_1 g_2 - 1)}}$$

Fig.6 shows the dependence of resonator transmittance T= 1-1/M^2 upon g_2 for g_1 =1.5. One can see that when g_2 changes from g_2 <0.67 (the case of stable resonator) to 1.0 the transmittance T varies from 0 and 0.93.

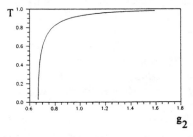

Fig. 6. Transmittance T of laser cavity vs g_2 in case g_1 =1.5.

So, changing the curvature of the flexible mirror surface by applying cosine voltage to the control electrodes one could change the transmittance of the telescopic unstable resonator with the same frequency and modulation of the laser output beam power could be obtained. The cosine voltage of frequency up to 10 kHz and amplitude up to 300V was applied for this purpose. Under control voltage of more than 1 kHz the period of the mirror surface oscillation could be compared with the life-time of the CO_2 laser active medium. Thus the Q-switch regime of laser generation could be obtained.

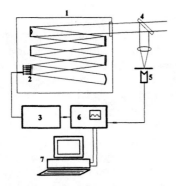

Fig. 7. Experimental setup. 1 - CW TL-5 laser; 2 - active mirror; 3 - generator; 4 - beam splitter; 5 - power meter; 6 - oscilloscope; 7 - IBM computer.

An experimental setup is presented on fig. 7 [5,6]. For detecting the change of output power of laser radiation a small part of the beam was directed by KCl beam splitter 4 to the power meter 5 with 100 ms response time. The signal from such detector came to the double channel digital oscilloscope 6 which digitized the amplitude of signal. The control signal from generator 3 was fed to the second oscilloscope channel. Data from oscilloscope was stored in the computer 7. The dependence between output beam power and control voltage amplitude and frequency was studied. The periodical laser generation regime with 100% power modulation depth was obtained at near-resonance frequencies - 3.8 and 7.6 kHz. The pulse peak power exceeded the average power threefold. The average power drop was insignificant (<10%). Fig. 8a, 8b demonstrates output beam pulses and control voltage pulses applied to deformable mirror at resonance frequencies. Here one can see that even 120 V at 3.8 kHz and 80 V at 7.6 kHz mirror control voltage is enough to get 100% power modulation.

Together with the power modulation the aperture of the output laser beam also changed from 35 to 65 mm at the plane where the convex resonator mirror was placed. At the same time changing the curvature of active mirror we managed to vary the beam convergence angle of industrial laser TL-5 in the range of 2 mrad. Such change of beam convergence by flexible mirror in resonator can be used in material processing for dynamic focusing of laser radiation. In different technological processes such as deep penetration welding or thick material processing the displacement of the laser focal spot along the beam axis is needed to concentrate energy in the developing area. This focal spot displacement usually was realized by moving the focusing lens along the laser beam axis or using extracavity active mirror[7]. The technique of change the beam convergence angle using flexible mirror within the resonator is simple and provides the focal spot displacement in a wide range with low deformation of the mirror surface.

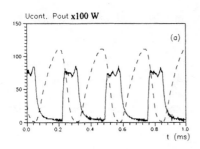

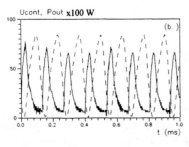

Fig. 8 Oscillograms of output beam power P_{out} and control signal U_{cont} at frequency 3.8 (a) and 7.6 (b) kHz.

———————— P_{out} x100, W

- - - - - - - - - U_{cont} , V

Thus the possibility to obtain the pulse-periodical generation and Q-switch regime of continuous pumped industrial laser was demonstrated. One of the main shortcoming of the proposed method of obtaining the Q-switch pulses is that the output radiation changes not only its power but also the parameters of the output radiation change: the size and curvature radius of the output beam are modified – due to the change of the resonator output phase of the beam is either plane (+80 Volts on the mirror) or spherical one (no voltage on the mirrors electrodes). And of course this is not very convenient to work with such beams and use them in the technological process.

The use of the intracavity active mirrors do not require any significant modification of the whole laser. At the same time it broaden the possible spheres of technological lasers application and can improve different technological processes. The proposed method is characterized by simplicity, does not involve replacement of laser power supply and enables to obtain the pulse peak power higher than CW power level.

3.2. THE USE OF THE CYLINDRICAL CORRECTOR.

In our first experiments with bimorph corrector described above we used Russian TL-5 laser. The measurement of this laser focused beam profile showed that along with the narrow kern, there exists rather high energy-intensive environment (pedestal)

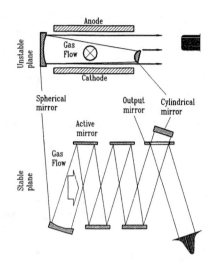
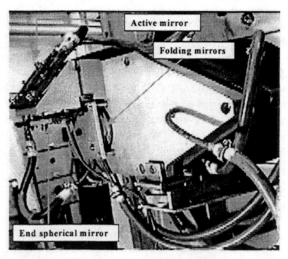

Fig. 9. Stable-unstable resonator with unilateral beam output. Modified scheme.

that contains more than a half of the total beam power. The total divergence of the beam is approaching to 1 mrad. The beam quality parameter of such laser beam is rather poor (~ 4).

This is why we have proposed and used a special scheme of the optical resonator [8,9]. It has the properties of an unstable telescopic resonator only in the plane perpendicular to the electrode walls, and it is stable in the plane parallel to these walls (Fig.8), as opposed to the known scheme of the stable - unstable resonator wherein the plane of the unstable resonator is parallel to the electrodes, and the plane of the stable resonator is perpendicular to the gas flow and electrodes.

The resonator includes two end mirrors, one of them is spherical and another is cylindrical .The generatix of the cylindrical mirror is parallel to the electrode walls. This resonator allowed to increase the far-field beam intensity by a factor of 2-3 in comparison with an ordinary telescopic resonator. One of the very important advantages of the «stable-unstable» resonator is a good space coupling of the resonator and GDC volumes. The geometry of the optical folding of this resonator takes into consideration the optimal utilization of gas discharge zones, being not included into the volume of the laser optical field. This became possible due to the flow of the excited gas [10].

Only the lowest gaussian mode of the stable resonator was selected by diaphragm installed inside the cavity. The principal mode size in our case was approximately 12 mm by $1/e^2$ level. Near-field beam intensity profiles of the stable-unstable resonator are presented on Fig. 10. The diaphragm aperture for one-mode generation is equal to 20 mm [11].

Fig. 10. Near-field beam intensity profile of the stable-unstable resonator. One mode generation.

The unstable resonator axis was located close to one of the electrode, the resonator caustic accounted for 50 mm, and the output beam size was 20 mm. As the practical applications require the compact beam, so we have to use the resonator with M=2 magnification. The output power of the laser employing the modified stable-unstable resonator was achieved to be 6 kW, i.e. no marked decrease in efficiency was practically observed.

As it is shown on Fig. 9, we substituted one of the folding mirrors with the cylindrical corrector. Applying sine voltage to the actuator at the frequency 2 kHz we obtained the modulation of the output radiation and received quasi Q-switch regime of generation – for the CW power 2 kW output intensity of the beam was 5 kW. The duration of the pulses was 5 μc.

The quality of the output beam, beam size and beam phase in the case of this kind of laser resonator and the use of active cylindrical corrector does not change because the modulation of the cavity parameters takes place only in the stable plane: in stable resonator the wavefront of the output beam coincides with the surface of the output mirror. And this surface in our experiments was constant. Some changes of the size of the beam in the stable plane inside cavity leads to the modulation of the intracavity losses but does not lead to the increase of the size of the output beam. In the unstable plane of the resonator we did not change any parameters of resonator and so, no deformations of laser beam appear.

4. CONCLUSION

This paper proved the possibility of obtaining Q-switch regime of CO_2 laser generation by means of intracavity adaptive mirrors. The advantage of the use of such mirrors is that they can be easily installed in laser camera without any changes of the laser resonator design.

ACKNOWLEDGMENTS

The authors would like to thank their colleagues A.Korotchenko and P.Romanov for their assistance during the joint research work and preparation of the experiment. This work was supported under the IPLIT-DERA Grant of Contract ELM1158 and NATO SfP Grant N974116.

REFERENCES

1. R.R.Stephens, R.C.Lind, "Experimental study of an adaptive-laser resonator", *Opt. Lett.* **3**, 79-80 (1978).
2. J.Bae, T.Nozokido, H.Shirai et. all "High peak power and high repetition rate characteristics in a current-pulsed Q-switched CO_2 laser with a mechanical shutter", *IEEE Journal of Quantum Electronics* **30**, N4, 887-892 (1994).
3. A.V.Kudryashov, V.I.Shaml'hausen, "Semipassive bimorph flexible mirrors for atmospheric adaptive optics applications", *Opt. Eng.* **35**(11), 3064-3073 (1996).
4. M.A.Vorontsov, G.M.Izakson, A.V.Kudryashov, G.A.Kosheleva, S.I.Nazarkin, Yu.F.Suslov, and V.I.Shmalgauzen, "Adaptive cooled mirror for the resonator of an industrial laser", *Sov. J. of Quantum Electron.* **15**, 888 (1985).
5. A.V.Kudryashov, V.V.Samarkin, "Control of high power CO2 laser beam by adaptive optical elements", *Opt. Comm.* **118**, 317-322 (1995).
6. *Laser Resonators: novel design and development.* Alexis Kudryashov, Horst Weber, editors, SPIE Press, 301 p. (1999).
7. H.Haferkamp et. all "Beam delivery using adaptive optics for material processing applications with high power CO_2 lasers," Paper presented at the Int. Symp. On Optical Tools for Manufacturing and Advanced Automation, 1993.
8. G.A.Abilsiitov, A.I.Bondarenko, V.S.Golubev. «Industrial lasers of the Research Center on Technological Lasers», *Kvantovaya Elektronika*, Vol. 17, pp. 672-677.
9. A.Borghese, R. Canevari, V. Donati and L. Garifo, « Unstable-stable resonators with toroidal mirrors», *Applied Optics*, **20**(20), pp.3547-3552 (1981).
10. A.M.Zabelin, "Fast-transverse flow laser with stable- unstable resonator", Patent RU #2092947 Int. CL HO1S 308.
11. A.M.Zabelin, A.V.Korotchenko, "CO_2 laser resonator: some new aspects in the design" in *Laser Resonators II*, Alexis Kudryashov, Editor, Proc. SPIE Vol. 3611, 317-322 (1999).

Stable resonators with graded-phase (aspherical) mirrors

Alan H. Paxton and William P. Latham

Air Force Research Laboratory/DEOB, Kirtland AFB, NM 87117-5776

ABSTRACT

Aspherical (graded-phase) mirrors [1-5] are potentially useful in stable resonators to obtain a desired intensity distribution at the output of a low-loss or moderate-loss resonator. The mode volume for single-mode operation can also be increased over that which can be obtained using a stable resonator. Much of the work on this concept to date has concentrated on resonators that have a super-gaussian mode at the output mirror. Here we explore some concepts involving other amplitude distributions that may be used to further enlarge the mode volume, although the output distribution may not be as desirable as the super gaussian.

1. INTRODUCTION

We will review the well-known concept [1-5] for the stable resonator with an aspherical mirror. The concept is most simply described in terms of a two-mirror resonator. For simplicity, let us consider the resonator to have a planar partially reflective mirror with constant reflectivity, through which the outcoupling occurs. Opposite this is a highly reflective nonplanar mirror. In a stable resonator, apertures can play a secondary role, and we can ignore perturbation of the mode due to aperture losses at the two ends of the resonator. In this approximation, we will show that the nonplanar mirror must transform the mode incident on it into its conjugate wave.

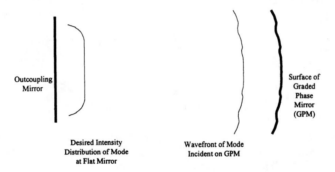

Figure 1. Sketch showing concept of stable resonator with graded phase mirror. The aspherical mirror conjugates the phase of the field incident on it.

The amplitude of a resonator mode reproduces itself upon a round-trip propagation. The complex mode amplitude, $u(r, z_p)$, at the planar mirror is transformed by propagation to $u(r, z_a)$ at the nonplanar mirror. Upon reflection from the nonplanar mirror, $u(r, z_a)$ is transformed to $u(r, z_r)$. Because we are considering a resonator mode, so propagation of $u(r, z_r)$ to the planar mirror and reflection transforms the field to $\eta u(r, z_p)$ where η is the resonator eigenvalue. Reflection from the planar mirror corresponds to multiplying the field by a single complex constant, propagation of $u(r, z_r)$ to the planar mirror without reflection results in amplitude $\rho u(r, z_p)$, where ρ is a complex constant. Therefore, the propagation of $u(r, z_r)$ from the nonplanar mirror to the planar mirror must be the time-reverse of the propagation of $u(r, z_p)$ from the planar mirror to the nonplanar mirror. For this to occur, $u(r, z_r)$ must be the complex conjugate of $u(r, z_a)$ multiplied by a constant that is independent of r. For a laguerre-gaussian or hermite-gaussian mode, which has parabolic phase (with π phase jumps for nonzero order modes), a spherical mirror transforms the mode to its conjugate wave at the nonplanar mirror. A mode with any other field amplitude function at the planar mirror requires the nonplanar mirror to have an aspherical surface.

A wave with phase $\phi(r)$ is conjugated by a mirror with surface profile

$$z(r) = -\lambda \phi(r, z_a)/2\pi + z_0, \tag{1}$$

In *Laser Resonators III*, Alexis V. Kudryashov, Alan H. Paxton, Editors,
Proceedings of SPIE Vol. 3930 (2000) • 0277-786X/00/$15.00

45

where z_0 is a small arbitrary constant and we use the convention

$$E = u(r, z)e^{i(kz-\omega t)} \qquad (2)$$

for the wave propagating in the positive z direction (from the planar mirror to the nonplanar one).

The prescription for designing a resonator is: (1) select the desired complex amplitude of the field at the planar mirror; (2) calculate the field at the position of the nonplanar mirror; (3) obtain the shape of the mirror surface according to Eq. 1. A sketch of a resonator with a graded-phase mirror is shown in Fig. 2. The resonator has apertures near each of the two mirrors.

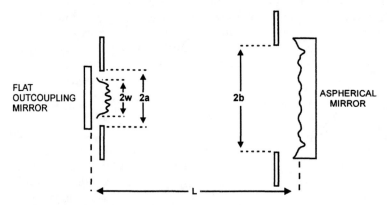

Figure 2. Sketch of resonator, showing field intensity near flat.

2. MODE AMPLITUDE FUNCTIONS

Much of the past work on this type of resonator has been focused the creation of modes that are super gaussian at the planar mirror. The mode shape is

$$u = e^{-(r/w_s)^n}, \qquad (3)$$

Where n is the order of the super-gaussian. For a Gaussian mode, n=2. A super-gaussian mode is desirable for some purposes because its intensity is more uniform than a Gaussian beam. A resonator for a super-gaussian mode with n=6 was studied theoretically and experimentally with a CO_2 gain medium (wavelength $\lambda = 10.6\mu$m) by Bélanger, Lachance, and Paré (Ref's 2 and 3). The super-gaussian beam radius was $w_s = 0.330$ cm and the resonator length was $L = 200$ cm. The mode intensity at the graded phase mirror (GPM) is shown in Fig. 3. The distribution peaks in the center and has a broad skirt. Our calculations showed that an aperture width of $b = 2.0$ cm led to the reproduction of an $n = 6$ super gaussian after a round-trip propagation, to a very good approximation. However, a value of $b = 1.2$ cm resulted in a 13% dip in the center of the intensity distribution. A discussion of aperture sizes necessary to obtain nearly the super-gaussian amplitude at the flat mirror is contained in Ref. 4.

Evidently, if the gain region is cylindrical and extends the entire length of the resonator, its radius must be much greater than the radius of the supergaussian beam. The gain medium may be appreciably shorter than the resonator, in which case the radius of the gain medium need not be much greater than the width of the supergaussian. Examples of very short gain media are flowing dye lasers and solid-state disk lasers. To generate a super-gaussian mode, the resonator must have a length that is about the same as the rayleigh range for the equivalent gaussian beam waist, w_0. For $n = 6$, $w_0 = .79w_z$ (see Ref. 3). This may still lead to a resonator that is of impractical length.

The wavefront (surface of constant phase) is also shown in Fig. 3. A parabolic fit that was derived according to the prescription of Ref. 4 is included for comparison. The maximum separation of the two curves is about 0.07λ.

Here, we wish to select an optical beam with planar phase at the outcoupling mirror. We want the beam to have significant phase perturbations at a relatively short propagation distance. The initial beam must have increased high spatial frequency content compared with a gaussian beam with a comparable width. As an example, we selected a beam amplitude

$$u(r, z_p) = e^{-(r/w_s)^n}[3 + \cos(2\pi r/\Lambda)], \qquad (4)$$

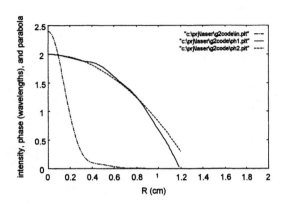

Figure 3. Intensity and wavefront of super-gaussian mode. Also shown is a parabola for comparison with the wavefront.

where $n = 6$, $w_s = 1.0$ cm, and $\Lambda = 0.183$ cm. We used wavelength $\lambda = 10.6 \mu m$ for easy comparison of our results with those of Bélanger et al. The intensity corresponding to the selected beam at the planar mirror is plotted in Fig. 4. We applied the Fox-Li iterative method to calcualte the resonator eigenvalue, *gamma* and the mode amplitude at the two mirrors. For purposes of the calculations, the reflectivities of the mirrors was set to 1, so the loss was entirely due to clipping at apertures. The calculational grid contained 512 points over a grid of width b, for all cases. The aperture radius at the planar mirror was $a = 1.1$ cm, the resonator length was $L = 200$ cm, and the aperture at the aspherical mirror was $b = 4.0$ cm, for one of the cases. The intensity and phase after a single propagation to the aspherical mirror are shown in Fig. 5. The resonator mode intensity at the planar mirror is shown in Fig. 6.

Table 1. Calculated values of $|\gamma|^2$ for resonators of various lengths.

L(cm)	Λ(cm)	$|\gamma|^2$, fundamental	$|\gamma|^2, l \neq 0$	l
200	4.0	0.9993	0.973	2
100	3.0	0.9995	0.966	1
30,40,50,60			$> .990$	> 0

Results of calculations for various resonator lengths are shown in Table 1. Values of $|\gamma|^2$ are given for the fundamental mode and for the lowest loss mode with azimuthal mode index $l \neq 0$. We did not calculate the eigenvalue for the second-to-lowest loss mode with $l = 0$.

The mode discrimination shown in Table 1 may be adequate for some applications.

A method that is similar to the one we have illustrated here was reported by Leger et al. in Ref. 6. The authors of that work, however, used a diffraction grating to introduce the high spatial frequency components. The gain medium was located between the planar mirror and the diffraction grating, and it was unnecessary to select a mode with intensity ripples at the planar mirror.

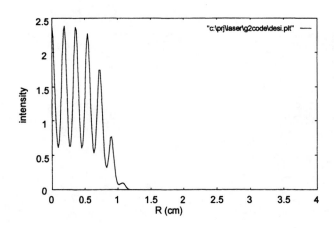

Figure 4. Intensity corresponding to the the beam with amplitude given by Eq. 4.

3. CONCLUDING REMARKS

The separation between the surface of the aspherical mirror and a nearly matching spherical or planar surface is less than 0.1 wavelengths for our example, as well as for the super-gaussian of Bélanger et al. Therefore, the operation of our example of a stable resonator with an aspherical mirror would be sensitive to phase aberrations of a similar magnitude.

References

1. P.-A. Bélanger and C. Paré, Optical Resonators Using Graded-Phase Mirrors, Opt. Lett., V. 16, pp. 1057-1059, 1991.
2. P.-A. Bélanger, R. L. Lachance, and C. Paré, Super-gaussian output from a CO_2 laser using a graded-phase mirror resonator, Opt. Lett. V. 17, pp. 739-742, 1992.
3. C. Paré and P.-A. Bélanger, Custom Laser Resonators Using Graded-Phase Mirrors, IEEE J. Quantum Electron. V. 28, pp. 355-362, 1992.
4. C. Paré and P.-A. Bélanger, Custom Laser Resonators Using Graded-Phase Mirrors, IEEE J. Quantum Electron. V. 30, pp. 1141-1148, 1994.
5. J. R. Leger, Diana Chen, and Kevin Dai, High modal discrimination in a Nd:YAG laser resonator with internal phase gratings, Opt. Lett., V. 19, pp. 1976-1978, 1994.

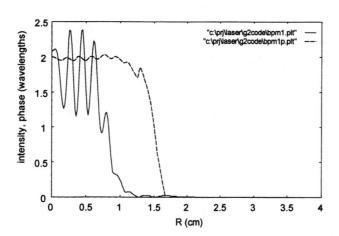

Figure 5. The intensity and phase at the aspherical mirror resulting from propagation of the field given by Eq. 4 to the aspherical mirror location.

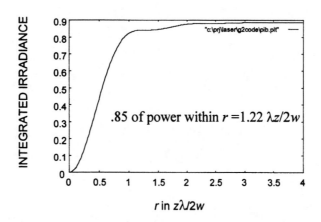

Figure 6. Integrated irradiance of the far-field pattern of the complex field amplitude of the resonator mode at the flat mirror, given by Eq. 4.

SESSION 3

Resonators Including Nonstandard Optical Elements

Lineselective Resonators with Variable Reflectivity Gratings (VRG) for Slab-Laser Geometry

Ralf Hocke[a] and Martin Collischon[b]

[a] Lehrstuhl fuer Hochfrequenztechnik,
Universitaet Erlangen Nuernberg, Cauerstr. 9, 91058 Erlangen, Germany
[b] Lehrstuhl fuer Optik, Physikalisches Institut Universitaet Erlangen Nuernberg,
Staudtstr. 7/B2, 91058 Erlangen, Germany

ABSTRACT

A diffusion cooled lineselective CO_2 laser with a variable reflectivity grating is presented in this paper. The lineselective CO_2 laser acts as a pump source for far infrared lasers. We use an rf-excited slab laser geometry to achieve a compact laser design due to the area power scaling in contrast to the conventional longitudinal DC excited lasers with length power scaling. The huge Fresnel number of a slab system normally leads to a higher order mode operation of the laser and uncontrolled transversal mode-hops. In order to reduce this effect, so called Variable Reflectivity Mirrors (VRM) were investigated for stable and unstable resonators in the past. Instead of a VRM we use a modified binary Littrow grating to achieve the same effect in a lineselective resonator setup. The reflectivity of a binary Littrow grating with constant depth and grating period depends on the width of the grating bars. The grating splits the incident power into the -1st and 0th diffraction order. The splitting ratio depends on the duty cycle of the grating. The -1st diffraction order is reflected back into the resonator whereas the 0th diffraction order is used to couple out the laser beam. Therefore the grating acts as lineselective, outcoupling and, due to the nature of VRM resonators, as a modeselective element as well. The gratings are realized by a microgalvanic process on copper substrates. Results of different resonator concepts (stable and unstable) with planar and convex gratings are presented.

Keywords: Lineselective Resonator, Diffusion Cooled CO_2 Slab Laser, Variable Reflectivity Grating, Binary Littrow grating, Apodization

1. INTRODUCTION

In recent years remote sensing of stratospheric OH concentration became more and more important for understanding the ozone cycle. Usually this is done with a heterodyne receiving system of 2.5 THz center frequency. The local oscillator, represented by a methanol laser, is pumped by a lineselective highly stable CO_2 laser[1].

These classical lineselective diffusion cooled CO_2 lasers usually are bulky and space-consuming systems. A more compact laser design, the so-called slab laser, was investigated in the last few years for high power generation. For this type of laser, the output power is area scaled and not length scaled as it is for cylindrical lasers[2]. An RF-excited lineselective slab laser is presented in Fig.1. The distance between the two electrodes is typically about 2 mm, whereas the width is about 20 mm. The typical length is in the range between 0.5 and 1 m.

These dimensions lead to waveguide propagation conditions in y-direction and free-space conditions in x-direction. As a result of the huge Fresnel number in x-direction the laser operates in higher order modes for optical stable resonator systems. Several different resonator concepts have been investigated in the last few years to suppress this effect[3-10].

In addition, it has to be taken into account that we have to use a Littrow grating in the resonator as lineselective element. In our resonator this element is utilized as lineselective and outcoupling element.

Further author information: (Send correspondence to Ralf Hocke)
Ralf Hocke: E-mail: ralf@lhft.e-technik.uni-erlangen.de
Martin Collischon: E-mail: martin.collischon@leister.com

In *Laser Resonators III*, Alexis V. Kudryashov, Alan H. Paxton, Editors,
Proceedings of SPIE Vol. 3930 (2000) ● 0277-786X/00/$15.00

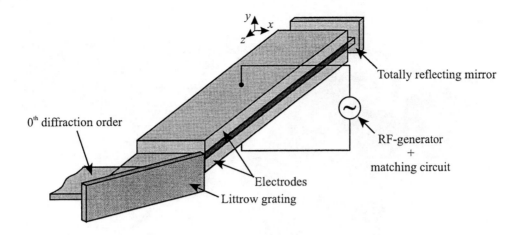

Figure 1: RF-excited lineselective slab laser.

2. BINARY LITTROW GRATINGS

For lineselective resonators of CO_2 lasers typically gratings with about 150 lines/mm are used. This means that the grating period p is about 7 μm. In this case and for an angle α of the incident wave of about 45° only two diffraction orders appear. This is shown in Fig. 2. The incident plane wave splits into two plane waves, i.e. the 0th and -1st diffraction order. The portions of power of these two diffraction orders are called the reflectivities R_0 and R_{-1} :

$$R_0 = \frac{P_0}{P_{IN}} \qquad\qquad R_{-1} = \frac{P_{-1}}{P_{IN}} \qquad\qquad (1)$$

The special case that the angle α holds :

$$\sin(\alpha) = \sin(\alpha_L) = \frac{\lambda}{2p} \qquad\qquad (2)$$

is called the Littrow configuration of the grating. For the given wavelength λ of the plane wave and the given grating period p the direction of the -1st diffraction order is opposite to the direction of the incident wave. The angle α is also called the Littrow angle α_L.

In resonators the -1st diffraction order is used as lineselective feedback into the resonator. For normal resonator configurations the 0th diffraction order is not utilized unlike in our resonator where it is used to couple out the laser beam. Additionally, our grating has a third function which will be explained in the following.

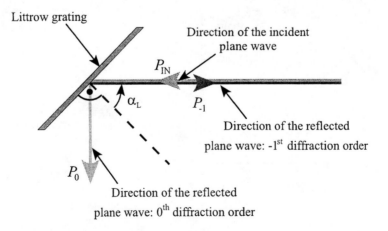

Figure 2: Littrow configuration of a diffraction grating.

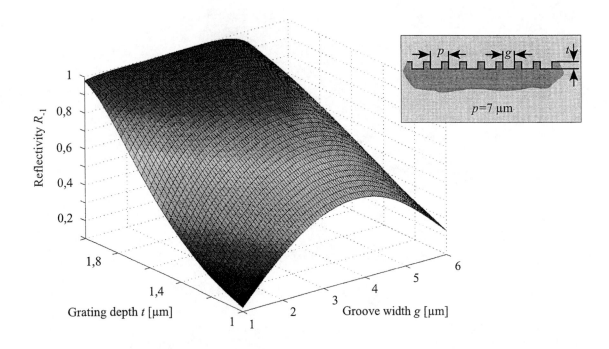

Figure 3: Diffraction efficiency of a binary Littrow grating.

The reflectivities R_{-1} and R_0 depend on the geometry of the grating surface. Different grating structures like the blazed grating or binary grating are known. In the following we concentrate on binary gratings as shown in Fig. 3. For theoretical calculations of these binary gratings we used the modal method[11], with the assumption that our substrate has an infinite conductivity. This means that the sum of R_0 and R_{-1} has to be 1 :

$$R_0 + R_{-1} = 1 \tag{3}$$

For a given grating period p the reflectivities are a function of the grating depth and the groove width, which is shown in Fig. 3. For a constant grating depth t of 1.4 μm and a grating period of 7 μm the reflectivities R_0 and R_{-1} are presented in Fig. 4a. The parameter groove width g allows us to realize a grating with areas of different

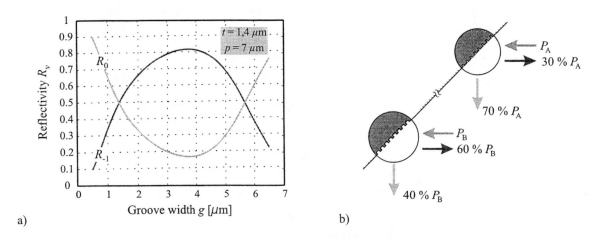

a) b)

Figure 4: a) Reflectivities of a binary Littrow grating of constant grating depth and period.
b) Binary Littrow grating with areas of different groove width.

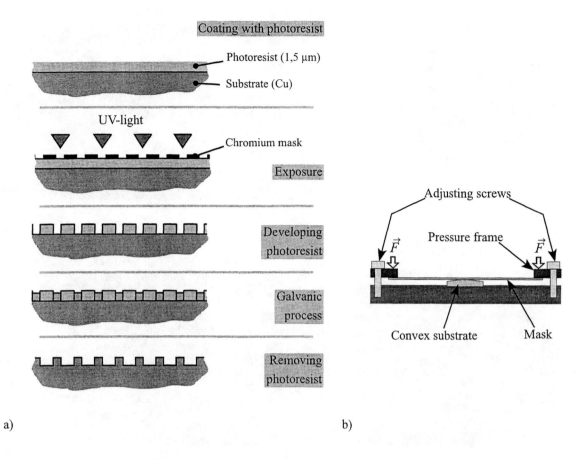

a)   b)

Figure 5: a) Photolithographic microgalvanic manufacturing process.
b) Mechanical setup for exposing convex substrates.

reflectivities as shown in Fig. 4b. The maximum reflectivity R_{-1} of about 83 % guarantees a sufficient feedback into the resonator. To increase this feedback a depth greater than 1.4 μm has to be chosen. Due to the photolithographic process the smallest structure to be realized was about 1 μm, which means that for $t = 1.4$ μm reflectivities R_{-1} between 30 and 83 % are possible.

The limitations of possible grating geometries with a constant grating depth t is due to the manufacturing process of the gratings, which is shown in Fig. 5a. We used a photolithographic microgalvanic process to realize the gratings. First, planar or convex copper mirrors were coated with photoresist and then exposed by UV-light through a chromium mask. The chromium mask itself was written with an e-beam-lithography. After developing the photoresist a galvanic process was used to fill the grooves between the resist with copper. At the end we removed the photoresist and we got the final copper grating.

For the purpose of exposing convex substrates we used the mechanical setup shown in Fig.5b. With the pressure frame and the adjusting screws we wrapped the mask round the surface of the convex substrate. This setup guarantees that we had no projection errors during the exposing process, which means that the unrolled grating has nearly a constant grating period.

Fig.6 presents some results of our gratings. Fig.6a shows the grating structure on a copper substrate with a diameter of 38 mm and a thickness of 8 mm. The area of the grating is about 14x10 mm^2. Fig.6b presents a SEM picture of the grating cross section. For this picture we covered the grating with nickel to protect the grating structure. Then the grating was cut in the middle and polished. We can see that the real structure is not a binary one. The shape of the grating bars is trapezoidal, due to the nature of the photolithographic process. Fig.6c and d show a top view of a grating for two different areas with the same period but different groove width. The surface roughness of the grating bars was measured with an atomic force microscope and is about ±60 nm.

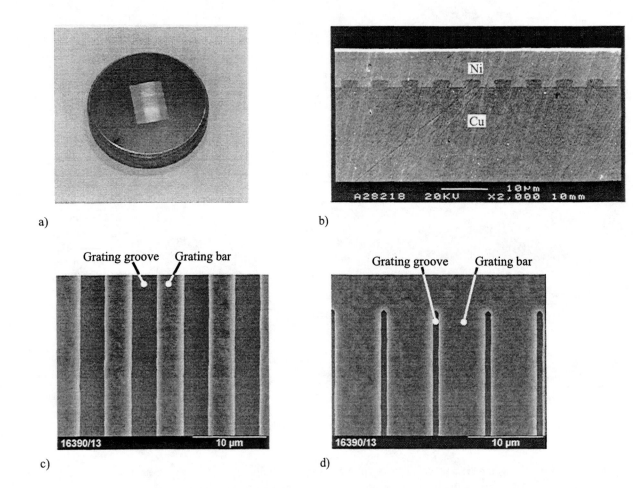

Figure 6: a) Binary Littrow grating on a copper substrate (diameter 38 mm).
b) Grating cross section (SEM picture).
c) Top view of the grating in the middle of the structure.
d) Top view of the grating at the edge of the structure.

3. RESONATORS WITH VARIABLE REFLECTIVITY GRATINGS

Different concepts to extract a beam with good beam quality, like unstable resonators or resonators with variable reflectivity mirrors (VRM) were investigated in the past. Such a VRM resonator is shown in Fig.7a. Due to the reduced off-axis reflectivity the losses increase for higher order modes in stable resonators. This means that higher order modes can be suppressed in a stable resonator with a variable reflectivity mirror and that monomode operation of the resonator is achieved by this concept. This effect is not limited to stable resonators so that unstable resonators also show monomode operation due to the missing hard aperture of the mirror.

Both VRM resonator concepts have the disadvantage that they are not lineselective. By duty cycle modulation of the binary gratings, shown in section 2, we can substitute the VRM in Fig.7a for a so-called Variable Reflectivity Grating (VRG) as shown in Fig.7b. Changing the groove width of the binary grating with constant grating period p enables us to realize for example a super-Gaussian reflection profile given by :

$$R(x) = R_{max} \cdot \exp\left\{-2\left|\frac{x}{w_p}\right|^n\right\} \tag{4}$$

R_{max} is the maximum reflectivity on the optical axis of the resonator. For a grating depth of 1.4 μm and a groove width of 3.5 μm we can realize $R_{max} = R_{-1} = 83\%$. The parameter n is the so-called super-Gaussian parameter.

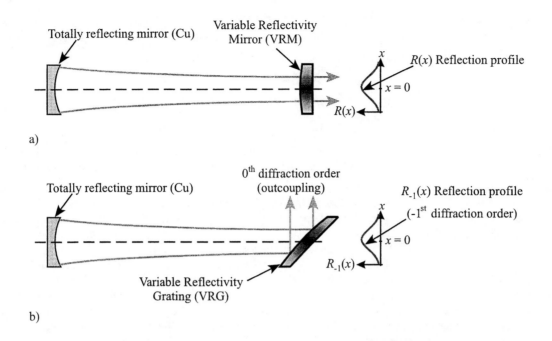

Figure 7: a) VRM resonator. b) VRG resonator.

For $n = 2$ we get a Gaussian profile with a profile width $2w_p$ and for $n \to \infty$ we get a hard aperture with a width of $2w_p$.

We realized two different resonator concepts, a stable and an unstable. The stable resonator had a concave totally reflecting mirror with a curvature of 10 m, a resonator length of 380 mm and a planar Littrow grating. For the reflection profile we have chosen $n = 3$, $w_p = 5$ mm and $R_{max} = 0.83$. The width of the slab was 12 mm. Fig.8a shows the simulated normalized field in the resonator (I), the outcoupled field (II) and the reflection profile for the -1st diffraction order (III). Here we considered a minimum reflection of about 42 % due to the minimum structure size of about 1-1.2 μm, which leads to the two shown terraces.

Fig.8b shows the measured beam profile on the 9P36 line of the CO_2 laser in a distance of 400 mm from the

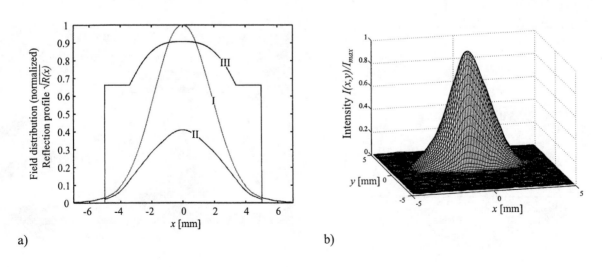

Figure 8: a) Simulated field and reflection profile of the stable resonator.
b) Measured beam profile of the stable resonator.

ν	$2W_{0\nu}$ [mm]	$Z_{r\nu}$ [m]	M_ν^2
x	5.39	2.36	1.02
y	2.05	0.33	1.05

Table 1: Measured beam parameters of the stable resonator ($\lambda = 9.7\ \mu$m)

grating. The measured beam parameters (measured with a Coherent Modemaster system) are given in Table 1, which fit the simulated parameters very well (beam waist diameter $2W_{0\nu}$, Rayleigh range $Z_{r\nu}$, beam parameter product M_ν^2).

To measure the reflection profile of our realized gratings we scanned the grating with a focused CO_2 laser beam and measured the reflectivities R_{-1} and R_0. The profile we measured was not the profile shown in Fig.8a (III). The measured reflection profile is a correlation function of the focused beam of the laser (Gaussian profile) and the reflection profile shown in Fig.8a (III). Due to this we got the ideal reflection profile R_{-1} *ideal* shown in Fig.9a (The ideal profile R_0 *ideal* is not shown). The direction $\tilde{x}$ is the projected x-direction on the surface of the grating. The measured profile R_{-1} fits the ideal profile very well, with the exception of the two terraces, where we got an absolut difference of about 15% between ideal and realized profile. This difference is due to two effects. One is the non-binary structure of our grating bars shown in Fig.6b. For a constant trapezoidal angle of the grating bars (which depends on the photolithographic parameters) we had the largest difference between the binary and the realized trapezoidal grating bars at the edge of the grating, where we find the smallest structures. This means that we can neglect this effect in the mid of the grating but not at the edge. The other effect is the larger slope of the reflectivity in Fig.3 for smaller structures. Due to this, manufacturing tolerances have more effect on the smaller structures.

The total efficiency of the grating ($R_{-1}(\tilde{x}) + R_0(\tilde{x})$) is better than 97 %. The resulting losses are a result of the surface roughness of the grating bars and the finite electrical conductivity of copper.

Fig.9b shows the laser we used to measure the properties of our VRG resonator. The grating is mounted in a gimbal mount in the grating unit on the right side of the picture. The angle of incidence and therefore the selected wavelength is tuned with a micrometer screw. The laser beam is coupled out through a ZnSe window as the 0th diffraction order. The resonator is thermally stabilized by an invar frame and the laser-plasma is excited with a hybrid RF amplifier. The RF frequency is 40.68 MHz and the gas pressure of our 75 % He, 15 % N_2 and 10 % CO_2 gas mixture typically is 40 mbar. The laser operates with a slow gas flow to reduce the demands on the mechanical setup. Typical laser output powers of our system with these non optimized parameters (see section 4) are in the range of 1-3 Watt on the 9P36-line.

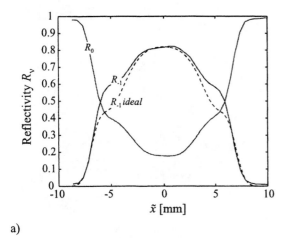

a)

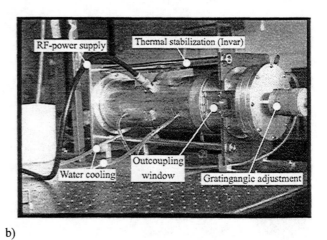

b)

Figure 9: a) Measured reflection profile of the grating for the stable resonator.
b) Lineselective laser

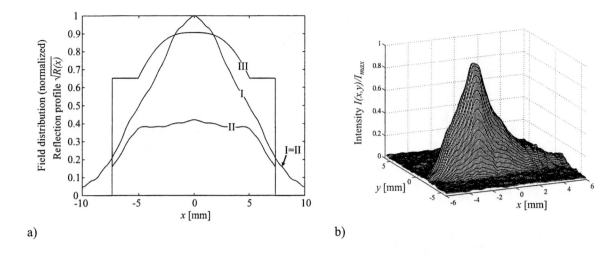

a)   b)

Figure 10: a) Simulated field and reflection profile of the unstable resonator.
b) Measured beam profile of the unstable resonator.

Our unstable resonator had a concave totally reflecting mirror with a curvature of 12 m, a resonator length of 380 mm, a super-Gaussian parameter $n = 3$, a maximum reflectivity of 83 % and a profile width $w_p = 7.3$ mm. The width of the slab was 20 mm. The grating was realized on a convex substrate with a curvature ϱ of 15 m. For a wavelength of 9.7 μm and a grating period $p = 7$ μm the effective mirror radius ϱ_{eff} for the -1st and the 0th diffraction order is 10.8 m :

$$\varrho_{eff} = \varrho \cdot \cos\left(\alpha_L\right). \tag{5}$$

For the chosen wavelength of 9.7 μm we therefore get an unstable resonator with a VRG. Fig.10a shows the simulated normalized field in the resonator (I), the outcoupled field (II) and the reflection profile for the -1st diffraction order (III). Fig.10b shows the measured beam profile on the 9P36 line of the CO_2 laser in a distance of 400 mm of the grating. The measured profile does not fit the simulated profile, nevertheless laser activity was possible at the selected laser line. The discrepancy between simulated and measured profile may result from the assumption of a cold resonator (resonator without active media) in our simulation tool.

Laser activity on the 10P20 line of the CO_2 laser was not achieved with our laser setup. For this wavelength the effective mirror radius for the grating is about 9.8 m. Due to this, the diffraction losses of the unstable resonator for the longer wavelength are larger than for the shorter wavelength and laser activity is impossible.

Both resonator concepts show monomode operation even if we tune the length of the resonator with a piezo transducer. The typical mode-hop behaviour of a lineselective resonator could not be observed. The resonator always oscillates in the same mode and has an output power characteristic as shown in Fig.11. Fig.11a shows the tuning characteristic of the stable and Fig.11b of the unstable resonator. During the tuning procedure the mode pattern was observed by means of a pyroelectric camera and we still found no change in the intensity pattern. This is a clear indication of monomode operation without mode-hops, which is very important for pump lasers, because the optimum pump frequency is normally not the mid frequency of the CO_2 laserline.

4. CONCLUSION

We realized different binary gratings with a spatial inhomogeneous reflection profile through a photolithographic microgalvanic process. The overall efficiency of the gratings is better than 97 %. With these gratings we realized a stable and unstable resonator in a slab laser, which both show monomode operation.

The effect that the simulated fields of the unstable resonator do not fit the measured fields is due to the assumption of a cold resonator in our simulation tool. At the moment we are engaged in optimizing a nonlineselective slab-laser.

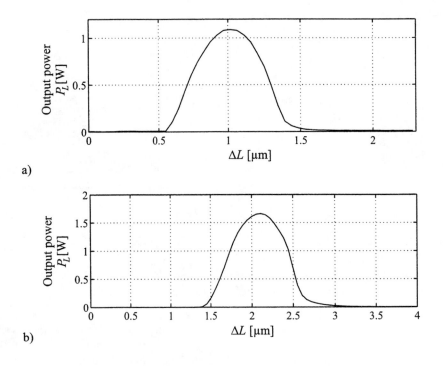

a)

b)

Figure 11: Laser power depending on the resonator length. a) stable, b) unstable resonator.

With those parameters we will redesign the gratings and the unstable resonator with a simulation tool which includes the active media.

The concept to use a VRG for lineselective resonators is a good solution for the problem of extracting a diffraction limited beam out of a slab cavity.

ACKNOWLEDGMENTS

This work was supported by the DFG (Deutsche Forschungsgemeinschaft). The authors wish to thank Prof. H. Brand, Prof. J. Schwider and Prof. L.-P. Schmidt for the useful discussions and support and all persons who have been involved in this work. Especially we wish to thank the mechanical workshop of the Lehrstuhl für Hochfrequenztechnik and Mr. Müller from Schlötter Galvanotechnik.

REFERENCES

1. K. Huber, R. Engelbrecht, R. Hocke, M. Raum, H. Brand, S. Martius, and L.-P. Schmidt, "A broadband, low noise heterodyne receiver for stratospheric measurements at 2.5 THz," in *Proceedings of IGARSS 99*, 1999.

2. K. Abramski, A. Colley, H. Baker, and D. Hall, "Power scaling of large-area transverse radio frequency discharge CO_2 lasers," *Applied Physics Letters* **54**, pp. 1833–1835, May 1989.

3. R. Nowack, H. Opower, H. Krüger, W. Haas, and N. Wenzel, "Diffusionsgekühlte CO_2-Hochleistungslaser in Kompaktbauweise," *Laser und Optoelektronik* **23**(3), pp. 68–81, 1991.

4. L. Serri, C. Maggi, L. Garifo, S. D. Silvestri, V. Magni, and O. Svelto, "Multikilowatt transverse flow CO_2 laser with super-gaussian mirror," *Conference Proceedings "Quantum Electronics and Plasma Physics"*, SIF, *Bologna*, pp. 59–63, 1991.

5. B. Sterman, A. Gabay, and S. yatsiv, "Narrow gaussian CO_2 laser meam in an M mode of a rectangular sheath cavity," *SPIE CO_2 Lasers and Applications II* **1276**, pp. 98–105, 1990.

6. N. Generalov, N. Solovyov, M. Yakimov, and V. Zimakov, "Beam quality improvement by means of unstable resonator with variable reflectivity output coupler," *SPIE* **3267**, pp. 144–155, 1998.

7. C. Pare and P. Belanger, "Optical resonators with graded-phase mirrors," *SPIE* **3267**, pp. 226–233, 1998.

8. M. Morin and M. Poirier, "Graded reflectivity mirror unstable laser resonator design," *SPIE* **3267**, pp. 52–65, January 1998.

9. Y. Takenaka, Y. Motoki, and J. Nishimae, "High-power CO_2 laser using gauss-core resonator for 6-kW large-volume TEM_{00} mode operation," *IEEE Journal of Quantum Electronics* **32**, pp. 1299–1305, August 1996.

10. H. Brand, R. Hocke, M. Collischon, and P. Kipfer, "Optische Flachkanalresonatoren mit räumlich inhomogen reflektierenden Littrow-Gittern, insbesondere für linienabstimmbare Bandleiterlaser." Patent, 1996.

11. J. Andrewartha, J. Fox, and I. Wilson, "Resonance anomalies in the lamellar grating," *Optica Acta* **26**(1), pp. 69–89, 1979.

Laser resonators with helical optical elements

V.E. Sherstobitov[*], A.Yu. Rodionov

Research Institute for Laser Physics, SC "Vavilov State Optical Institute",
12, Birzhevaya line, 199034, St.Petersburg, Russia.

ABSTRACT

The features of fundamental modes in large-aperture unstable resonators with helical mirrors are for the first time to our knowledge investigated via computer simulation. It is shown that a large-aperture laser can oscillate at a single transverse mode with axially symmetric intensity and nonzero topological charge determined by the charge of the helical coupling mirror. Energy efficiency of such a laser does not differ from that for conventional unstable cavity.

Keywords: Unstable resonator, helical mirror, topological charge, Laguerre-Gaussian beam.

1. INTRODUCTION

Since the pioneering work of Vaughan and Willets [1] where for the first time a laser beam with a helical wavefront was demonstrated experimentally, such beams have been constantly attracting the attention of researchers in a large number of laboratories (see for instance the reference in [2-7]).

The interest to investigation of such beams is explained by a phase singularity on the beam axis and unusual features of the beams, that can be used in some applications in science and technology.

Helical beams are described by complex amplitude of the form

$$E(\rho,\varphi) = E_0(\rho)e^{in\varphi} . \tag{1}$$

Here ρ - the transversal coordinate, φ - azimuthal angle in a cylindrical coordinate system, associated with the beam axis and $E_0(\rho) \equiv 0$ at $\rho = 0$. Here $n = 0, \pm 1, \pm 2...$ is the so-called topological charge of the beam that is defined by the phase gradient circulation around a closed loop that encloses a zero of the field [3]. The phase singularity of helical beams (1) at the beam axis manifests itself in existence of the so-called optical vortex on the wavefront of the beam. The circulation of the phase gradient on any loop around the beam axis $\rho = 0$ is equal to $2\pi n$. The beam intensity distribution of such beams has axially symmetric structure.

The well-known example of such beams is the TEM*$_{01}$ "doughnut" laser mode [1] corresponding to a helical wave with $n = 1$ and having the annular intensity profile.

A more general example of helical beams is given by Laguerre-Gaussian resonator modes [8]

$$E_{lp}(\rho,\varphi) = const\left(\frac{\sqrt{2}\rho}{w}\right)^l L_p^l\left(\frac{2\rho^2}{w^2}\right)e^{\pm il\varphi}e^{-\frac{ik\rho^2}{2q}} . \tag{2}$$

Here $\frac{1}{q} = \frac{1}{R} - \frac{i\lambda}{\pi w^2}$, R - is the radius of curvature of the wavefront, λ-is the wavelength, w is the distance at which the intensity of the fundamental Gaussian beam is e^{-2} times that on the axis and L_p^l is the generalized Laguerre polynomial. Here $p = 0,1,2,3..$ and $l = 0,1,2...$ are the radial and angular mode numbers. The mentioned above "doughnut" mode corresponds to $l = 1, p = 0$ in (2).

One of the most interesting features of helical beams is that such beams can exhibit a nonzero angular momentum even for linear polarization of radiation. As was shown in [4, 6] each photon in a helical beam with the Laguerre-Gayssian amplitude (2) carries $\mp l\hbar$ angular momentum if its charge of singularity $n = \pm l$. This momentum in distinction to momentum associated with circular polarization (which is equal to $\pm \hbar$) is referred to as "orbital angular momentum" [4]. The angular momentum of helical beams can be transferred to birefringent optical elements that reverse the spirality of the

---

[*] Correspondence: e-mail: sherstob@ilph.spb.su; Telephone: (812) 328 5894; Fax: (812) 328 5891

62

In *Laser Resonators III*, Alexis V. Kudryashov, Alan H. Paxton, Editors,
Proceedings of SPIE Vol. 3930 (2000) ● 0277-786X/00/$15.00

beam [6] or to small absorbing particles making the latter to rotate [9]. Helical beams can be used for trapping small particles where such beams with zero intensity on the beam axis have some advantages [10] in comparison with Gaussian ones because the dark central spot reduces the scattering forces due to back reflections from the top of the particle. All these features can be important in biological applications for manipulation of cells, bacteria and their component parts [11].

A very interesting application of helical beams is connected with generation of optical vortex solitons that are stationary structures that can be formed by helical beams in defocusing nonlinear optical medium [12]. Investigation of optical vortex solitons in different nonlinear media (see reference in [7]) can provide new potential applications of laser beams in various fields of science and technology. In a number of applications high-power helical beams are of interest and the task of their generation is one of central problems in current research.

Usually helical beams are produced by diffraction of conventional laser beams on a computer-generated hologram [13,4]. This hologram should be just a recording of the interference pattern between a required helical beam and some simple reference beam. This is a rather flexible technique for producing singular beams. However the efficiency of transformation of the incident beam into the required helical beam is usually rather low and besides the intensity of the beam is limited by a damage threshold of the hologram material for the reading-out radiation.

Another approach is based on conversion of high-order Hermite-Gaussian beams into the required Laguerre-Gaussian beams with the use of a cylindrical lens converter [16] and has been demonstrated experimentally in [16, 6, 17]. However this approach requires application of special techniques to provide oscillation of the laser on the single Hermite-Gaussian mode of the required index which can be a problem in a high-power large-aperture laser.

An alternative conversion technique [18] uses a special spiral phase plate introducing a helical phase shift into a TEM_{00} beam at the output of a laser. The spiral phaseplate is a transparent plate with a thickness $h(\varphi) = h_0 + h_1\varphi/2\pi$ increasing proportionally to the azimuthal angle φ around a point in its center so that the height of the step after one turn around the center is equal to h_1. Being placed into an immersion liquid such a plate produces at its output a phase shift $\Phi(\varphi) = \dfrac{\Delta n h_1\varphi}{\lambda}$ where Δn is the difference in the refractive indices of the plate and the immersion and λ-the radiation wavelength in vacuum. By changing the temperature of the immersion liquid it was possible in [18] to adjust the phase shift introduced into the beam and obtain $\Phi(\varphi) = n\varphi$. It was sufficient for transformation of a Gaussian beam with the wavelengh λ into a helical beam with the topological charge n. Such approach to generation of helical beams can be easily implemented in laboratory [18]. However its use is limited because of a typically low power of the fundamental TEM_{oo} mode and thermal distortion in the spiral cell which is especially significant for transformation at a large n.

It is well-known that Laguerre-Gaussian modes (2) can be generated in lasers with stable resonators [1,14]. However the modes with azimuthal factors $e^{-il\varphi}$ and $e^{il\varphi}$ differing by the sign of the azimuthal angle φ are degenerate due to azimutal symmetry of the equation for eigenmodes of the cavity. As a result these modes have equal losses and oscillate usually simultaneously with the same frequency. The intensity pattern of the output beam should exhibit in this case a $\cos^2 n(\varphi - \psi)$ azimuthal modulation. It is necessary to note that the phase ψ can vary in time rapidly so that the average intensity pattern can have an annular shape. Nevertheless the orbital angular momentum of such a beam can be equal to zero. Sometimes the vortex wave with the required sign of the charge can be selected in the laser output due to a special feedback in the cavity [14] or due to some lack of the cavity symmetry (e.g. scattering of radiation on a helical structure in a ring cavity [15]). However such techniques are rather complicated or not properly controlled.

The goal of this paper is to investigate a possibility to generate helical beams with nonzero orbital angular momentum within linear laser resonators. This is done by computer simulation of the fundamental mode in a cavity comprising a helical mirror (HM) instead of one of the cavity mirrors.

The helical mirror is a mirror with an azimuthal profile producing a phase shift $\Phi(\varphi) = n\varphi$ in the reflected beam with the required wavelength λ. Such a profile can be generated by diamond turning or with the use of optical coating techniques.

It is shown that introducing HMs into the unstable cavity can provide oscillation of a large-aperture laser on a fundamental mode with a nonzero orbital angular momentum.

2. TRANSFORMATION OF GAUSSIAN BEAM BY HELICAL MIRRORS

A schematic profile of a HM is presented in Fig 1. The mirror "depth" linearly grows as a function of azimuthal angle φ so that the step height is equal to $n\lambda/2$ after one turn around the mirror axis. The profile can be applied on a conventional concave or convex mirror so that the reflected beam will have an azimuthal phase profile applied on the converging or diverging wavefront.

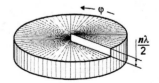

Fig.1. Helical mirror

In the following Sections we consider properties of laser cavities in which one mirror (or two) is replaced by HM. Here we obtain analytical expressions for spacial and angular characteristics of helical beams obtained by reflection of a Gaussian beam from a HM. These data are of interest in design and analysis of an unstable cavity with HM.

Let us consider reflection of a converging Gaussian beam with complex amplitude

$$E_0(\rho) = E_0 e^{-\frac{\rho^2}{w^2}} \cdot e^{\frac{ik\rho^2}{2R}} \tag{3}$$

from the HM that introduces a phase factor $e^{in\varphi}$. Here $k = 2\pi/\lambda$, ρ -as before is a transversal coordinate, R-radius of curvature of the incident beam, w-is the radius at that the beam intensity of the incident beam drops to e^{-2} of its maximum.

According to Huygens-Fresnel principle complex amplitude $E(\rho',\varphi')$ in a plane located at a distance L from the HM is given by the integral

$$E(\rho',\varphi') = \frac{iE_0}{L\lambda} e^{-\frac{ik\rho'^2}{2L}} \int\limits_0^\infty \int\limits_0^{2\pi} e^{-\frac{\rho^2}{B^2}} e^{in\varphi} \exp\left[\frac{ik\rho\rho'}{L}\cos(\varphi - \varphi')\right] \rho \, d\rho \, d\varphi \quad , \tag{4}$$

where

$$\frac{B}{w} = \left[1 - \frac{ikw^2}{2L}\left(1 - \frac{R}{L}\right)\right]^{-1/2} \tag{5}$$

Using a standard procedure of integration over φ we find

$$E(\rho',\varphi') = 2\pi i^{(n+1)} \frac{E_0}{L\lambda} e^{-\frac{ik\rho'^2}{2L}} e^{in\varphi'} \int\limits_0^\infty e^{-\frac{\rho^2}{B^2}} J_n\left(\frac{k\rho\rho'}{L}\right) \rho \, d\rho \, , \tag{6}$$

where $J_n(x)$ – Bessel function of the order n. Integration over ρ can be done with the use of integral representation for degenerate hypergeometrical function $M(r,s,x)$ (see formula 11.4.28 in [19])

$$\int\limits_0^\infty e^{-a^2 t^2} t^{\mu-1} J_\nu(bt) dt = \frac{\Gamma\left(\frac{\nu}{2} + \frac{\mu}{2}\right)\left(\frac{b}{2a}\right)^\nu}{2a^\mu \Gamma(\nu+1)} M\left(\frac{\nu}{2} + \frac{\mu}{2}, \nu+1, -\frac{b^2}{4a^2}\right) \, , \tag{7}$$

where $\mathrm{Re}(\mu + \nu) > 0; \mathrm{Re}\, a^2 > 0$.

Using (7) and the known formula (see in [19]) for relation between hypergeometrical functions of negative and positive arguments, we find

$$E(\xi',\varphi') = i^{(n+1)} \frac{E_0 k w^2}{2L} \frac{\Gamma\left(\frac{n}{2} + 1\right)}{\Gamma(n+1)} \left(\frac{B}{w}\right)^2 e^{-i\xi'^2 \frac{2L}{kw^2}} e^{-\frac{\xi'^2 B^2}{w^2}} \left(\frac{B}{w}\xi'\right)^n M\left(\frac{n}{2}; n+1; \left(\frac{B}{w}\xi'\right)^2\right) e^{in\varphi'}, \tag{8}$$

where $\xi' = \frac{wk\rho'}{2L}$ - dimensionless transversal coordinate in the plane of observation. As is seen from (8) the beam reflected from the HM is really a helical beam with the azimuthal dependence of $e^{in\varphi'}$.

The intensity distribution $EE^* = I_n$ at a distance L can be easily found to yield:

$$I_n(\xi',\varphi') = \left(\frac{kw^2}{2L} E_0\right)^2 \frac{\left[\Gamma\left(\frac{n}{2} + 1\right)\right]^2}{(n!)^2} \cdot \left[1 + \left(\frac{kw^2}{2}\right)^2\left(\frac{1}{R} - \frac{1}{L}\right)^2\right]^{-\frac{n+2}{2}} \cdot \xi'^{2n}$$

$$\cdot \exp\left\{-2\xi^2\left[1 + \left(\frac{kw^2}{2}\right)^2\left(\frac{1}{R} - \frac{1}{L}\right)^2\right]^{-1}\right\} \cdot \left|M\left(\frac{n}{2}; n+1; \frac{\xi'^2}{1 \pm \frac{ikw^2}{2}\left(\frac{1}{R} - \frac{1}{L}\right)}\right)\right|^2 \tag{9}$$

It is evident that for $n=0$ (9) is readily transformed into the well-known intensity distribution of a convergent Gaussian beam

$$I_0(\xi',\varphi') = \left(\frac{kw^2 E_0}{2L}\right)^2 \left[1+\left(\frac{kw^2}{2}\right)^2 \left(\frac{1}{R}-\frac{1}{L}\right)^2\right]^{-1} \exp\left\{-2\xi'^2\left[1+\left(\frac{kw^2}{2}\right)^2\left(\frac{1}{R}-\frac{1}{L}\right)^2\right]^{-1}\right\} \quad (10)$$

if we take into account that $M(0,s,x)\equiv1$.

As is seen from (8) the beam reflected from the HM differs from the Laguerre-Gaussian functions that are the eigenmodes of the free space propagation operator. It means that this beam does not propagate as a pure mode but changes the form of its intensity distribution as a function of distance L. The reason for such a behavior of the beam is that its phase distribution is changed in the process of reflection but the amplitude distribution remains the same. Such a beam should be decomposed into a series of different Laguerre polynomials with the same azimuthal index (see [18] where such a decomposition was made for a spiral phaseplate). It should be noted however that discrepancy in intensity distribution just after the reflection is especially large at the beam axis where all the higher-order Laguerre-Gaussian functions are equal to zero. At the same time namely the near-axis part of the reflected beam suffers the maximum phase perturbation because the azimuthal phase gradient is maximum here and goes to infinity on the beam axis. As a result the radiation should be rapidly carried out from the near-axis area at some distance from the HM. The intensity distribution in central part of the beam that is of interest in resonator applications of HMs should in this case rapidly approach to that described by the Laguerre-Gaussian mode with the same azimuthal index.

This is illustrated by the gray -scale intensity profile of a converging Gaussian beam with $n=0$ shown as a function of the distance L after reflection from the HM with $n=1$ (Fig 2a).

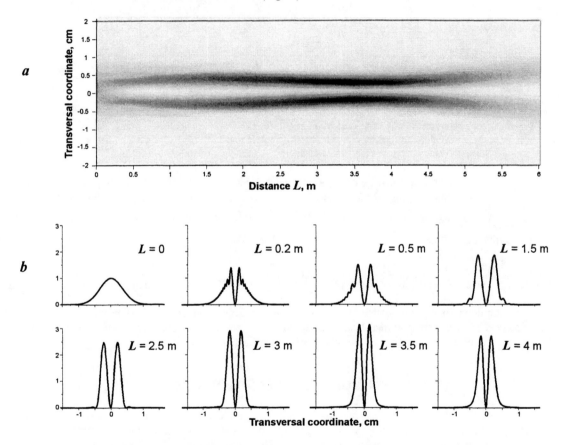

Fig. 2. Evolution of convergent Gaussian beam with $w=0.707$ cm, $\lambda=10.6$ μm, R=4 m after reflection from a helical mirror with $n=1$.
a)   Gray-scale intensity
b)   Beam intensity distribution at different distances L normalized to initial Gaussian beam intensity.

The profile was calculated both via integration of (2) with the use of FFT procedure or directly using Eq. (9) for intensity as a function of L. The hyper-geometrical function $M(r,s,x)$ was calculated with the use of the well-known decomposition in x^k series [19]. Both approaches gave the same results. As is seen from the Figure the radiation is actually carried out rapidly from the near-axis area just after the reflection and the beam (to be more correct, its central part, carrying the bulk of energy) starts propagating in a self-reproducing manner. The beam cross-sections for the case shown in Fig.2a are given in Fig.2b for several distances L. The intensity ripples noticeable at small distances from the HM are explained by components described by higher-order Laguerre polynomials L_p^l with $l=1$, $p\neq0$. They have a large beam divergence and decay rapidly along the propagation path.

The angular characteristics of the beam obtained via reflection of a Gaussian beam with a plane wavefront and the beam waist w_0 from HMs with different topological charges n are shown in Fig.3.

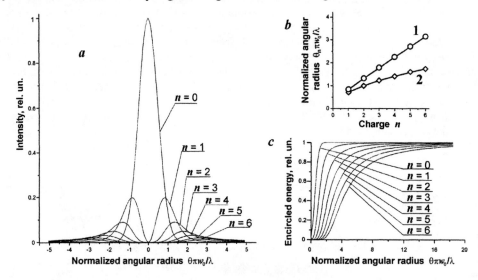

Fig. 3. Angular characteristics of a Gaussian beam after reflection from helical mirrors with different n
a) Angular intensity distribution b) Angular radius of intensity maximum, 1 - numerical results, 2 - similar data for Laguerre–Gaussian beams. c) Encircled energy as a function of normalized radius

All the curves describing the angular intensity distributions (Fig.3a) are normalized on the same power in the incident Gaussian beam (its angular distribution is also shown in the Figure by curve $n=0$). The dependence of the angular radius θ_n corresponding to position of a maximum intensity in the far field of the reflected beam on the charge n is shown by curve 1 in Fig.3b. It is approximated by linear dependence $\theta_n \approx \dfrac{\lambda}{\pi w_0}(0.38+0.42\left|n\right|)$ for $n=1,2,3\ldots$. Note that this dependence differs from that for Laguerre-Gaussian modes where $\theta_n = \dfrac{\lambda}{\pi w_0}\cdot\sqrt{\dfrac{\left|n\right|}{2}}$ (curve 2). As has been already told it is explained by the difference in the amplitude distribution of the beams. Note that the beam divergence of the beam reflected from HM with $n=1$ differs not very significantly from that of the initial Gaussian beam. At the 0.8 energy level it exceeds the latter approximately two times only. Such beams can be useful in various applications.

3. RESONATORS WITH HELICAL MIRRORS

If we replace one of the mirrors in an unstable cavity by a HM, we can expect that under certain conditions the fundamental mode in such a cavity would be a helical wave with the topological charge controlled by parameters of the HM and geometry of the cavity. It would allow generating high-power helical laser beams in large-aperture lasers using HMs of small diameter.

3.1 Transformation of the charge by optical elements

Taking in mind carrying out analysis of such a possibility let us remind first how the topological charge of helical beams is transformed by different optical elements that can be used in a laser cavity. Following the commonly accepted nowterminology [3] we shall consider a topological charge of the helical beam to be positive for the helical wave similar to a right screw relatively to the propagation direction and negative in the opposite case.

When a helical beam with the right screw is reflected from a mirror (Fig.4a) a normal to the mirror component of the wave vector changes its sign for the opposite and the tangential component remains the same. It is evident that as a result of the reflection the right helical wave is transformed into a left one, i. e. the topological charge n changes for $(-n)$ (Fig.4b)

Another situation is realized when the helical beam is reflected in the opposite direction by a roof-reflector (Fig.4c). Due to two reflections suffered by the beam its charge remains the same.

Similar property is exhibited by a combination comprising a cylindrical lens and a mirror placed in the focal plane of the lens normally to the beam axis (Fig.4d). Besides changing the direction of the incident helical beam for the opposite such a system produces transformation of its cross-section that is equivalent to reflection in the horizontal plane. This reflection results in additional change of the charge sign so that the reflected beam has the same charge as the incident one. It is clear that a linear combination comprising two confocal cylindrical lenses following one another will change the charge of the beam passing through such a combination to the opposite.

An interesting case of the helical beam transformation is the reflection from a phase-conjugate mirror (PCM). In this case the wavefronts of the incident and reflected beams coincide for any angles of incidence and the direction of propagation changes for the strictly opposite. As a result the right screw remains the same, i.e. the topological charge of the beam does not change (Fig.4e). Experimentally this feature of helical beams was demonstrated recently in [20].

Let us consider now transformation of the topological charge when the beam is reflected by a HM similar to that shown in Fig 1. Its "depth" grows linearly with φ when we look along the direction of the incident beam and go counter-clockwise (Fig. 4f). The step is equal to $m = \dfrac{\lambda}{2}$ at $\varphi = 2\pi$. In the incident helical wave with a positive charge the wavefront time lag also grows with φ. It is clear that in the process of reflection this time lag will increase due to additional time delay introduced by the mirror. So the charge m created by the mirror will be

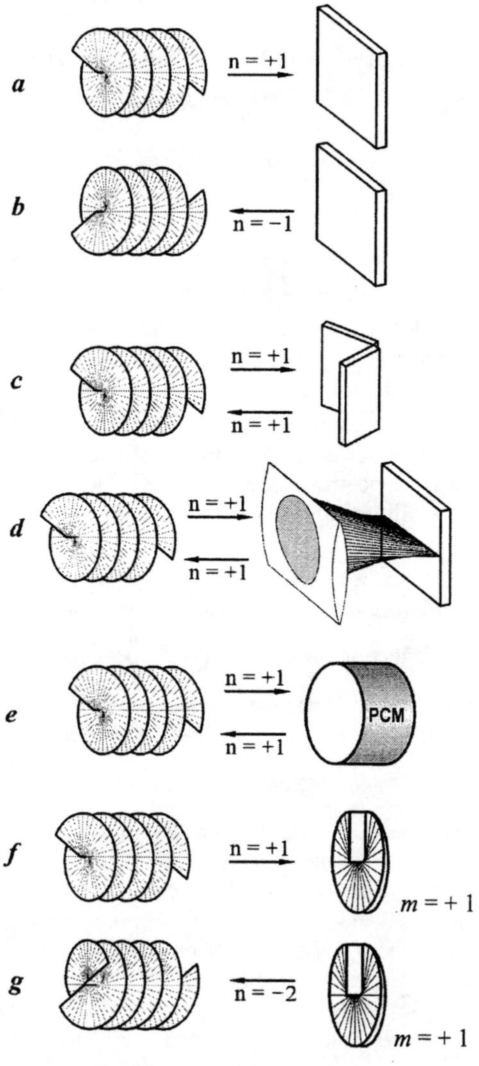

Fig.4. Transformation of the beam charge by different optical elements (see text)

added to the positive charge n of the incident beam. We shall call such a HM "positive" and a mirror with the opposite screw – "negative" and associate with such mirrors topological charges $(+|m|)$ and $(-|m|)$ respectively.

Taking into account that the sign of the charge is changed by the reflection we find for the charge of the reflected beam $n_{ref} = -(n+m)$ (see Fig. 4g where $n=1$, $m=1$).

Depending on the number of reflections suffered by the beam in a round trip over the cavity and topological charges of HMs within the cavity we can realize oscillating modes with different topological charges.

3.2 Charge reproduction in different resonators

Let us consider a cavity comprising a traditional mirror and a HM with an arbitrary charge m $(m = \pm 1, \pm 2 ...)$. Evolution of the beam charge n for two passes in the cavity is shown in Fig. 5a.

The equation for reproduction of the topological charge of steady-state oscillating modes has the form

$$n = n + m \qquad\qquad (11)$$

It has a trivial solution $m=0, n=0$ corresponding to a conventional cavity that does not comprise any HMs. It means that for $m \neq 0$ oscillation of helical mode in such a cavity is impossible.

In a cavity with a roof reflector instead of a conventional mirror (Fig. 5b) reproducing of a mode with a nonzero topological charge is possible. In this case the equation for the mode charge is of the form

$$n = -(n+m) \qquad (12)$$

and has the only solution

$$n = -\frac{m}{2} \qquad (13)$$

Taking into account that n is an integer we find that for each even $m = 2p$ there exists the only solution $n = -p$ for the charge of a spiral beam incident on the HM. Note that the reflected beam has the charge $-(n+m) = -p$. Thus, it looks like the HM «unwinds» the incident helical wavefront with the charge n to the same charge but having the opposite sign. Simultaneously it changes its sign additionally as a conventional reflecting mirror. As a result we have for both directions of propagation in the cavity the beam with the same topological charge $n = -p$ equal to half the charge of the HM taken with the opposite sign.

It is necessary to note that according to equation (13) conventional beams with $n=0$ can not reproduce their charge at each round trip along the cavity. However if we take n in Fig. 5b to be arbitrary (in particular $n=0$) and consider a possibility to reproduce a beam charge after two round trips we find the condition

$$n = -[-(n+m)+m] \equiv n \qquad (14)$$

Thus for two round trips any arbitrary charge (positive or negative) can be reproduced in the cavity shown in Fig.5b, and in particular $n=0$. Note, however, that the beam with $n=0$ (for instance, incident on the helical mirror) will have after the reflection the charge m (not $\frac{m}{2}$ as in the previous case). As a result this beam will have a larger beam divergence during the next round trip (see Fig. 3) and consequently larger losses in the cavity.

Similar consideration can be done for the cavity comprising the HM and a PCM (Fig. 5c). As in the previous case such a cavity can oscillate at a single azimuthal mode with $n = -\frac{m}{2}$ if we require the charge reproducing for each round trip.

However it is clear that for PCM of infinite size it can also oscillate at modes with any arbitrary topological charge and any tilts of the beam with respect to the axis, if we take into account a possibility to reproduce the charge after two round trips. Just as in conventional cavities with PCMs in order to provide single mode oscillation in this case it is necessary to place a mode selector into the cavity. For modes with $n\neq0$ such a selector can be implemented as a stop with an annular slot placed in the focal waist of the beam inside the cavity. The size of this annular slot and its width can be estimated from the plots of Fig. 3 for each particular n.

Let us turn now to a cavity comprising two HMs with the charges m_1 and m_2 as shown in Fig.6d. For the charge n of the wave propagating to the right in the Figure we find

$$n = -[-(n+m_2)+m_1] \qquad (15)$$

Equation (15) has nontrivial solutions ($n=\pm1\pm2...$) only for $m_1=m_2=m$. Note, that the charges of the waves propagating in the opposite directions are generally speaking different for arbitrary m and equal respectively to n and $-(n+m)$. Due to symmetry of the cavity there are sets of such modes differing by the direction of the mode with the charge n.

For even $m = 2p$, where p is a signed integer, a mode exists for which the two oppositely propagating beams have the same topological charge $n = -p$, equal to half the charge of each mirror taken with the opposite sign. Along with this mode there exists a set of modes for which one of the beams can have the charge $n = -p+k$, where $k=\pm1\pm2...$ At the same time

Fig. 5. Evolution of the beam charge in different cavities with helical mirrors (HM) (see text)

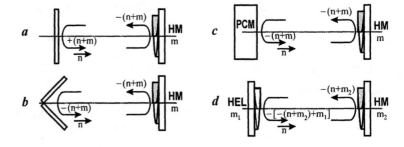

68

the beam of the opposite direction shall have the charge $-(n+m)=-p-k$. Again due to symmetry of the cavity for each $k{\neq}0$ we have two different beams propagating in the same directions. They have the topological charges $n=-p\pm|k|$ and transform into each other after each reflection. Thus, for instance, if $m=-2$ and $p=-1$, then the mode having the same charge for both directions will be the mode with $n=1$. Besides this beam the beams with $n=1\pm|k|$ can bounce in the cavity (for instance, with $n=0$ and $n=2$ for each direction). In this case the beam with $n=0$ transforms after the reflection into the beam with $n=2$ and vice versa. A similar pair is formed by the beams with $n=-1$ and $n=3$. Relative losses of all these modes depend on specific geometry of the cavity. As a rule the minimum losses should have the mode with $n=-\dfrac{m}{2}$ for the both opposite beams in the cavity.

4. COMPUTER SIMULATION OF UNSTABLE RESONATORS WITH HELICAL MIRRORS

In this section we present the results of computer simulation of unstable cavities in which one of the resonator mirrors is replaced by a helical mirror (HM). These results confirm feasibility of generation of helical beams in a laser with a large-aperture unstable cavity.

The steady-state configuration of the fundamental mode in the cavity was found by a standard iteration approach. It used FFT procedure for calculation of the beam propagation between the mirrors of the resonator. We used a 1024x1024 mesh, which enabled the calculation of the fundamental mode with the required accuracy $\sim10^{-4}$ at equivalent Fresnel numbers $N_{eq} \leq 10$. As an initial field distribution in the iteration procedure we used the waves with various topological charges and constant amplitude over cross-section of the cavity. When it was necessary to investigate the process of fundamental mode build up from the background noise we used as the initial field a set of such waves comprising several waves with different topological charges. We calculated the modes as in "empty" resonators so in resonators containing a gain medium. In the latter case the gain medium was changed for a thin amplifying layer with the increment $g(\rho,\varphi)L$ per one trip, where g is defined by equation

$$g(\rho,\varphi)L = \frac{g_0 L}{1+I(\rho,\varphi)} \quad . \tag{16}$$

here g_0 – small-signal gain, L- the cavity length, I- the total intensity of two oppositely propagating beams, normalized on the saturation intensity. The amplifying layer was located near the plane of the rear reflector of the unstable resonator.

The calculations were performed in parallel for conventional unstable cavities and their helical analogues designed with the use of the concept shown in Fig. 5b of the previous Section.

Here we present some results of the computer simulation obtained for a positive branch confocal unstable cavity with parameters typical for a large-aperture pulsed CO_2 laser and for its helical modification.

The confocal unstable resonator comprised a fully reflecting rear concave mirror (R_1=8m) and a coupling convex mirror (R_2=4m) (see Fig. 6a), placed at a distance $L=\dfrac{1}{2}(R_1 - R_2)$ of two meters.

Magnification of the cavity $M = R_1/R_2$ was equal to M=2. Both mirrors were spherical so that geometrical loss of the fundamental mode per round trip [21] was defined by $\delta_g = 1-1/M^2$.

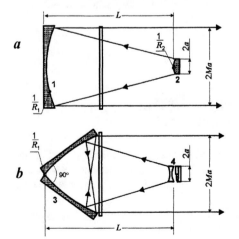

Fig. 6. Schematic representation of a laser with confocal unstable resonator (*a*) and resonator with helical mirror (*b*)

1- concave mirror (curvature $1/R_1$)

2- convex mirror (curvature $1/R_2$)

3- roof-reflector (equivalent curvature $1/R_1$)

4- convex mirror (curvature $1/R_2$) with helical azimuthal profile (*m*=-2)

The unstable helical resonator is shown in Fig 6b. It was formed by a roof–reflector equivalent to a concave mirror with the curvature radius R_1=8m (such a reflector can be composed, for instance, of two concave cylindrical mirrors) and a convex coupling mirror with radius of curvature R_2=4m. The coupling mirror had on its surface an azimuthal relief with m=–2 (the right screw with the step $h_1 = \lambda$).

Calculations were performed for the wavelength λ=10.6 μm and diameters of the coupling mirror $2a$ in the range of 2÷4 cm. The edge of the coupling mirror used in calculations was considered to be "smoothed" in order to damp convergent waves in the cavity [21, 22] and to decrease the amplitude of the intensity ripples in the beam cross-section and consequently to reduce small-scale nonlinear distortions in the gain medium [23].

Fig. 7 shows the calculation results for spatial characteristics of the fundamental mode in resonators of both configurations for the limiting case of "smoothing". In this case the reflectivity of the coupling mirror as a function of transverse coordinate was Gaussian so that reflected the beam intensity dropped to e^{-2} level at a distance a=1.5 cm from the axis. For the plots presented in Fig. 7 the small signal increment $g_0 L$ =2.

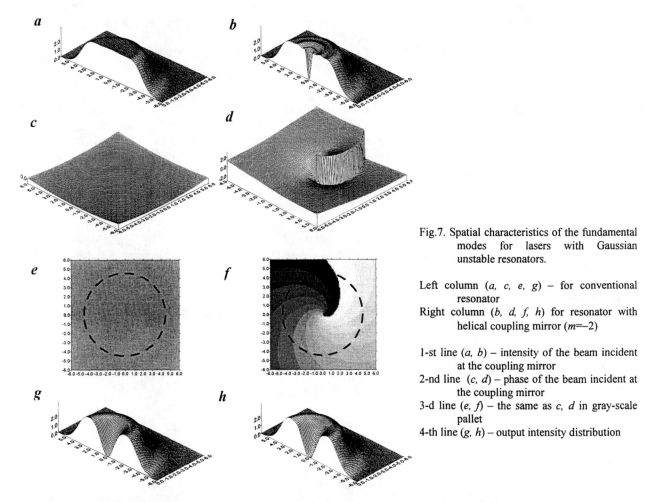

Fig.7. Spatial characteristics of the fundamental modes for lasers with Gaussian unstable resonators.

Left column (*a, c, e, g*) – for conventional resonator

Right column (*b, d, f, h*) for resonator with helical coupling mirror (*m*=–2)

1-st line (*a, b*) – intensity of the beam incident at the coupling mirror
2-nd line (*c, d*) – phase of the beam incident at the coupling mirror
3-d line (*e, f*) – the same as *c, d* in gray-scale pallet
4-th line (*g, h*) – output intensity distribution

As the initial field distribution we used in these calculations a set of helical waves with a constant amplitude over the cavity cross-section (equal for all the waves) and five different topological charges (*n*=0,1,2,3,4).

Transition to a steady-state mode of propagation was observed after 10–20 iterations for the relative accuracy of 10^{-4}. In the conventional unstable resonator it was the fundamental Gaussian mode with *n*=0 [22] and a slightly flattened at the top intensity distribution due to gain saturation (Fig. 7a). The phase profile (Fig. 7c) exhibited a slight sphericity also caused by saturation. (For empty cavity the phase should be constant over the cross-section of the cavity [22] for the plane wave propagating to the coupling mirror within the cavity).

For the helical resonator the fundamental mode had a form of a helical wave with *n*=1 for both opposite directions in full agreement with the results of the previous Section. Its intensity distribution was similar to that for the mode with *n*=0 in the conventional resonator but had a drop to zero on the axis (Fig.7b). The phase profile (Fig.7d) changed linearly with φ from 0 to 2π which is the evidence of $n=1$. The helical behavior of the phase is readily seen from the gray-scale pattern (see Fig. 7f) of the same phase distribution. Some "spirality" of the "rays" seen in Fig. 7f is explained by small sphericity of the wavefront connected, as in the previous case, with gain saturation. The intensity distributions at the output of the laser were very close to each other for both resonators (Figs. 7g and 7h).

A more practical situation was simulated in the same cavities but with the coupling mirrors limited in size and equal to $2a=4$cm in diameter. Reflectivity of these mirrors was equal to 100% over the entire surface of the mirror except a narrow ring near the mirror edge, where it dropped linearly to zero over a zone with the width $a/2N_{eq}$ [22] equal to 1 mm. Here

$$N_{eq} = \frac{a^2}{R_2\lambda}$$ is the equivalent Fresnel number [24] which is a measure of the angle between the wavefronts of the

convergent and divergent waves in the unstable cavity calculated near the edge of the coupling mirror in the geometrical approximation [21]. As it was noted in [21] the linear "smoothing" of the reflectivity can be replaced in practice by making a "saw-tooth" profile of the coupling mirror rim. Such an approach has been successfully used in experiments [23] on damping LIMP-effects in an electron-beam sustained CO_2 laser with an unstable resonator.

Fig. 8 shows the calculation results similar to those presented in Fig. 7 but obtained for the linearly "smoothed" mirror edge. The number of iterations necessary in this case was larger than before because of more complicated structure of the field and amounted to 30-40. However we see that the situation has not changed significantly. The fundamental mode in the conventional unstable cavity with the "smoothed" coupling mirror is very close to the well-known solution obtained in the geometrical approximation. It shows almost constant intensity (Fig. 8a) and phase distributions (Figs. 8c, 8e) over the coupling mirror surface and Fresnel ripples in the periphery part of the beam.

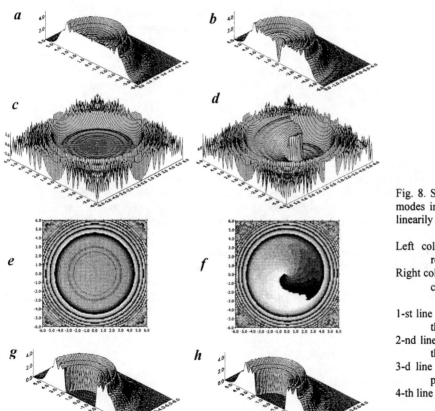

Fig. 8. Spatial characteristics of the fundamental modes in lasers with unstable resonators having linearly "smoothed" coupling mirrors.

Left column (a, c, e, g) – for conventional resonator
Right column (b, d, f, h) for resonator with helical coupling mirror (m=–2)

1-st line (a, b) – intensity of the beam incident at the coupling mirror
2-nd line (c, d) – phase of the beam incident at the coupling mirror
3-d line (e, f) – the same as c, d in gray-scale pallet
4-th line (g, h) – output intensity distribution

The fundamental mode in the helical cavity is presented as before by a helical wave with $n=1$ having practically the same intensity distribution as $n=0$ mode in the conventional resonator but differing by the central dip in intensity on the axis (Fig. 8b). Its phase distribution (Fig. 8d and 8f) has again a spiral form characteristic for a wave with the topological charge $n=1$.

As in the Gaussian resonator the output intensity distributions of the fundamental modes practically coincide with each other for the conventional and helical unstable cavities (Figs. 8g and 8h). Such a coincidence in the intensity distributions resulted in practically the same outputs with a 1% accuracy in both these resonators. The loss δ_n per round trip for the mode $n=0$ in the conventional resonator and $n=1$ in the helical one also coincided. For the "smoothed" mirror case they were $\delta_0=0.784$ and $\delta_1=0.781$ which is very close to the geometrical loss $\delta_g=0.75$. For Gaussian resonator the losses also where almost equal ($\delta_0=0.841$ and $\delta_1=0.849$) and a little larger due to saturation effects.

It is necessary to note that in spite of the coincidence of intensity distributions at the output of the conventional and helical resonators the beam divergence diagrams are quite different (Fig. 9) for $n=0$ and $n=1$ modes.

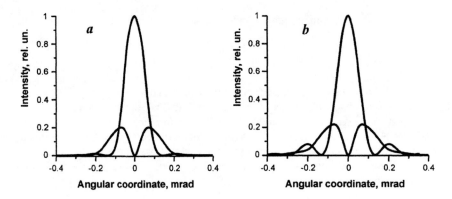

Fig.9. Far-field intensity distributions at the output of laser with conventional (1) and helical (2) unstable resonator
a) with Gaussian coupling mirror, *b*) with linearly "smoothed" coupling mirror

One of the beams ($n=0$) has the central maximum in the far-field, and the other ($n=1$) has a form of annulus. In the Gaussian case (Fig. 9a) they resemble the corresponding distributions shown in Fig. 3. For the "smoothed" mirror case (Fig. 9b) each plot differs a little from the corresponding one for the Gaussian resonator but this difference is not very significant.

5. CONCLUSION

Thus we have analyzed by computer simulation a possibility of generation of helical beams in large-aperture lasers with unstable resonators. It was shown that for some modifications of linear unstable resonators replacing the coupling mirror by a helical mirror can result in generation of the large-aperture helical beam at the output of the laser.

We investigated characteristics of the fundamental mode in loaded unstable helical resonators with the Gaussian reflectivity coupling mirror and the "smoothed" coupling mirror having tapered reflectivity near the mirror edge. Both resonators showed good selective properties and provided oscillation at a single helical mode with the topological charge determined by the charge of the coupling mirror.

Comparison of calculation results with data obtained for conventional resonators with similar parameters shows that transfer to helical modification of the cavity should not result in any decrease in the laser efficiency.

Large-aperture lasers with helical unstable resonators can find their applications in scientific research for investigation of vortex solitons in nonlinear media or manipulation of small particles and atomic beams. They can be also used in technology for applications where the ring structure of the helical beams or their nonzero orbital momentum is of importance.

ACKNOWLEDGMENTS

The authors are indebted to V.I. Kuprenyuk for discussions and to I.M. Kozlovskaya and A.M. Kokushkin for their technical assistance in preparation of the manuscript.

REFERENCES

1. J.H.Vaughan and D.V.Willets, "Interference properties of a light beam helical wave surface" Opt. Comm. **30**, pp. 263-267, 1979.
2. P. Coullet, L. Gil and F. Rocca "Optical vortices", Opt. Comm. **73**, pp.403-408, 1989.

3. I.V. Basistiy, M.S. Soskin and M.V. Vasnetsov, "Optical wavefront dislocation and their properties, Opt. Comm. **119,** pp. 604-612, 1995.

4. N.R. Heckenberg, R.Mc. Duff, C.P. Smith, H. Rubinsztein-Dunlop, M.J. Wegener. "Laser beams with phase singularities", Optical and Quantum Electronics **24,** s951-s962, 1992.

5. E. Abramochkin and V. Volostnikov, "Spiral-type beams", Opt. Comm. **102,** pp. 336-350, 1993.

6. L. Allen, M.V. Beijersbergen, R.J.C. Spreeuw, J.P. Woerdman, "Orbital angular momentum of light and the transformation of Laguerre-Gaussian laser modes", Phys. Rev. A. **45,** n. 11, pp.8185-8189, 1992.

7. D. Rozas, C.T. Law, G.A. Swartzalander, "Propagation dynamics of optical vortices", JOSA B, **14,** pp. 3054-3065, 1997.

8. H. Kogelnik and T. Li, "Laser Beams and Resonators", Appl. Opt. **5,** p. 1550, 1966.

9. H.He, M.E.J. Friese, N.R. Heckenberg and H. Rubinsztein-Dunlop, "Direct observation of transfer of angular momentum to absorptive particles from a laser beam with a phase singularity", Phys. Rev. Lett. **75,** n.5, pp.826-829, 1995

10. N.R. Heckenberg, T.A. Neiminen, M.E.J. Friese, H. Rubinsztein-Dunlop, "Trapping microscopic particles with singular beams", SPIE **3487,** pp.46-52, 1998.

11. S.M. Block, Optical tweezers: "A new tool for biophysics in Noninvasive Techniques in Cell Biology", pp. 375-402, Wiley-Liss, Inc., 1990.

12. G.A. Swartzlander, Jr., and G.T. Law, "Optical vortex solitons observed in Kerr nonlinear media", Phys. Rev. Lett., **69,** pp. 2503-2506, 1992.

13. V.Yu. Bazhenov, M.V. Vasnetsov and M.S. Soskin, "Laser beams with screw wavefront dislocations", JETP Lett., **52,** n.8, pp. 1037-1039, 1990

14. C. Tamm and C.O. Weiss, "Bistability and optical switching of spatial patterns in a laser", JOSA B, **7,** n.6, pp. 1034-1038, 1990

15. N.N. Rosanov, "On formation of radiation with wavefront dislocations", Optika i spectroscopia, **75,** №4, pp.861-867, 1993.

16. E. Abramochkin, V. Volostnikov, "Beam transformations and nontransformed beams", Opt.Comm., **83,** pp.123-135, 1991.

17. E. Abramochkin, V. Volostnikov, "Spiral-type beams optical and quantum aspects", Opt. Comm., **125,** pp. 302-323, 1996.

18. M.W. Beijersbergen, R.P.C. Coerwinkel, M. Kristensen, J.P. Woerdman, " Helical-wavefront laser beams produced with a spiral phaseplate", Opt.Comm., **112,** pp.321-327, 1994.

19. Handbook of Mathematical Functions. Ed. By M. Abramowitz and I. Stegun, Nat. Bureau of Standards, Appl. Math. Series, 1964.

20. I.G. Marienko, M.S. Soskin, M.V. Vasnetsov, " Phase conjugation of wavefronts containing phase singularities", SPIE, **3487,** pp.39-41, 1998.

21. Yu.A. Anan'ev "Unstable resonators and their applications", In: "Kvantovaya electronika", Ed. N.G. Basov, №6, p. 3, 1971

22. Yu.A. Anan'ev, V.E. Sherstobitov, «The influence of edge effects on properties of unstable cavities». In: «Kvantovaya electronika», Ed. N.G .Basov, №3, p.82, 1971 (in Russian)

23. S.A. Dimakov, S.I. Zavgorodneva, L.V. Koval'chuk, A.Yu. Rodionov, V.E .Sherstobitov, V.P.Yashukov, "A study of spatial characteristics of the radiation from a CO_2 EIL with intracavity apodization", Kvantovaya electronika, **17,** №3, pp. 291-295, 1990.

24. A.E. Siegman, R. Arrathoon, "Modes in unstable optical resonators and lens waveguides", IEEE of Quant. Electr., QE-3, p.156, 1967.

SESSION 4

Characterization of Laser Beams

Measuring laser beam parameters, phase and spatial coherence using the Wigner distribution

Bernd Eppich[a], Sandra Johansson[b], Holger Laabs[c], Horst Weber[d]

[a,b,d]Laser- und Medizin-Technologie gGmbH, 12489 Berlin, Germany
[c,d]Technische Universität Berlin, 10623 Berlin, Germany

ABSTRACT

Detailed laser beam characterization is essential for the proper choice of lasers source according to the respective application as well as for the optimization of optical systems. Since most lasers generate partially coherent beams, intensity and phase distributions are not enough to describe them. In addition the knowledge of the coherence distribution is necessary. So far different setups have been used to measure phase and coherence distribution with limited accuracy. Here we demonstrate a new measurement procedure which is based on the retrieval of the Wigner distribution, from which all relevant information can be derived. The setup is very simple and the results seems to be fairly accurate.

Keywords: laser beam characterization, beam parameters, phase space distribution, phase retrieval, coherence retrieval

1. THE WIGNER FUNCTION FOR PARTIALLY COHERENT BEAMS

Originally, the Wigner distribution function has been introduced by Eugene Wigner in 1932 as a phase space distribution in quantum mechanics[1]. In 1968 a similar distribution has been introduced in theoretical optics by Walther as a link between coherence theory and classical radiometry[2]. A detailed description of the Wigner distribution for partially coherent optical beams has then been given by Bastiaans[3,4,5]. Although a significant number of applications have been found for the Wigner distribution in the field of theoretical optics, it didn't found its way into experimental (practical) optics. Only within the last few years, coming along with an increasing interest in beam characterization and standardization efforts, first experimental results regarding the measurement of the Wigner distribution have been published[6,7]. In the following we will give a brief introduction of the Wigner distribution as far as it is necessary for this paper. More details can be found elsewhere[3,4,5].

Within the paraxial approximation and neglecting polarization effects, a partially coherent optical beam can be described by the scalar, complex field distribution $E(\bar{r},t)$. Assuming a temporally stationary process, the so-called mutual coherence function Γ at a transverse plane z is only a function of time differences:

$$\Gamma(\bar{r}_{\perp,1},\bar{r}_{\perp,2},t_1-t_2)=\Gamma(\bar{r}_{\perp,1},\bar{r}_{\perp,2},t_1,t_2)=\left\langle E(\bar{r}_{\perp,1},t_1)\cdot E^*(\bar{r}_{\perp,2},t_2)\right\rangle, \tag{1}$$

where $\bar{r}_\perp = \begin{pmatrix} x \\ y \end{pmatrix}$ is the transverse spatial vector and z the position in the direction of propagation of the beam.

Although, the brackets in (1) denote an ensemble average, it is usually not wrong to consider it as an time average over a sufficiently long span. The temporal Fourier transform of the mutual coherence function is known as the cross spectral density function:

$$\Gamma(\bar{r}_{\perp,1},\bar{r}_{\perp,2},\omega)=\int\Gamma(\bar{r}_{\perp,1},\bar{r}_{\perp,2},\tau)e^{i\omega\tau}\,d\tau \tag{2}$$

For quasi-monochromatic beams, i.e. beams with a small bandwidth compared to the mid frequency, the temporal-frequency variable ω can be neglected. This will be assumed throughout the remainder of this paper.

---

[a] e-mail: b.eppich@lmtb.de

In *Laser Resonators III*, Alexis V. Kudryashov, Alan H. Paxton, Editors,
Proceedings of SPIE Vol. 3930 (2000) • 0277-786X/00/$15.00

Given the cross spectral density in one transverse plane the beam is completely characterized and its propagation through first order optical systems (combinations of perfect lenses and free propagation) can be calculated. Starting from the generalized Kirchhoff diffraction integral for coherent fields[8]

$$E_o(\vec{r}_{\perp,o}) = \int E_i(\vec{r}_{\perp,i}) K(\vec{r}_{\perp,i}, \vec{r}_{\perp,o}) d\vec{r}_{\perp,i} \text{ with } K(\vec{r}_{\perp,i}, \vec{r}_{\perp,o}) = \frac{ik}{2\pi\sqrt{\det(\mathbf{B})}} e^{-ikL} e^{-i\frac{k}{2}(\vec{r}_{\perp,i}\mathbf{B}^{-1}\mathbf{A}\vec{r}_{\perp,i} - 2\vec{r}_{\perp,i}\mathbf{B}^{-1}\vec{r}_{\perp,o} + \vec{r}_{\perp,o}\mathbf{D}\mathbf{B}^{-1}\vec{r}_{\perp,o})}$$
(3)

where k is the wavenumber and $\mathbf{A}$, $\mathbf{B}$, $\mathbf{C}$, and $\mathbf{D}$ are the 2×2 submatrices of the symplectic[10] 4×4 geometrical-optical system matrix

$$S = \begin{pmatrix} \mathbf{A} & \mathbf{B} \\ \mathbf{C} & \mathbf{D} \end{pmatrix}$$
(4)

one easily obtains the propagation law for the cross spectral density through first order systems due to

$$\Gamma_o(\vec{r}_{o,1}, \vec{r}_{o,2}) = \int K(\vec{r}_{i,1}, \vec{r}_{o,1}) \Gamma(\vec{r}_{i,1}, \vec{r}_{i,2}) K^*(\vec{r}_{i,2}, \vec{r}_{o,2}) d\vec{r}_{i,1} d\vec{r}_{i,2}$$
(5)

where $\Gamma_i(\vec{r}_{\perp,1}, \vec{r}_{\perp,2})$ and $\Gamma_o(\vec{r}_{\perp,1}, \vec{r}_{\perp,2})$ is the cross spectral density in the input plane and output plane, respectively.

The measurable (time averaged) intensity distribution is given by

$$I(\vec{r}_\perp) = \Gamma(\vec{r}_\perp, \vec{r}_\perp).$$
(6)

Although the cross spectral density is sufficient for the complete description of partially coherent beams, there are a couple of other distributions having some additional advantages. The Wigner distribution is one of them. It is defined by a certain spatial Fourier transform of the cross spectral density

$$F(\vec{r}_\perp, \vec{q}_\perp) = \int \Gamma(\vec{r}_\perp + \tfrac{1}{2}\vec{r}_\perp', \vec{r}_\perp - \tfrac{1}{2}\vec{r}_\perp') e^{-i\vec{q}_\perp^T \vec{r}_\perp'} d\vec{r}_\perp'.$$
(7)

Obviously, the Wigner distribution has the same informational content as the cross spectral density. Its outstanding importance results from the following properties of the Wigner distribution. The measurable intensity distribution is given by integration over the variable $\vec{q}_\perp = \begin{pmatrix} u \\ v \end{pmatrix}$:

$$I(\vec{r}_\perp) = \int F(\vec{r}_\perp, \vec{q}_\perp) d\vec{q}_\perp$$
(8)

The propagation through first order systems becomes simply

$$F_o(\vec{r}_\perp, \vec{q}_\perp) = F_i(\mathbf{D}\vec{r}_\perp - \mathbf{B}\vec{q}_\perp, -\mathbf{C}\vec{r}_\perp + \mathbf{A}\vec{q}_\perp),$$
(9)

where $F_i(\vec{r}_\perp, \vec{q}_\perp)$ is the Wigner distribution at the input plane and $F_o(\vec{r}_\perp, \vec{q}_\perp)$ the resulting Wigner distribution at the output plane.

Considering $\vec{q}_\perp$ as an angular (spatial frequency) variable, the equations (8) and (9) corresponds to the behavior of a bundle of geometric-optical rays passing a first order system. Although this analogy is limited[9], it is often helpful to consider the Wigner distribution as a geometrical-optical ray density, giving the amplitude of a ray passing through the point $\vec{r}_\perp$ and having the direction $\vec{q}_\perp$. Hence, the far field of the beam expressed as a function of the angular coordinates can be obtained by integrating the Wigner distribution over the spatial variables:

$$I_q(\vec{q}_\perp) = \int F(\vec{r}_\perp, \vec{q}_\perp) d\vec{r}_\perp$$
(10)

2. BEAM PARAMETERS DERIVED FROM THE WIGNER DISTRIBUTION

Although the four dimensional Wigner distribution as well as the cross spectral density give complete beam characterization, it is often desirable to get specific information in a more condensed form, according to the application under consideration. All of these beam parameters and distributions can be derived from the Wigner distribution or the cross spectral density. For some of them it is preferable to use the Wigner distribution, for others the cross spectral density usually delivers more accurate results. In this section we will present a brief description of some useful quantities and how they can be obtained.

1. Second order moments

The centered second order moments of the Wigner distribution enable a very quick and easy description of the gross propagation properties of a beam through first order optics. From a maximum of ten parameters the beam diameter behind optical systems can be predicted. The moments of the Wigner distribution are defined as

$$\left\langle x^n y^m u^p v^t \right\rangle = \frac{\int F(\vec{r}_\perp, \vec{q}_\perp) x^n y^m u^p v^t \, d\vec{r}_\perp \, d\vec{q}_\perp}{\int F(\vec{r}_\perp, \vec{q}_\perp) \, d\vec{r}_\perp \, d\vec{q}_\perp} \tag{11}$$

whereas the centered moments are given by

$$\left\langle x^n y^m u^p v^t \right\rangle_c = \frac{\int F(\vec{r}_\perp, \vec{q}_\perp)(x - \langle x \rangle)^n (y - \langle y \rangle)^m (u - \langle u \rangle)^p (v - \langle v \rangle)^t \, d\vec{r}_\perp \, d\vec{q}_\perp}{\int F(\vec{r}_\perp, \vec{q}_\perp) \, d\vec{r}_\perp \, d\vec{q}_\perp}. \tag{12}$$

Of particular interest are the ten centered second order moments, i.e. $n + m + p + t = 2$. The pure spatial moments $\left\langle x^2 \right\rangle_c$, $\left\langle xy \right\rangle_c$, and $\left\langle y^2 \right\rangle_c$ can directly be calculated from the power density distribution as

$$\left\langle x^n y^m \right\rangle_c = \frac{\int I(\vec{r}_\perp)(x - \bar{x})^n (y - \bar{y})^m \, d\vec{r}_\perp}{\int I(\vec{r}_\perp) \, d\vec{r}_\perp} \tag{13}$$

and are closely related to the transverse beam extension. Hence, they have been used for the definition of the beam diameter in the recent international draft standard ISO/FDIS 11146. The ten centered second order moments can be arranged in a 4×4 matrix

$$\mathbf{P} = \begin{pmatrix} \left\langle x^2 \right\rangle & \left\langle x y \right\rangle & \left\langle x u \right\rangle & \left\langle x v \right\rangle \\ \left\langle x y \right\rangle & \left\langle y^2 \right\rangle & \left\langle y u \right\rangle & \left\langle y v \right\rangle \\ \left\langle x u \right\rangle & \left\langle y u \right\rangle & \left\langle u^2 \right\rangle & \left\langle u v \right\rangle \\ \left\langle x v \right\rangle & \left\langle y v \right\rangle & \left\langle u v \right\rangle & \left\langle v^2 \right\rangle \end{pmatrix} \tag{14}$$

resulting in a simple propagation law for first order optics

$$\mathbf{P}_o = \mathbf{S} \cdot \mathbf{P}_i \cdot \mathbf{S}^T \tag{15}$$

which enables a quick and easy prediction of the beam diameters. Furthermore, two invariant quantities can be derived from this propagation law and the symplecticity[10] of the system matrices. One of them is the determinant of the matrix. This quantity is equivalent to the beam propagation factor defined in ISO/FDIS 11146. The other one has recently found by Nemes[11] and others and is given by

$$T = \left(\left\langle x^2 \right\rangle \left\langle u^2 \right\rangle - \left\langle xu \right\rangle^2 \right) + \left(\left\langle y^2 \right\rangle \left\langle v^2 \right\rangle - \left\langle yv \right\rangle^2 \right) + 2 \left(\left\langle xy \right\rangle \left\langle uv \right\rangle - \left\langle xv \right\rangle \left\langle yu \right\rangle \right). \tag{16}$$

This parameter is closely related to the intrinsic astigmatism of the beam as described by Nemes.

2. Phases

The complex valued cross spectral density can be uniquely decomposed into the following real valued terms:

$$\Gamma\left(\vec{r}_{\perp,1},\vec{r}_{\perp,2}\right)=\sqrt{I\left(\vec{r}_{\perp,1}\right)}\sqrt{I\left(\vec{r}_{\perp,2}\right)}\,\kappa\left(\vec{r}_{\perp,1},\vec{r}_{\perp,2}\right)e^{i\,\Phi\left(\vec{r}_{\perp,1},\vec{r}_{\perp,2}\right)} \tag{17}$$

where $I\left(\vec{r}_{\perp}\right)=\Gamma\left(\vec{r}_{\perp},\vec{r}_{\perp}\right)$ is the power density distribution, $\kappa\left(\vec{r}_{\perp,1},\vec{r}_{\perp,2}\right)$ is the coherence distribution (next section), and $\Phi\left(\vec{r}_{\perp,1},\vec{r}_{\perp,2}\right)$ might be called the **mutual phase distribution**. From equation (6) and (17) it follows

$$\kappa\left(\vec{r}_{\perp},\vec{r}_{\perp}\right)=1 \tag{18}$$

and

$$\Phi\left(\vec{r}_{\perp},\vec{r}_{\perp}\right)=0\ . \tag{19}$$

For coherent fields the coherence distribution equals unit everywhere and the mutual phase distribution can be expressed as

$$\Phi\left(\vec{r}_{\perp,1},\vec{r}_{\perp,2}\right)=\varphi\left(\vec{r}_{\perp,1}\right)-\varphi\left(\vec{r}_{\perp,2}\right)\ . \tag{20}$$

Here, $\varphi\left(\vec{r}_{\perp}\right)$ is the familiar phase distribution of the complex field, a two dimensional quantity. In general, for partially coherent beams the four dimensional mutual phase distribution can not be decomposed in this way. The mutual phase might then be considered as the average phase relation between any pair of points, as could be measured, e.g. by Young's double hole interference experiment. Note, that for partially coherent beams in general the mutual phase can not be completely compensated by phase plates, while this is always possible for coherent beams. Hence, usually it is not necessary to know the complete phase information. E.g., it can be shown that the beam propagation factor M^2 of a beam only depends on the Poynting vector distribution[12] $\vec{S}\left(\vec{r}_{\perp}\right)$ which is given by a certain partial derivation of the mutual phase weighted by the intensity:

$$S_x\left(\vec{r}_{\perp}\right)=\frac{1}{2}I\left(\vec{r}_{\perp}\right)\left(\frac{\partial}{\partial x_1}-\frac{\partial}{\partial x_2}\right)\Phi\left(\vec{r}_{\perp,1},\vec{r}_{\perp,2}\right)\Bigg|_{\vec{r}_{\perp,1}=\vec{r}_{\perp,2}}\ ,\quad S_y\left(\vec{r}_{\perp}\right)=\frac{1}{2}I\left(\vec{r}_{\perp}\right)\left(\frac{\partial}{\partial y_1}-\frac{\partial}{\partial y_2}\right)\Phi\left(\vec{r}_{\perp,1},\vec{r}_{\perp,2}\right)\Bigg|_{\vec{r}_{\perp,1}=\vec{r}_{\perp,2}}\ . \tag{21}$$

For coherent beams this simplifies to

$$\vec{S}\left(\vec{r}_{\perp}\right)=I\left(\vec{r}\right)\vec{\nabla}\varphi\left(\vec{r}_{\perp}\right)\ . \tag{22}$$

As mentioned above, any beams having the same intensity, coherence and Poynting distribution do have the same beam propagation factor and the same second order moments even if their mutual phase distributions differ. Once again, in the case of partially coherent beams in general the Poynting vector distribution can not be completely compensated by phase elements. Instead, it could be asked for which phase element minimizes the Poynting distribution:

$$\int\vec{S}\left(\vec{r}_{\perp}\right)-I\left(\vec{r}_{\perp}\right)\vec{\nabla}\varphi_c\left(\vec{r}_{\perp}\right)d\vec{r}_{\perp}=\min\ . \tag{23}$$

The phase shift distribution $\varphi_c\left(\vec{r}_{\perp}\right)$ of this element may define the **coherent phase distribution**, i.e. that part of the mutual phase that can be compensated by phase elements. This is the smallest valuable amount of phase information. Only in the case of completely coherent fields the mutual phase, the Poynting distribution, and the coherent phase are equivalent.

3. Coherence

Equation (17) also defines the coherence distribution $\kappa\left(\vec{r}_{\perp,1},\vec{r}_{\perp,2}\right)$. It expresses the coherence between any pair of points. From the definition it follows immediately

$$0\leq\kappa\left(\vec{r}_{\perp,1},\vec{r}_{\perp,2}\right)\leq 1 \tag{24}$$

where the upper bound means complete coherence and the lower bound means no coherence. For some applications a high degree of coherence is desirable, i.e. for the generation of any kind of interference pattern. For other applications coherence is disturbing, e.g. if homogeneous power density profiles are necessary. From the coherence distribution some other coherence parameters can be derived. The transverse coherence length is a measure for the average size of coherent regions within the beam cross section:

$$L_c = \sqrt{\frac{\int |\Gamma(\vec{r}_{\perp,1}, \vec{r}_{\perp,2})|^2 \cdot |\vec{r}_{\perp,1} - \vec{r}_{\perp,2}|^2 \, d\vec{r}_{\perp,1} \, \vec{r}_{\perp,2}}{\int |\Gamma(\vec{r}_{\perp,1}, \vec{r}_{\perp,2})|^2 \, d\vec{r}_{\perp,1} \, \vec{r}_{\perp,2}}} \quad . \tag{25}$$

A global degree of coherence can be defined by

$$K = \frac{\int |\Gamma(\vec{r}_{\perp,1}, \vec{r}_{\perp,2})|^2 \, d\vec{r}_{\perp,1} \, d\vec{r}_{\perp,2}}{\left(\int \Gamma(\vec{r}_\perp, \vec{r}_\perp) d\vec{r}_\perp \right)^2} = \lambda \frac{\int h(\vec{r}_\perp, \vec{q}_\perp)^2 \, d\vec{r}_\perp \, d\vec{q}_\perp}{\left(\int h(\vec{r}_\perp, \vec{q}_\perp) d\vec{r}_\perp \, d\vec{q}_\perp \right)^2}, \quad 0 \le K \le 1 \; . \tag{26}$$

The global degree of coherence is invariant under propagation through passive and lossless systems. It is unit for coherent beams and vanishes for completely incoherent beams.

3. MEASUREMENT OF THE WIGNER FUNCTION

1. One transverse dimension

Due to the uncertainty relation, the Wigner distribution cannot be measured point by point. To understand the actual measurement procedure we will first reduce the complexity of the problem considering beams with only one transverse dimension. The Wigner distribution is then a function of one spatial coordinate x and one angular coordinate u. Its propagation through first order optical systems is given by

$$F_o(x, u) = F_i(D x - B u, -C x + A u) \tag{27}$$

where the 2×2 matrix $\mathbf{S} = \begin{pmatrix} A & B \\ C & D \end{pmatrix}$ defines the ray propagation through the optical system. As in the two dimensional case, the power density distribution is obtained by integration over the angular variable:

$$I(x) = \int F(x, u) \, du \tag{28}$$

Hence, given a Wigner distribution $F(x, u)$ the power density distribution behind a first order system is obtained by

$$I(x, \mathbf{S}) = \int F(Dx - Bu, -Cx + Au) \, du \tag{29}$$

which can also be expressed as

$$I(x, \mathbf{S}) = \int F(x', u') \delta(x - Ax' - Bu') \, dx' \, du' \tag{30}$$

using Dirac's delta function. Fourier transforming both sides of eq. (30) with respect to the positional coordinate x results in

$$\int I(x, \mathbf{S}) e^{ikxw} \, dx = Z(B w, A w) \tag{31}$$

where

$$Z(s, w) = \int F(x, u) e^{ik(wx - us)} \, dx \, du = \int \Gamma\left(x + \frac{s}{2}, x - \frac{s}{2}\right) e^{ikwx} \, dx \tag{32}$$

is the so-called ambiguity distribution[14], a two dimensional Fourier transform of the Wigner distribution. From equation (31) it can be seen, that the Fourier transform of any power density distribution behind first order optical systems can be considered as central slices of the ambiguity distribution in the source plane (fig. 1). Hence, the ambiguity distribution can be scanned slice by slice by measuring the power density distributions behind various optical systems. The next step in the reconstruction of the Wigner distribution is then to re-sample the ambiguity distribution into a Cartesian coordinate system in order to enable the two dimensional Fourier transform into the Wigner space. Instead of creating the Wigner distribution, the cross spectral density may be calculated by an one dimensional Fourier transform of the ambiguity distribution with respect to the variable w. It should be noted that the whole reconstruction algorithm for the Wigner distribution is essentially an inverse Radon transform, since the power density distributions behind first order systems can be considered as angular projections of the initial Wigner distribution which is mathematical a forward Radon transform[11].

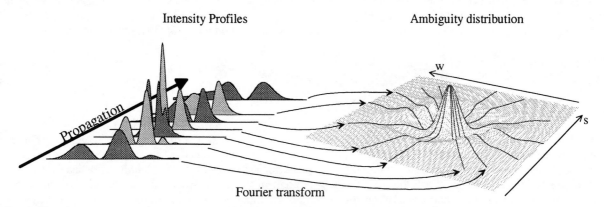

Fig. 1: Retrieval of the ambiguity distribution from measured intensity profiles. The Fourier transform of each profile gives a slice of the ambiguity distribution. For further processing the values between the slices have to be interpolated.

2.   Measurement Setup

The simplest optical system for scanning the ambiguity distribution is free propagation over the distance z,

$$\mathbf{S} = \begin{pmatrix} 1 & z \\ 0 & 1 \end{pmatrix}.$$

(33)

Using equation (31) one obtains

$$\int I(x,z)e^{ikxw}\,dx = Z(zw,w).$$

(34)

The number and positions necessary for an optimum sampling of the ambiguity distribution obviously depends on the ambiguity distribution itself. But it can be shown that in general best results are obtained if the power density profiles are acquired within a range of about ±3 Rayleigh length around the beam waist. The distances between the z positions should increase with the distance from the waist. If the beam waist is not accessible it usually has to be created by an appropriate lens. Hence, the required measurement procedure matches almost perfectly to the one which is described in the ISO draft ISO/FDIS 11146, where it is proposed for the determination of the beam propagation factor.

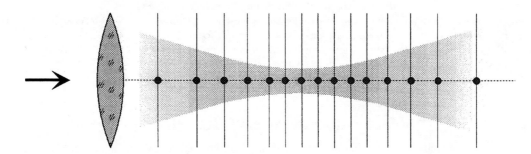

Fig. 2: Measurement setup for optimum sampling of the ambiguity distribution. Intensity profiles within approximately ±3 Rayleigh length around the beam waist are needed. In the waist region the measurement position should lie closer.

3. Two transverse dimensions

Real beams actually do have two transverse dimensions. Hence, their ambiguity distribution is four dimensional. Extending equation (31) to two transverse dimensions shows that the two dimensional Fourier transform of each intensity distribution gives a two dimensional central slice of the ambiguity distribution. It can be shown, that stigmatic optical systems, i.e. combinations of spherical lenses and free propagation, having a ray propagation matrix of the form

$$\mathbf{S} = \begin{pmatrix} A & 0 & B & 0 \\ 0 & A & 0 & B \\ C & 0 & D & 0 \\ 0 & C & 0 & D \end{pmatrix} \tag{35}$$

are not capable for a complete sampling of the ambiguity distribution. In fact, using stigmatic systems only a three dimensional subspace of the ambiguity distribution would be scanned. Although it is in principal possible to use a variety of astigmatic systems for a complete scan, this seems not to be practical. Thus, we propose a solution which takes into account a certain loss of information, but allows the same simple measurement setup as described above. The approach is based on angular projections of the four dimensional Wigner distribution. Defining the azimuthal angle α as shown in figure 3 the projected Wigner distribution $F_\alpha(a, w)$ is defined as

$$F_\alpha(a, w) = \int F(\vec{r}_\perp, \vec{q}_\perp) \delta(a - x\cos\alpha - y\sin\alpha) \delta(w - u\cos\alpha - v\sin\alpha) d\vec{r}_\perp \, d\vec{q}_\perp \tag{36}$$

The integration of the projected Wigner distribution with respect to the angular coordinate w delivers the angular projection of the intensity distribution:

$$I_\alpha(a) = \int I(\vec{r}_\perp) \delta(a - x\cos\alpha - y\sin\alpha) d\vec{r}_\perp = \int F_\alpha(a, w) dw \ . \tag{37}$$

The propagation of the projected Wigner distribution through stigmatic first order systems is exactly the same as the propagation of the two dimensional Wigner distribution:

$$F_o(x, u) = F_i(D x - B u, -C x + A u) \ . \tag{38}$$

Hence, the same reconstruction algorithm and measurement setup can be applied to the projected Wigner distribution. The advantage is that all the input data for the reconstruction of **all** Wigner projections can be acquired at once by only one single caustic scan as described above. The angular projections of the Wigner distribution form a three dimensional subspace of the complete four dimensional Wigner distribution. Hence, some information are lost as mentioned above. Nevertheless it can be shown, that most relevant quantities are conserved. An example is the phase distribution. Although the complete mutual phase distribution cannot be retrieved from the projections, the **coherent phase distribution** (as defined in eq. 23) can be obtained even for partially coherent beams with no symmetries. For beams which are assumed to be stigmatic and having no twisted phase a single projection is enough for a complete description of the beam. For simple astigmatic beams two orthogonal projections are sufficient.

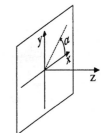

Fig. 3: Definition of azimuthal angle.

4. NUMERICAL SIMULATIONS

Several numerical simulations have been performed to test the stability and sensitivity of the reconstruction algorithm against noise. We calculated the propagation of the incoherent superposition of a TEM$_{03}$ and TEM$_{04}$, both suffering from an off axis gaussian phase distortion with a maximum distortion of $\lambda/10$. To the intensity distribution statistical noise of various levels has been added before they have been used as input data of the reconstruction algorithm. We investigated the dependency of the beam propagation factor M^2, the global coherence K, and the global transverse coherence length L_c on the noise level (given in percentage of the maximum intensity value) using a simulated scan of 21 intensity profiles. In addition we reconstructed the phase distribution in the reference plane. Figure 4 shows how weakly the beam parameters depend on the statistical noise. A noise level of 5% is quite high, since the input data are usually projections of CCD camera images and this operation reduces the relative noise level. Nevertheless even with a noise level of 10% the deviation from the theoretical values are below 3% for all parameters. Figure 5 shows that the reconstruction of the phase profile is more sensitive to noise. Up to a noise level of 3% the maximum deviations from the known phase distributions are about $\lambda/100$. Higher noise delivers significant deviations. But these deviations can be reduced if the number of intensity profiles is simultaneously increased as shown in figure 6, where the number of profiles has been doubled at a noise level of 5%.

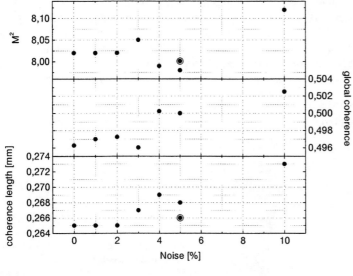

Fig. 4: Calculated global beam parameters at various noise levels. They are quite insensitive to noise.

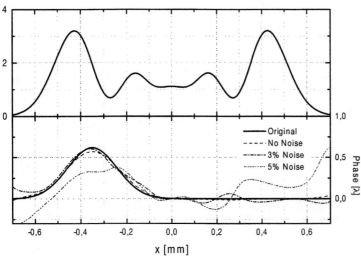

Fig. 5: Reconstructed phase profiles at various noise levels. At noise level less than 5% the deviations to the known phase profile is less than $\lambda/100$. At a noise level of 5% significant deviations occur.

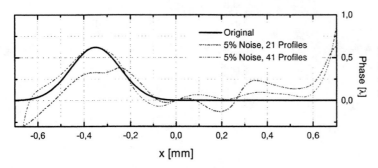

Fig. 6: Reconstructed phase distribution at a noise level of 5% and two different numbers of intensity profiles used. Enlarging the number of profils improves the accuracy.

5. EXPERIMENTAL EXAMPLE

To demonstrate the potential use of the new method we used it to investigate the beam properties of a special fiber pumped Nd:YAG laser. Fig. 7 shows a scheme of the laser setup. The HR coated backside of the laser crystal together with an external spherical mirror forms a stable resonator. Pumping occurs within a small region through the backside of the laser crystal by a diode via a fiber. The off axis pumping position can be adjusted by a screw. The overlap between the gain profile and the different Gauss-Hermite modes determines the number and distribution of excited modes, which are mutual incoherent. Hence, by variation of the pumping position and power the beam propagation factor M^2 as well as the global degree of coherence K will change. Open questions were, if the setup allows excitation of single higher order Gauss-Hermite modes or not, and if the pumping mechanism causes significant phase distortions. To answer these questions we measured the Wigner distribution at various pumping positions and successively derived the desired information from it. Figure 8 shows the beam propagation factor M^2 and the global coherence K as a function of the beam position. Even at the pumping positions where the coherence has its maximum, it is significantly below unit indicating that more than just one single mode is excited. Between these pumping positions the global coherence drops down by approximately a factor of two indicating a higher number of excited modes. The two different symbols in the plots indicate two independent measurements demonstrating the accuracy of the method.

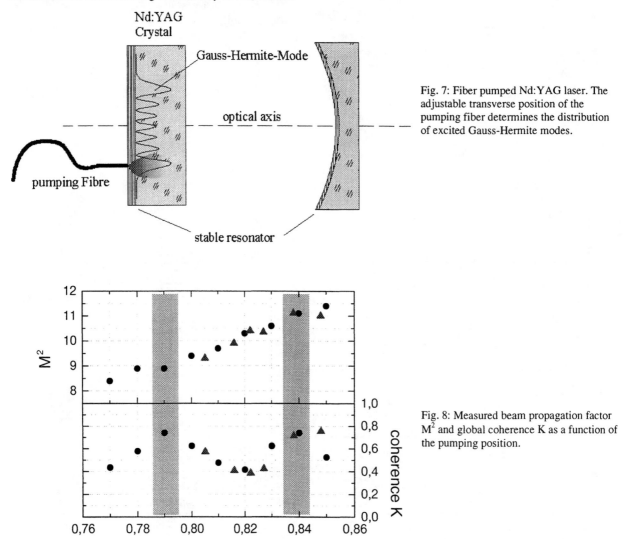

Fig. 7: Fiber pumped Nd:YAG laser. The adjustable transverse position of the pumping fiber determines the distribution of excited Gauss-Hermite modes.

Fig. 8: Measured beam propagation factor M^2 and global coherence K as a function of the pumping position.

In order to give an impression of the accuracy of the phase retrieval capabilites we repetitively measured the Wigner distribution at the same pumping conditions and derived the phase profile. The upper part of figure 9 shows the intensity profile in the reference plane. In the lower part the retrieved phase profiles (given in wavelength) of two independent measurements are shown. The phase destortions are very weak, less then $\lambda/100$. Nevertheless, it was possible to repetitively reproduce this measurement with a very high accuracy, confirming our numerical investigations.

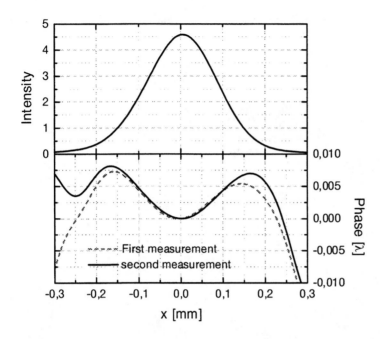

Fig. 9: Phase retrieval based on two independent measurements. Although the phase distortions are very weak, they could be nicely reproduced.

6. SUMMARY

It has been shown how the Wigner distribution can be retrieved from a couple of measured intensity profiles. The setup is very simple and cheap. It corresponds to the setup for the measurement of the beam propagation factor M^2 as described in the draft standard ISO/FDIS 11146. The complete procedure takes only a few minutes and has the potential for automation. The reconstruction algorithm is transparent, robust, and sufficiently fast (retrieval of a Wigner distribution with 200×200 sampling points takes less then ten seconds).

From the Wigner distributions all relevant beam information, phase and coherence distributions can be derived. Numerical simulations and first experiments have indicated that the new method is relatively insensitive to noise and provides fairly accurate results.

All these properties may favorite it for many beam characterization tasks even in industrial environments.

7. ACKNOWLEDGMENTS

This work was supported by the German Federal Ministry of Science and Research and by the city of Berlin.

8. REFERENCES

1. E. Wigner, "On the quantum correction for thermodynamic equilibrium", *Phys. Rev.* **40**, pp. 749-759, 1932
2. A. Walther, "Radiometry and Coherence", *Opt. Soc. Am.* **58**, pp. 1256-1259, 1968
3. M.J. Baastians, "Wigner distribution function and its application to first order optics", *J. Opt. Soc. Am.* **69**, pp. 1710-1716, 1979
4. M.J. Baastians, "The Wigner distribution function for partially coherent light", *Optica Acta* **28(9)**, pp. 1215-1224, 1981
5. M.J. Baastians, "Application of the Wigner distribution function to partially coherent light", *J. Opt. Soc. Am.* **A3**, pp. 1227-1238, 1986

6. D.F. McAlister et al., "Optical phase retrieval by phase-space tomography and fractional-order Fourier transforms", *Optic Letters* **20**, pp. 1181-1183, 1995

7. B. Eppich, "Measurement of the Wigner distribution function based on the inverse Radon transformation", in *Beam Control, Diagnostics, Standards, and Propagation*, L.W. Austin, A. Giesen, D.H. Leslie, H. Weichel,, Vol. 2375, pp. 261-268, SPIE, 1995

8. S.A. Collins, "Lens-System Diffraction Integral written in terms of Matrix Optics", *J. Opt. Soc.* **A1**, pp. 1168-1177, 1970

9. A.T. Friberg, "On the existence of a radiance function for finite planar sources of arbitrary state of coherence", *J. Opt. Soc. Am.* **69**, p. 192-198, 1979

10. R.K. Luneburg, *Mathematical theory of optics*, University of California Press, 1964

11. C.M. Vest, "Formation of images: Radon and Abel transforms", *J. Opt. Soc. Am.* **64**, pp. 1215-1218, 1974

12. D.Paganin, K.A. Nugent, "Noninterferometric phase imaging with partially coherent light", Phys.Rev.Lett. **80**, pp. 2586-2589, 1998

13. G. Nemes, A.E. Siegman, "Measurement of all ten second-order moments of an astigmatic beam by the use rotating simple astigmatic (anamorphic) optics", *J. Opt. Soc. Am.* **11**, pp. 2257-2264, 1994

14. A. Papoulis, "Ambiguity function in Fourier optics", *J. Opt. Soc. Am.* **64**, pp. 779-788, 1974

15. R. Castaneda, "On spatial coherence beams", OKTIK 109, pp. 77-83, 1998

Multi-Gaussian Beams - A Super-Gaussian Alternative

Anthony A. Tovar[*]

Physics and Engineering Programs, Eastern Oregon University, La Grande, OR 97850-2899

ABSTRACT:

The multi-Gaussian beam shape is proposed as a model for aperture functions and laser beam profiles which have a nearly flat top, but whose sides decrease continuously. Beams and apertures of this type represent a simple, elegant, and intuitive alternative to super-Gaussian beams. The design of lasers that have these beams as its eigenmodes is discussed. Such laser resonators would have increased energy extraction from the laser amplifier.

Keywords: Gaussian beams, propagation, diffraction, super-Gaussian beams, laser resonators, optical systems, high power lasers

1. INTRODUCTION

A top-hat laser beam profile is highly sought in a wide variety of applications. However, this top-hat profile, represented by a circ function, destabilizes as it propagates, leading to an undesirable ringing phenomena. This is well known from the study of diffraction of a plane wave by a circular hole. The oft-used Gaussian-profiled beam has the advantage of retaining its smooth shape as it propagates through free space. The disadvantage of the Gaussian shape is that the low intensity at the sides of the beam does not make it a good enough approximation to the desired top-hat shape. For example, the Gaussian shape is much less efficient at extracting energy from a laser amplifier than the top-hat.

A popular alternative to the Gaussian and the circ profiles is the super-Gaussian profile. However, the evaluation of the free space propagation of the field with this profile cannot be performed in a closed form, and numerical techniques are required. This difficulty would be alleviated if one could write the beam profile as a sum of Gaussian beams, as the propagation characteristics of Gaussian beams are well known. With this and acoustics applications in mind, Wen and Breazeale proposed a beam profile that consists of a sum of superimposed complex Gaussian [1]. Their proposed beams have the following form:

$$f_N(x) = \sum_{k=1}^{N} A_k \exp(-B_k x^2) \tag{1}$$

They obtained the complex A_k and B_k coefficients by a computer optimization. This superposition was also used as a model for an aperture function to obtain the propagation characteristics of Bessel, Gaussian, and similar beams through apertures [2].

Another alternate to super-Gaussian functions involves writing the profile as a sum of polynomial-Gaussian functions of the form [3], [4]

$$f_N(x) = \exp(-x^2) \sum_{k=0}^{N} \frac{x^{2k}}{k!} \tag{2}$$

These "flattened Gaussian beams" can be rewritten as a superposition of Laguerre-Gaussian beams, whose propagation characteristics are also well known.

In this study, a new class of beams, known as multi-Gaussian beams [5], is proposed which have the advantages of the super-Gaussian, while having diffraction characteristics which are analytically solvable. The multi-Gaussian beams consist of a small sum of finite-width Gaussian beams side-by-side each of which represents an intuitive component of the entire beam. Unlike the Wen and Breazeale beams, all of the Gaussians have the same width, phase curvature, and absolute phase. Unlike the flattened Gaussian beams, each of the multi-Gaussian beam components can be traced individually without resort to further series expansion.

2. THE MULTI-GAUSSIAN SHAPE

The multi-Gaussian shape is composed of a sum of Gaussian function components, each with the same $1/e$ radius, or spotsize, w. However, as shown in Fig. 1, each of the Gaussian function components is offset by some fixed amount. In this study, the spacing is chosen to be equal to the spotsize, w. The general formula for the shape of the proposed multi-Gaussian functions is

In *Laser Resonators III*, Alexis V. Kudryashov, Alan H. Paxton, Editors,
Proceedings of SPIE Vol. 3930 (2000) • 0277-786X/00/$15.00

$$MG(x) = \frac{\sum_{m=-N}^{N} \exp\left[-\left(\frac{x-mw}{w}\right)^2\right]}{\sum_{m=-N}^{N} \exp(-m^2)} \qquad (3)$$

The numerator represents the sum of off-axis Gaussian function components, and the denominator normalizes the function so its maximum value is unity. Reminiscent of a diffraction integral, several off-axis functions are added together. However, each of the Gaussian function components here is of finite size and spacing. The multi-Gaussian shape may be said to have an order, N. From Eq. (3) it can be seen that a multi-Gaussian function of order N is composed of $2N+1$ Gaussian function components. The summation in the denominator of Eq. (3) has a value between unity and approximately 1.7726, and is typically closer to the latter.

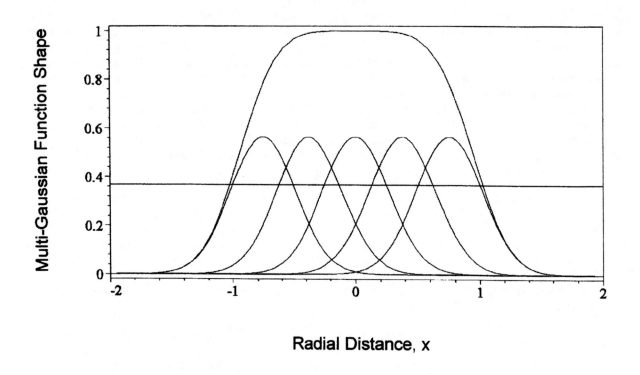

Fig. 1: An $N=2$ multi-Gaussian function is made up of a sum of $2N+1$ off-axis Gaussian function components.

If the multi-Gaussian shape were to represent the electric field amplitude of a flat-topped laser beam, then the complex electric field distribution at an input plane would be

$$E'(x) = E_0' \frac{\sum_{m=-N}^{N} \exp\left[-\left(\frac{x-mw}{w}\right)^2\right]}{\sum_{m=-N}^{N} \exp(-m^2)} \qquad (4)$$

This representation of the beam would only be valid at a location where there was no transverse phase variation. We refer to this location as the "waist plane" in analogy to Gaussian beam theory.

Outside the waist plane it is difficult to obtain a simple expression for the overall beam width in terms of the width of the Gaussian function components. However, one can obtain such an expression at the waist plane. If the width of the individual Gaussian terms is w and the width of entire multi-Gaussian beam is W, then they are related by [5]

$$w = \frac{W}{N + \sqrt{1 - \ln\left(\sum_{m=-N}^{N} \exp(-m^2)\right)}} \tag{5}$$

where, as before, N is the order of the multi-Gaussian beam.

While some beam models contain high spatial frequency local maxima, the multi-Gaussian functions, like the super-Gaussian functions are smooth. Fig. 2 contains plots of super-Gaussian and multi-Gaussian functions of different order. For purposes of comparison, both the super-Gaussian and the multi-Gaussian functions are chosen to have the same maximum value and function width. Qualitatively the plots are very similar. Both can provide excellent models for a hard aperture. While the diffraction of a super-Gaussian apertured Gaussian beam requires numerical solution of a diffraction integral, expressions for the diffraction of multi-Gaussian apertured Gaussian beams can be obtained analytically.

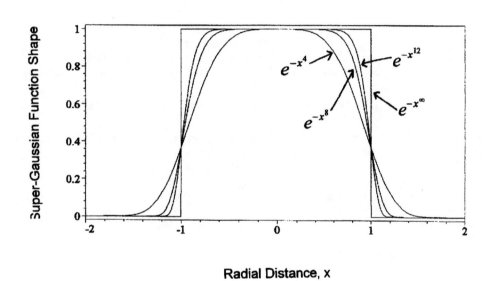

Fig. 2a: Different order super-Gaussian functions and the circ function.

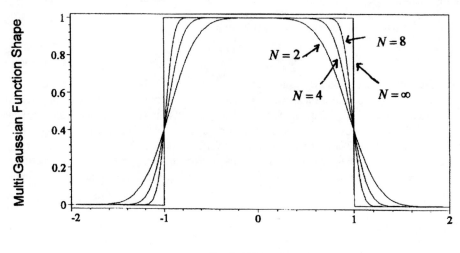

Radial Distance, x

Fig. 2b: Different order multi-Gaussian functions and the circ function. While they are very similar to super-Gaussian functions, they are much easier to deal with analytically.

3. MULTI-GAUSSIAN BEAMS IN OPTICAL SYSTEMS

Because formulas governing the propagation of Gaussian beams is well-known, it can be shown that the electric field of a multi-Gaussian beam propagating through an ABCD optical system is [5]

$$E'_{MG,2}(x) = E'_{0,2} \frac{\sum_{m=-N}^{N} f_m \exp\left[-\left(\frac{x - mw_1 A}{w_1(A^2 + B^2/z_{01}^2)^{1/2}}\right)^2\right] \exp\left[-i\frac{\beta_0}{2}\left(\frac{AC + BD/z_{01}^2}{A^2 + B^2/z_{01}^2}\right)\left(x - \frac{mw_1 B/z_{01}^2}{AC + BD/z_{01}^2}\right)^2\right]}{\sum_{m=-N}^{N} \exp(-m^2)} \quad (6)$$

where

$$f_m \equiv \frac{\exp\left[\frac{i}{2}\tan^{-1}\left(\frac{B}{Az_{01}}\right)\right]}{\left(A^2 + B^2/z_{01}^2\right)^{1/4}} \exp\left[-im^2\left(\frac{BC/z_{01}}{AC + BD/z_{01}}\right)\right] \quad (7)$$

These equations may be used to study the propagation of multi-Gaussian beams through optical systems that are represented by complex *ABCD* matrices. An *ABCDGH* matrix is used to characterize a misaligned complex optical system, and these formulas can be generalized to include the propagation characteristics of multi-Gaussian beams in these types of systems. Similarly, the one-dimensional beam cross-section can easily be extended to the full two dimensions.

3.1. Diffraction of multi-Gaussian beams
The above results may be applied to the propagation of multi-Gaussian beams in free space by using the ABCD matrix,

$$M_{free\ space} = \begin{pmatrix} 1 & z \\ 0 & 1 \end{pmatrix}, \quad (8)$$

for a length z of free space. Substituting this matrix for *A, B, C,* and *D* above yields

$$E'_{MG,2}(x) = E'_{0,2} \frac{\sum_{m=-N}^{N} f_m \exp\left[-\left(\frac{x-mw_1}{w_1(1+z^2/z_{01}^2)^{1/2}}\right)^2\right] \exp\left[-i\left(\frac{z}{z_{01}}\right)\left(\frac{x-mw_1}{w_1(1+z^2/z_{01}^2)^{1/2}}\right)^2\right]}{\sum_{m=-N}^{N} \exp(-m^2)} \tag{9}$$

where

$$f_m = \frac{\exp\left[\frac{i}{2}\tan^{-1}\left(\frac{z}{z_{01}}\right)\right]}{\left(1+z^2/z_{01}^2\right)^{1/4}} \tag{10}$$

The intensity distribution of a 2nd order multi-Gaussian beam is shown in Fig. 3 for several different propagation distances. The 2nd order beam retains the flat-topped shape for more than a Rayleigh length making it desirable as a beam mode for a laser resonator.

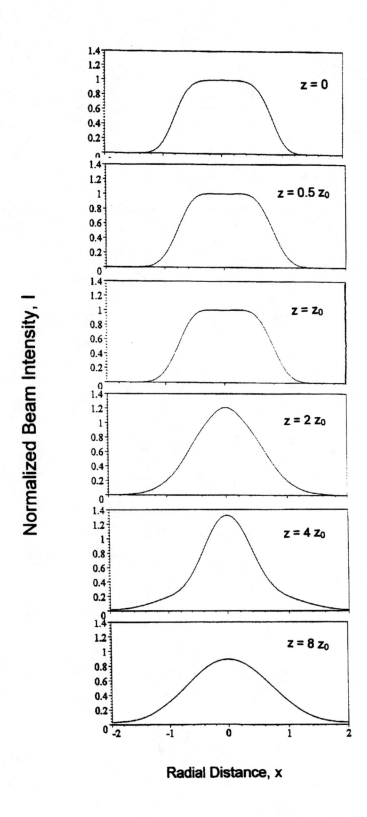

Fig. 3:   Intensity distribution of a multi-Gaussian beam as it propagates through free space.

3.2. Multi-Gaussian beams in the far-field

When a Gaussian beam at its waist is allowed to propagate for a very long distance, the radius of phase curvature of the beam approaches infinity. In this far-field regime, a multi-Gaussian beam will again obtain a form similar to Eq. (3). The expression for the diffraction angle becomes [5]

$$\theta_{diffract} = \tan^{-1}\left[\frac{2\lambda}{\pi W_1}\left\{N + \left[1 - \ln\left(\sum_{m=-N}^{N}\exp(-m^2)\right)\right]^{1/2}\right\}^2\right] \tag{11}$$

Those familiar with Gaussian beam theory will note that this formula reduces to the convention Gaussian beam formula when $N = 0$. Recalling that the summation in Eq. (11) is nearly independent of beam order, the diffraction angle becomes approximately quadratic in beam order for small beam divergences.

4. DISCUSSION

A Gaussian beam propagating in free space will have an infinite radius of phase curvature (i.e. flat phase fronts) at the beam's waist and at infinity. Under these conditions, the corresponding multi-Gaussian beam is flat-topped. A circ function can be written as an infinite order multi-Gaussian beam. The conclusion would seem to be that a circ profile remains a circ profile in the far field which appears to contradict the well-known result that such a beam has nulls in the far field.

While a flat-topped multi-Gaussian beam certainly changes beam shape as it propagates and becomes flat-topped again in the far field, this apparent contradiction can be explained by a limitation on the appropriate size of a multi-Gaussian beam which exists because of the paraxial approximation. In a flat-topped multi-Gaussian beam with an initial beam width of W_0, the width of each of the corresponding Gaussian beam components is

$$w_0 = \frac{W_0}{N + \sqrt{1 - \ln\left(\sum_{m=-N}^{N}\exp(-m^2)\right)}} . \tag{12}$$

The paraxial approximation was used in establishing the propagation characteristics of the Gaussian beam components with the consequence that the spotsize of any Gaussian must be on the order of or greater than the beam's wavelength. Applying this condition to Eq. (12) yields

$$N + \left[1 - \ln\left(\sum_{m=-N}^{N}\exp(-m^2)\right)\right]^{1/2} < \frac{W_0}{\lambda} \tag{13}$$

The summation in Eq. (13) ranges from unity to approximately 1.7726, and for purposes of this calculation, the bracketed quantity in Eq. (13) is on the order of unity so that the maximum beam order is

$$N_{max} \cong \frac{W_0}{\lambda} - 1 \tag{14}$$

To focus a beam to a wavelength across, the maximum order from Eq. (14) is zero and the multi-Gaussian beam reduces to a pure Gaussian beam. In the opposite limit, a circ beam has $N = \infty$ which can only occur if the beam size, W_0, is also infinite. So an infinite width circ beam remains an infinite width circ beam. These results cannot be applied to a circ beam of finite extent. However, in many applications the width of the desired beam is many wavelengths across. In such regimes, a large order multi-Gaussian beam (or aperture) may be used which is an excellent approximation to the circ function.

REFERENCES

[1]   J. J. Wen and M. A. Breazeale, "A diffraction beam field expressed as a superposition of Gaussian beams," J. Acoust. Soc. Am. **83**, pp. 1752-1756 (1988).

[2]   D. Ding and X. Liu, "Approximate description for Bessel, Bessel-Gauss, and Gaussian beams with finite aperture," J. Opt. Soc. Am. A **16**, pp. 1286-1293 (1999).

[3]   F. Gori, "Flattened Gaussian beams," Opt. Commun. **107**, pp. 335-341 (1994).

[4]   V. Bagini, R. Borghi, F. Gori, A. M. Pacileo, M. Santarsiero, D. Ambrosini, and G. S. Spagnolo, "Propagation of axially symmetric flattened Gaussian beams," J. Opt. Soc. Am. A **13**, pp. 1385-1394 (1996).

[5]   A. A. Tovar, "Propagation of flat-topped multi-Gaussian beams," J. Opt. Soc. Am. A, submitted.

---

*Correspondence: Email: mailto:superdad@eou.edu; WWW: http://physics.eou.edu/~superdad; Telephone: 541 962 3310

Characterization of excimer lasers for use with optical integrators

George N. Lawrence

Applied Optics Research, 3023 Donee Diego Dr., Escondido, CA 92025, glad@aor.com

Ying Lin

ABSTRACT

Excimer lasers are the preferred source of illumination for photolithographic systems. Lenslet array optical integrators are very frequently used to achieve good uniformity of illumination at the plane of the mask — the target to be recorded by the photolithographic system[1-6]. Lenslet arrays break the incident beam up into subapertures and the resulting beams are overlapped at the plane of the target with variable incident angles. The overlapping of the tilted beams segments results in coherent interference effects leading to fine-scale intensity modulation called orange peel We show that the ideal method of characterization of excimer lasers for this application is the Young's double hole test with hole spacing set to the lenslet element separation. This method to be superior to other common measures of laser quality such as angular divergence and M^2.

Keywords: Photolithography, excimer lasers, metrology, optical homogenizers, lenslet arrays

1. Introduction

Commercial excimer lasers are usually characterized in terms of angular divergence or M^2 value[7-10]. While these measures of performance provide a very approximate indication of performance in a photolithographic system using a lenslet array homogenizer, we can, in fact, show that the Young's double hole fringe visibility test provides an exact measure how well a given laser will suppress orange peel. Orange peel is a term describing the fine scale nonuniformity that appears in photolithographic applications as described below. We have used exact laser and diffraction modeling to calculate performance of excimer lasers through the complete optical system including pulsed laser gain, line narrowing prisms and gratings, apertures and aberration, lenslet arrays, and partial coherence effects[11-13]. We use the commercial code GLAD for laser modeling[14]. However the particular issue of best testing method may be understood from well-understood principles of Fourier optics and statistical optics, as we outline them here.

Consider a simplified view of a photolithographic system as illustrated in Fig. 1. This system is based on the classical Kohler illumination concept. An excimer laser is the illumination source. A lenslet array creates an array of point sources which are then imaged by a condenser lens into the pupil of the relay lens. Underfilling of the relay lens by a controlled amount yields a specified level of partial coherence that is helpful in achieving sharply resolved edges. The beams from the various lenslet elements overlap exactly at the plane of the target — the location for the mask. The overlap of the beams creates a very flat envelope to the illumination. Consider the initial gaussian-like beam envelope of the excimer laser (indicated by the dashed line). Because of beam overlap, the irradiance at the target plane is nearly flat (indicated by the dashed line at the target plane). With the Kohler design the condenser system and the relay system are, to a large extend, separated in function and may be understood by considering each system separately.

Figure 2 shows the condenser system separately. The figure indicates the excimer beam in terms of its speckled instantaneous form with smooth gaussian-like envelope. The lenslet array creates a corresponding array of point sources. These points are recollimated by the large condenser lens and overlap precisely at the target plane. To the extent that the beams from neighboring lenslet have some low level of mutual coherence, the overlap of the beams results in fine-scale interference at a low level of modulation. Figure 3 illustrates the overlap of two adjacent beam to create an interference pattern. The period of the interference pattern may be on the order of 20 microns depending on the focal lengths of the lenslet elements and the condenser lens. We would like to suppress this orange peel down to a few percent uniformity or less.

In *Laser Resonators III*, Alexis V. Kudryashov, Alan H. Paxton, Editors,
Proceedings of SPIE Vol. 3930 (2000) • 0277-786X/00/$15.00

95

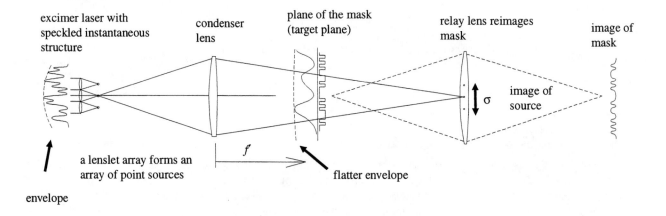

Fig. 1. A representative photolithographic system in a Kohler illumination configuration. An excimer laser is the source of illumination. It has an instantaneous speckle pattern but a smooth time-averaged envelope. A lenslet array (indicated by 3×3 array) forms an array of point sources. The array of point sources is imaged by the condenser lens into the pupil of the relay lens. The degree of filling of the pupil of the relay lens is the σ value and determines the degree of partial coherence in the image of the mask formed at the photographic recording material on the extreme right. The envelope of the overlapped distribution is much flatter, although there is fine microstructure created.

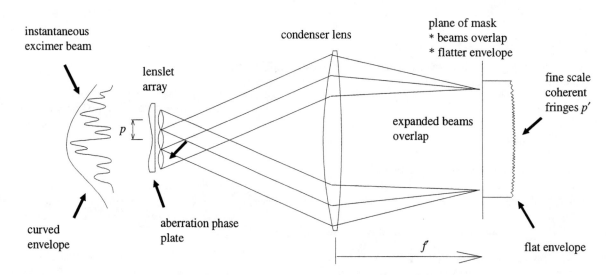

Fig. 2. Condenser subsystem. An excimer beam is incident on a lenslet array. All beams overlap at the rear focal plane of the large lens. This is the plane of the mask (target plane). The overlap of the beam produces a much flatter envelope. The beams from each lenslet are essentially collimated and have different angles when they overlap. A randomizing phase plate may be used for reduction of spatial coherence. Fine interference fringes are produced, with a period p' on the order of 100's of microns. These fine fringes must be suppressed down below a few percent so as to not be visible in the final exposure. We must have the spatial coherence width smaller than the lenslet element-to-element separation p — the characteristic period of the lenslet array — to suppress the coherent fringes.

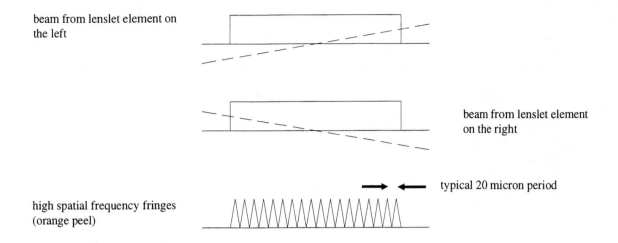

beam from lenslet element on the left

beam from lenslet element on the right

typical 20 micron period

high spatial frequency fringes (orange peel)

Fig. 3. The overlapping pupils from the lens array elements arrive at the target plane with angles corresponding to the angles of the chief rays. High spatial frequency fringes are produced. The modulation is reduced when the autocorrelation size is less than the lenslet diameter size.

The orange peel due to residual coherence between adjacent lenslet elements, may be reduced by reducing the spatial coherence of the laser. Excimer lasers tend to run with a very high number of spatial modes and, therefore, have very low spatial coherence widths. In photolithography, it is necessary to have a very narrow spectral line width to avoid problems of chromatic aberration. The excimer lasers are equipped with line narrowing elements in the laser cavity such as prisms, gratings, and etalons. Reduction of the line width tends to force an increase in spatial coherence width of the laser. See Sengupta for a good explanation of a typical line-narrowing concept[15]. While it is possible to preserve a narrow spatial coherence while decreasing the line width, this requires a more expensive design. It is important, therefore, to precisely determine the degrees of spatial coherence. If the spatial coherence width is too large, the orange peel will be objectionable. If the spatial coherence is made too narrow, the laser will be unnecessarily expensive.

2. Reduction of spatial coherence width

The lenslet array homogenizer lends itself optical analysis. Figure 4 illustrates the distribution that will be observed at the target plane if the laser source has perfect coherence. The case illustrated shows the results of an 8×8 lenslet array. For perfect coherence a field of super-modulated peaks will be created due to the multiple beam interference of a the beams from all lenslet elements. This is the worst possible case of orange peel. We can reduce the spatial coherence readily and cheaply by adding an aberration plate to the lenslet array. Figure 5 shows two types of aberration plates: different piston terms for each lenslet element or a continuously varying aberration plate. Both types of aberration plates will work well. Detailed diffraction analysis that we have done shows that such aberration plates help to reduce the level of orange peel below what can be achieved with a given laser. In our view, the aberration plate should always be included in the design, but we have not found any mention of the use of such aberration plates with lenslet arrays in the literature on application of lenslet homogenizers to photolithography.

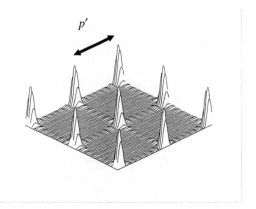

Fig. 4. Super-modulation fine structure created by an 8×8 array of lenses and perfect coherence. Typical period may be 20 microns.

The aberration plate, by itself, can reduce the super-modulated multiple beam interference pattern of Fig. 4 to a uniformity of about 100% — similar to the modulation of a field of random speckle. This nominal 100% uniformity was achieved with the laser being perfectly coherent. It should be noted that the homogenizer acts as an optical integrator to improve the irradiance uniformity of any type of input beam that may be applied. As such, we can not specifically tune the aberration plate to the incident beam envelope as the envelope will be varying. Hence a random aberration plate will work well. If we knew the laser envelope was always the same, we could design specific aberration plates to reduce the uniformity somewhat below the 100% level of the random plate — down to values of 70% or lower.

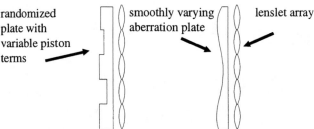

Fig. 5. Two form of aberration plates used in conjunction with a lenslet array to decrease spatial coherence.

While we find that adding fixed aberration is always helpful, it is not sufficient to achieve the 1% or lower uniformity levels that are desirable. Figure 7 illustrates the orange peel when a laser with moderate spatial coherence is added to the system with the aberration plate in place. To make the pattern easily visible for this paper we used a somewhat large spatial coherence width such that the uniformity is about 5% in Fig. 7, rather than the 1% or comparable level that would be desirable for photographic application.

2. Measuring the spatial coherence

The orange peel is primarily the result of interference between neighboring lenslet elements, as is illustrated in Fig. 3. Of course interference between elements separated by two or more element spacings can occur, but such higher order interference is generally less than nearest-neighbor interference and, with proper design, will be negligible. The orange peel and any point A' in the target plane is the result of complimentary points A_1, A_2, etc. in the pupils of the lenslet elements (see Fig. 8). The target plane is nearly, although not exactly, in a conjugate plane of the input plane of the lenslet array. Hence, points in the lenslet pupil are approximately imaged into corresponding points in the target plane.

We may, therefore, measure the tendency of the laser to form orange peel simply measuring the correlation between point A_1 and A_2. These are points separated by the lenslet spacing. If the statistical properties of the laser are constant over the aperture then any pair of points

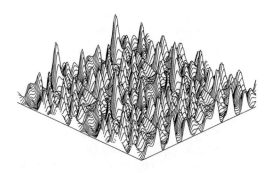

Fig. 6. Fine-scale nonuniformity with a perfectly coherent beam and an aberration plate. — about 100% uniformity achievable.

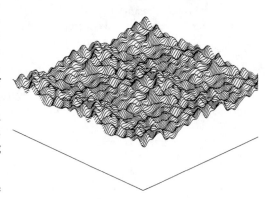

Fig. 7. Addition of dynamic effects of laser speckle to reduce uniformity to about 5%.

separated the same lenslet spacing will have the same mutual coherence. Constant statistical properties are referred to as stationary statistics. In fact, both theory and experiment Lin and Buck show that in excimer lasers the statistics are not stationary so we may need to consider averages over the aperture of the interference between point pairs. This is a straightforward extension of the principles set forward below.

The spatial coherence of two points A_1 and A_2 is directly found from the classical Young's double hole fringe visibility test[17-18]. Figure 9 illustrates this test. The laser speckle field due to multiple mode beating is illustrated on an instantaneous basis by a speckle field. Figure 9 shows a set of narrowly separated holes giving relatively high fringe visibility and a set

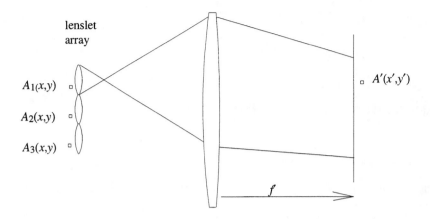

Fig. 8. The points $A_1(x,y)$, $A_2(x,y)$, etc. at the same relative position in each lenslet element overlap at a common image point $A'(x',y')$. The fringe visibility at the point A' at the mask may be computed by considering only the time-averaged mutual coherence of the points A_1, A_2, etc. separated by distance p.

of widely spaced holes giving low fringe visibility. In the photographic application, we will need very low coherence between the points corresponding to very low fringe visibility values — on the order of 1% or less. To be useful, the excimer laser must have sufficiently narrow spatial coherence width that for A_1 and A_2 separated by the lenslet spacing the fringe visibility is on the order of 1% or less.

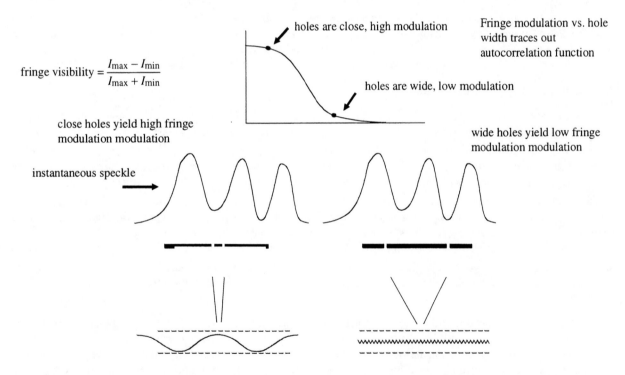

Fig. 9. Young's double hole experiment shown with closely spaced holes (left) and widely spaced holes (right). Samples of the instantaneous speckled form are shown. With time averaging, the fringe visibility assumes the form as shown in the lower illustration. The fringe visibility function is the same as the spatial coherence function due strictly to the time-varying speckle modulation.

The fringe visibility function is the same as the spatial coherence function when only the dynamic effects of multiple mode beating are considered. The fringe visibility of two-hole interference is completely independent of fixed aberration. As we can easily add as much fixed aberration as we like with the inexpensive aberration plate, we wish to measure the speckle-induced coherence function of the laser, as these dynamic effects are most beneficial and can not readily be added in other ways (without costly moving elements or electro optic effects).

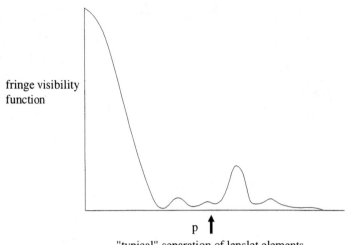

fringe visibility function

p

"typical" separation of lenslet elements

Fig. 10. A "typical" spatial coherence function similar to that measured by Kwata, et. al. The bumpy side band structure makes the level of orange peel very sensitive to the exact lenslet spacing. The $1/e^2$ width tells us little about the spatial coherence at large point separation values of the type indicated by the point p.

We could vary the separation of the two holes and measure the full fringe visibility function to determine the dynamically-based spatial coherence function. However we only need to know this function at one specific point on the curve corresponding to the lenslet element separation. If the exact lenslet element spacing is unknown then one can measure the function over a range of values. It is seen that the simple Young's hole visibility test measures exactly what we want.

3. Typical spatial coherence functions

Early experiments by Kawata, et. al. showed that the spatial coherence function may be quite "bumpy' in the skirts of the distribution and it exactly in the skirts that the laser and lenslet array are designed to operate so that low modulation values are achieved[7]. A "typical" example of such a distribution is illustrated schematically in Fig. 10. Theory and numerical modeling show us that such bumpy spatial coherence functions are created by clipping apertures in the laser and in the optical train. Such clipping apertures are essential to proper operation of the spectral line narrowing system in lasers with grating/prism control[15]. Given the likelihood of peaks and valleys in the spatial coherence function, it is most important to determine precisely where the lenslet separation p falls among the peaks and valleys. Figure 10 illustrates schematically some possible conditions. If the spatial coherence function were a simple gaussian function (no peaks and valleys), then it could be readily characterized by the width of the gaussian function. Single-number measures of performance such as angular divergence and M^2 would be sufficient for the gaussian spatial coherence function, provided, of course, that a means is found to separate the contribution from fixed aberrations in the laser from the dynamic effects which are what we want to measure.

Fig. 11 also illustrates more realistic spatial coherence functions with peak and valleys. If the lenslet separation p falls on a dip in the curve, orange peel will be greatly suppressed. It is quite possible that if the lenslet array spacing is increased (the figure on the left) the correlation may actually increase, contrary to our expectation from the simple gaussian model. Given a bumpy sidelobe structure, the "width" of the function is not very useful. Rather than measure the width at some relative value on the curve such as the $1/e^2$ point (a horizontal measure), we should measure the height of the curve at the specific point p (a vertical measurement).

three spatial coherence functions with the same $1/e^2$ width

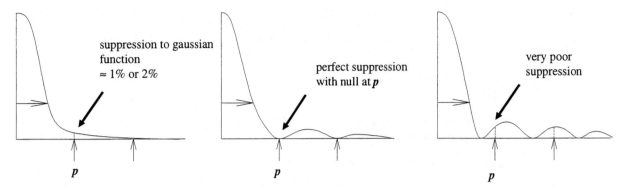

Fig. 11. Comparison of spatial coherence function shapes. A strictly gaussian shape may be characterized by the width of the gaussian function, but such a smooth function is not usual because of the strong effect of clipping apertures. Peaks and valleys in the wings of the coherence function are more typical. The system will be operated with the coherence function smaller than the lens spacing so that the fringe visibility function is on the order of a few percent or less, corresponding to a lens separation p. If p falls on a null, the orange peel will be greatly suppressed (center figure). If p falls near a peak (right figure), the orange peel may be quite obvious. Depending on the exact nature of the side lobe structure, reducing the spatial coherence at the $1/e^2$ width might actually cause the orange peel to get worse.

Fortunately the Young's double hole test is easy to do, requiring only a plate with two holes appropriately spaced — perhaps 100 microns to 200 microns apart. No lens is need and the fringe visibility measurement may be made in any suitable plane behind the holes. The principle difficulty is that one must measure low contrast fringes, albeit of relatively large spatial period.

4. Conclusions

Analysis of the lenslet array shows that the Young's double hole fringe visibility test is ideally suited to characterizing the ability of an excimer laser to suppress orange peel. The test may be made in the wings of the spatial coherence function at the distance corresponding to the lenslet element separation. The Young's double hole test is insensitive to fixed aberration in the excimer laser, as it should be, since the dynamic speckle-induced spatial coherence narrowing is the most important property of the excimer laser. Fixed aberrations of the laser contribute somewhat to orange peel suppression, but we may easily add as much aberration as we like cheaply and easily with an aberration plate in the system, so such fixed aberrations of the laser should not be counted in the spatial coherence measurement. To do so would lead to an overly optimistic view of the level of orange peel suppression.

Both angular divergence and M^2 do not differentiate between fixed aberrations and the dynamically-induced effects and therefore over predict orange peel suppression for a laser having some fixed aberrations such as coma, spherical aberration, and other aberrations. More importantly these single-number measures can not address the large fluctuations in performance that depend on how the lenslet element separation p falls among the peaks and valleys of the wings of the spatial coherence function.

5. References

1. K. A. Valiev, L. V. Velikov, G. S. Volkov and D. Yu. Zaroslov, " The optimization of excimer lasers radiation characteristics for projection lithography, " Proc. of 1989, Inter. Symp. on MicroProcess Conference,37-42 (1989).

2. Yoshiharu Ozaki, Kiichi Takamoto, "Cylindrical fly's eye lens for intensity redistribution of an excimer laser beam", Appl. Opt. **28**,106-110 (1989).

3. Zhou, Chongzi; Lin, Dajian; Yao, Hanmin, "Calculation and simulation of intensity distribution of uniform-illumination optical systems for submicron photolithography," SPIE Vol. 301, pp 652-657 (1997).

4. C.-Y. Han, Y. Ishi,and K. Murata, "Reshaping collimated laser beams with Gaussian profile to uniform profiles," Appl. Opt. **22**,3644-3647 (1983).

5. X. Deng, X. Liang, Z. Chen, W. Yu and R. Ma, "Uniform illumination of large targets using a lens array," Appl. Opt. **25**, 377-381 (1986).

6. Kenji Nishi, Yokohama and Naomasa Shiraishi, Kawasaki; "Illumination optical apparatus and method having a wavefront splitter and an optical integrator," US patent, no. 5815249.

7. Shintaro Kawata, Ikuo Hikima, Yukata Ichihara, and Shuntaro Watanabe, "Spatial coherence of KrF excimer lasers," Appl. Opt. **31**, 387-396 (1992).

8. B. A. See, "Measuring Laser divergence," Optics and Laser Technology, **29**, 109-110 (1997).

9. Anthony E. Siegman and Steven W. Townsend, "Output beam propagation and beam quality from a multimode stable-cavity laser," IEEE J. Quantum Electron. **29**,1212-1217 (1993).

10. David L. Wright and Steven Guggenheimer, "Status of ISO/TC 172/SC9/WG1 on standardization of the measurement of beam widths, beam divergence, and propagation factor," SPIE Proc. Vol. 1834, 2-17 (1992).

11. W. Goodman, *Introduction to Fourier Optics*, McGraw-Hill, New York, pp 86-88 (1968).

12. Gaskill, *Linear Systems, Transforms, and Optics*, Academic Press, p139 (1976).

13. W. Goodman, *Statistical Optics*, John Wiley & Sons, New York, p321 (1985).

14. GLAD is a commercial laser and physical optics computer program, Applied Optics Research, www.aor.com, glad@aor.com.

15. Uday K. Sengupta, "Krypton fluoride excimer laser for advanced microlithography," Optical Engineering, **32**, 2410-2420 (1993).

16. Ying Lin and Jesse Buck, "Numerical modeling of the excimer beam," SPIE Vol. 3677, 700-710 (1999).

17. L. G. Nazarova,"Measurement of the degree of laser coherence by Young's method," UDC 621.375.9:535, 403-405(1970).

18. Kiichi Takamoto, "Young's interference fringes with multiple-transverse-mode laser illumination", J. Opt. Soc. Am. A, **6**, 1137-1141 (1989).

SESSION 5

Solid State Laser Resonators

Self-starting 100W-average-power laser with a self-adaptive cavity

O.L. Antipov[+][a], A.S. Kuzhelev[a], D.V. Chausov[a], A.P. Zinov'ev[a], O.N. Eremeykin[a]
A.V. Fedin[++][b], A.V. Gavrilov[b], S.N. Smetanin[b]

[a] Institute of Applied Physics of the Russian Academy of Science,
603600, 46 Uljanov Str., Nizhny Novgorod, Russia
[b] Kovrov State Technological Academy, Division of Laser Physics and Technology,
601910, 19 Mayakovsky Str., Kovrov-city of Vladimir region, Russia

ABSTRACT

The numerical and experimental investigation of a self-starting Nd:YAG laser oscillator with a cavity completed by population gratings induced in the laser crystal by generating beams themselves are reported. The spatio-temporal characteristics of the laser comprising three Nd:YAG amplifiers and a saturable absorber were investigated. The generation of single mode beam with an average power of as large as 100 W and high quality was achieved.

Key words: laser oscillator; Nd:YAG laser crystals; cavity with nonlinear dynamic mirrors; refractive index and gain gratings; self-adaptive resonator; high average power; good beam quality

1. INTRODUCTION

The cavity of laser oscillators can be completed by holographic gratings of refractive index and gain which are induced in active laser media by generating beams themselves. Such kind of the self-starting laser oscillator based on laser crystals has been demonstrated recently.[1-8] An important advantage of the lasers of a new class is the self-adaptive property of cavity provided by a nonlinear dynamic mirror which is, in fact, a self-pumped phase conjugate mirror. The adaptive property is attractive for creation of high average-power lasers with good-quality beam.

Another interesting property of the new-kind laser oscillator is self-Q-switching of the cavity, which is caused by switching of the nonlinear mirror induced by generated optical beams. Such self-Q-switching mechanism provides for long coherence length of the generated beam that starts initially from broadband spontaneous emission. The single-longitudinal-mode generation in the self-starting laser was reported.[6-9]

The self-starting laser oscillators with dynamic cavity of two types have been recently demonstrated. The first type of the oscillator based on the laser crystals (LaCs), such as Nd:YAG and Ti:S, incorporates a nonreciprocal transmission element in the cavity formed by the gain grating (GG).[2,4,5] Another laser oscillator with reciprocal cavity was reported to be formed by a moving resonant refractive index grating (RIG) which accompanies the population grating (PG) induced in a Nd:YAG crystal by generating beams.[6-9] The latter type of the laser has demonstrated the capability for generating beams with high average power (as large as 60W) and near-diffraction-limited quality.[7,8]

In this paper we present new results of our theoretical and experimental investigation of the self-starting lasers based on the Nd:YAG crystals with the reciprocal cavity completed by the dynamic holographic grating. We also demonstrate the capability of these lasers to generate the beams with the 100W average power and the good quality.

[+] E-mail: antipov@appl.sci-nnov.ru; Tel.: +7(8312)352642; Fax: +7(8312)363792

[++] E-mail: kasapr@ks.ru; Tel.: +7(092)3232879; Fax: +7(092)3232533

104

In *Laser Resonators III*, Alexis V. Kudryashov, Alan H. Paxton, Editors,
Proceedings of SPIE Vol. 3930 (2000) ● 0277-786X/00/$15.00

2. NUMERICAL INVESTIGATIONS OF THE SELF-STARTING LASER OSCILLATOR

2.1. Theoretical model of the oscillator

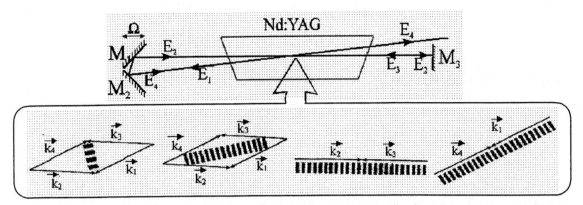

Fig. 1. Schematic of a self-starting generator with a loop cavity. M_1 is the vibrating mirror that gives the frequency shift Ω to reflected waves; M_2 and M_3 are linear mirrors; $k_1,...k_4$ are wave vectors of the interacting optical waves $E_1,...E_4$ in the cavity; the dashed area indicates PGs inside the LaC.

The principle of operation of the self-starting laser oscillator is connected with the two-wave mixing and four-wave mixings, in the laser amplifier, of optical waves starting initially from the amplified spontaneous emission (ASE). Four optical ASE waves with complex amplitudes $E_1,...E_4$ can interfere with each other (Fig. 1). The interference fields will induce the population gratings (PGs) which can be accompanied both by the gain gratings (GGs) and the refractive index gratings (RIGs). These gratings completed the cavity of the laser-oscillator.

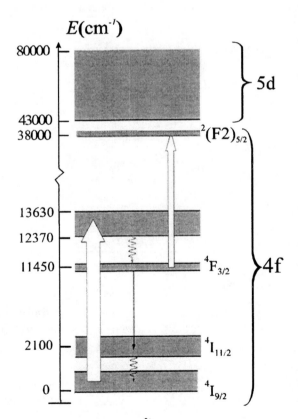

Fig. 2. Energy levels of Nd^{3+} ions in Nd:YAG

The characteristic pulse duration of the generating waves was measured in the previous experiments to about microsecond.[6-9] Therefore, when analyzing the self-starting conditions of the scheme it is necessary to take into consideration the temporal dynamics of the gratings. It is well known that the temporal behavior of the GGs is discribed by the kinetic equation for population of the high metastable level of the working laser transition, $^4F_{3/2}$ (Fig. 2). The RIG is induced mainly due to population or depopulation of a higher-lying level $^2F(2)_{5/2}$ of the 4f electron shell of Nd^{3+} ions.[10,11] This energy level, having large polarizability in near-IR and visible light, determines in practice the total inertial change of the Nd^{3+} ion polarizability under optical pumping. In other words, the population (or depopulation) of the energy level $^2F(2)_{5/2}$ with a life time of about 3 μs explains the resonant refractive-index changes of the Nd:YAG crystal. Therefore, this level must be taken into account in an explanation of the nonlinear interactions in the pumped Nd:YAG crystals at the microsecond temporal intervals.[10,11]

The numerically calculated model presented herein takes into consideration two metastable levels: $^4F_{3/2}$ and $^2F(2)_{5/2}$ (Fig. 2). We supposed that the $^2F(2)_{5/2}$ level can be populated only by pump-induced transition from the metastable level $^4F_{3/2}$ which is the highest level of the working laser transition. The $^2F(2)_{5/2}$ level is assumed to be depopulated by radiative transitions. In this model changes in population of the $^2F(2)_{5/2}$ level follow changes in population of the $^4F_{3/2}$ level with a delay time of about 3 μs. The equations for the average-in-space population of the $^2F(2)_{5/2}$ level (M_0) and the population grating amplitudes (M_{ij}) are:

$$\frac{\partial M_0}{\partial t} + \frac{M_0}{T_2} = \frac{N_0}{T_2} U_p(t), \quad \frac{\partial M_{ij}}{\partial t} + \frac{M_{ij}}{T_2} = \frac{N_{ij}}{T_2} U_p(t),$$ (1)

where N_0 and N_{ij} are the average-in-space population and the gratings of population of the $^4F_{3/2}$ level, respectively; T_2 is the longitudinal relaxation time of the $^2F(2)_{5/2}$ level, which is normalized to the relaxation time of the working transition (T_1); U_p is the pump velocity, $U_p(t) = (e^{-t/t_{off}})^2(1 - e^{-t/t_{on}})^2$ (t_{off} = 300 μs, t_{on} = 200 μs).

The changes in population of the $^2F(2)_{5/2}$ and $^4F_{3/2}$ levels determined the resonant changes of the refractive index of the Nd:YAG crystals, which can be described by the following expression:

$$\Delta n^e(v) = \frac{2\pi F_L^2}{n_0}(N \cdot \Delta p_n + M \cdot \Delta p_m),$$ (2)

where $F_L = (n^2+2)/3$ is the factor of a local field (Lorentz factor), n_0 is the linear index of refraction, Δp_n and Δp_m are the polarizability difference of the active Nd^{3+} ions in the ground state ($^4I_{9/2}$) and in the excited $^4F_{3/2}$ and $^2F(2)_{5/2}$ levels, respectively.

Therefore, the refractive-index changes consist of the average-in-space component (~N_0 and M_0) and the PG component (~ N_{ij} and M_{ij}). The resonant RIG induced in the laser crystal (LaC) by optical beams takes part in the formation of the nonlinear cavity.

The numerically calculated set of equations for the amplitudes of the generating waves E_1-E_4 in the plane wave (and plane polarization) approximation is

$$\mu\frac{\partial E_1}{\partial t} + \frac{\partial E_1}{\partial z} = (\sigma_1 N_0 + i\beta_2 M_0)E_1 + (\sigma_1 N_{12} + i\beta_2 M_{12})E_2 + (\sigma_1 N_{13} + i\beta_2 M_{13})E_3 +$$

$$+ (\sigma_1 N_{14} + i\beta_2 M_{14})E_4 + \sigma_0 N_0 F_1(z,t)\exp(2\pi i\varphi_1(z,t)),$$

$$\mu\frac{\partial E_2}{\partial t} - \frac{\partial E_2}{\partial z} = (\sigma_1 N_0 + i\beta_2 M_0)E_2 + (\sigma_1 N_{12}^* + i\beta_2 M_{12}^*)E_1 + (\sigma_1 N_{23} + i\beta_2 M_{23})E_3 +$$

$$+ (\sigma_1 N_{13} + i\beta_2 M_{13})E_4 + \sigma_0 N_0 F_2(z,t)\exp(2\pi i\varphi_2(z,t)),$$

$$\mu\frac{\partial E_3}{\partial t} + \frac{\partial E_3}{\partial z} = (\sigma_1 N_0 + i\beta_2 M_0)E_3 + (\sigma_1 N_{13}^* + i\beta_2 M_{13}^*)E_1 + (\sigma_1 N_{23}^* + i\beta_2 M_{23}^*)E_2 +$$

$$+ (\sigma_1 N_{12} + i\beta_2 N_{12})E_4 + \sigma_0 N_0 F_3(z,t)\exp(2\pi i\varphi_3(z,t)),$$

$$\mu\frac{\partial E_4}{\partial t} - \frac{\partial E_4}{\partial z} = (\sigma_1 N_0 + i\beta_2 M_0)E_4 + (\sigma_1 N_{14}^* + i\beta_2 M_{14}^*)E_1 + (\sigma_1 N_{12}^* + i\beta_2 M_{12}^*)E_3 +$$ (3)

$$+ (\sigma_1 N_{13}^* + i\beta_2 M_{13}^*)E_2 + \sigma_0 N_0 F_4(z,t)\exp(2\pi i\varphi_4(z,t)),$$

where μ is the wallk-off time in the LaC, $\mu = l/c$, c is the light velocity in the rod; all distances, times and intensities are normalized to the rod length (l), longitudinal relaxation time of the working transition (T_1), and saturation intensity (I_s), respectively; $\sigma_1 = \sigma_0(1+i\beta_1')$, σ_0 is the cross section of the working transition; β_1' is the ratio of the real part of resonant susceptibility to the imaginary part for the $^4F_{3/2}$ level, $\beta_1' = 4\pi^2 F_L^2 \Delta p_n /(\sigma_0 n_0 \lambda)$; β_2 is the nonlinearity coefficient determined by the polarizability of the $^2F(2)_{5/2}$ level, $\beta_2 = 4\pi^2 F_L^2\Delta p_m/(n_0\lambda)\equiv\sigma_0\beta_2'$; $F_1(z,t),...F_4(z,t)$ are amplitudes of the distributed Langevin's noise sources for the spontaneous polarization;[12] $\varphi_1(z,t),... \varphi_4(z,t)$ are phases of the Langevin's noise sources; $F_i(z,t)$ and $\varphi_i(z,t)$ are the random delta-correlated functions.

106

In our analysis we suppose that the generation of the quasi-monochromatic optical waves in the novel laser is a results of the self-consistent growing of the broadband ASE and narrowing of its frequency band. The process of self-organization of the generated beam was modeled by taking into consideration both the narrow-band generating waves $E_1...E_4$ describing by the set of equations (3) and the broadband ASE waves $E^S_1,...E^S_4$, co-propagating with the generating waves. The latter waves do not generate but they have two main functions: first, they saturate the amplification coefficient of the amplifiers; second, they interfere with each other, providing additional induction of the PGs of the $^4F_{3/2}$ level. It is possible to show that among the PGs induced by the broadband ASE there are the PGs with the same grating vectors as the PGs induced by the generating waves. So, both the waves can jointly induce the PGs.

For simplicity we supposed that the ASE waves $E^S_1,...E^S_4$ are described by the same the set of equations (3) with replacement of the amplitudes $E_1,...E_4$ to the amplitudes $E^S_1,...E^S_4$ and with the use of a new noise source $F^S_1,...F^S_4$. It was assumed that the braodband spontaneous emission noise has an effective gain ($\alpha_{ef} = \sigma_{ef}N_0$) that was less than the amplification coefficient of the narrow-band generating waves in the center of the luminescence line.

For changes of population of the $^4F_{3/2}$ level the following equations was used

$$\frac{\partial N_0}{\partial t} + N_0 = N_p(t) - N_0 \cdot I_\Sigma \tag{4}$$

$$\frac{\partial N_{12}}{\partial t} + N_{12} = -N_0(E_1 E_2^* + E_3 E_4^* + E_1^S E_2^{S^*} + E_3^S E_4^{S^*}) - N_{12} \cdot I_\Sigma,$$

$$\frac{\partial N_{13}}{\partial t} + N_{13} = -N_0(E_1 E_3^* + E_2 E_4^* + E_1^S E_3^{S^*} + E_2^S E_4^{S^*}) - N_{13} \cdot I_\Sigma$$

$$\frac{\partial N_{14}}{\partial t} + N_{14} = -N_0(E_1 E_4^* + E_1^S E_4^{S^*}) - N_{14} \cdot I_\Sigma, \tag{5}$$

$$\frac{\partial N_{23}}{\partial t} + N_{23} = -N_0(E_2 E_3^* + E_2^S E_3^{S^*}) - N_{23} \cdot I_\Sigma,$$

where I_Σ are the total intensity of the optical fields, $I_\Sigma = \sum_{j=1}^{4}\left(\left|E_j\right|^2 + \left|E_j^S\right|^2\right)$, $N_p(t)$ is the optical pump.

The initial conditions were as follows

$$E_1(z,t=0) = E_2(z,t=0) = E_3(z,t=0) = E_4(z,t=0) = 0$$
$$N_0(z,t=0) = N_{12}(z,t=0) = N_{13}(z,t=0) = N_{14}(z,t=0) = N_{23}(z,t=0) = 0 \tag{6}$$

The boundary conditions were

$$E_1(z = 0, t) = 0,$$
$$E_2(z = 1, t) = E_1(z = 1, t)r_{12}e^{i2kL_1}e^{i\Omega t}$$
$$E_3(z = 0, t) = E_2(z = 0, t)r_3 e^{i2kL_2}, \tag{7}$$
$$E_4(z = 1, t) = E_3(z = 1, t)r_{12}e^{i2kL_1}e^{i\Omega t},$$

where L_1 and L_2 are the distances between the end of the Nd:YAG rod and the mirrors M_1/M_2 and M_3, respectively; Ω is the frequency shift between the incident wave and the wave reflected from the vibrating mirror M_1. The zero boundary condition for the weakest wave E_1 indicate the absence of any linear reflection and scattering in the output of the scheme.

The sets of Eqs. (1), (4), (5) for the gratings together with the Eqs. (3) for the generating optical waves $E_1,...E_4$ and similar equations for the broadband ASE waves $E^S_1,...E^S_4$ with the initial and boundary conditions (6) and (7) described the operation of the self-starting oscillator.

2.2. Results of numerical calculation

It was calculated that the nonlinear generation of the narrow-band waves $E_1,...E_4$ occurs in the presence of the broad-band ASE waves $E^S_1,...E^S_4$, which induce the noise PG, and even without any "linear" diffuse reflection on the rod end and the internal scattering. The numerical calculations showed the existence of the threshold amplitude of the Langevin noise source for the broadband ASE wave E^S_1, $\varepsilon^S=<|F^S_1(z,t)|^2>$, which depends on the amplifier gain and the coefficients of nonlinearity β_1 and β_2 (Fig. 3). In fact, the threshold level of the ASE wave determines the amplitude of the noised PGs which is necessary for the self-starting generation.[9]

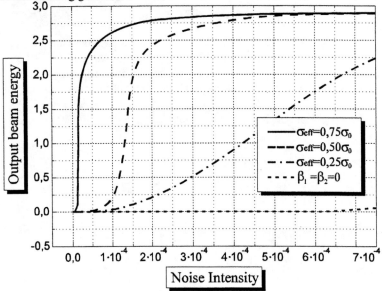

Fig. 3. Numerically calculated dependence of the output beam energy on the broadband noise intensity (ε^S) for different effective noise gain (σ_{eff})

Below the generation threshold only broadband noise amplification took place (curve 4 in Fig. 4), at the threshold the single-pulse generation was realized (curve 2), and beyond the threshold the pulse train was generated (curve 1 in Fig. 4).

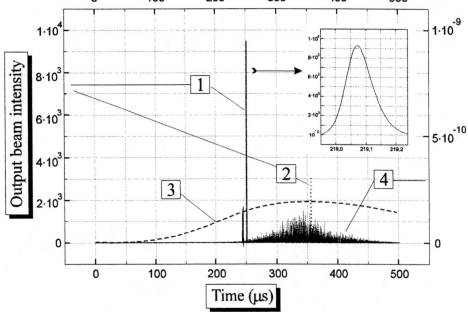

Fig. 4. Numerically calculated oscillograms of generated pulses (curves 1,2,4) and pulse shape of full luminescence (curve 3).

The threshold of the generation and the energy of the generated pulses were found to be dependent on the frequency detuning Ω (Fig. 5a), which indicates the influence of the mirror vibrations on the generation conditions.[9] The resonances of the frequency detuning corresponded to the optimal nonlinear increment of the weakest waves due to two-wave-mixing and four-wave-mixing energy transfer from the stronger waves.

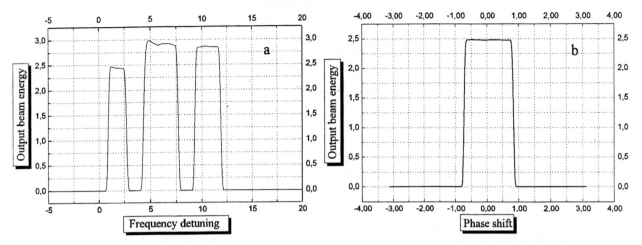

Fig. 5. Numerically calculated dependencies of the output energy on the frequency detuning (a) and phase shift (b).

It was calculated that energy of the generation pulse train depends on the absolute phase shift for the optical waves in the open space between the Nd:YAG rod ends and the mirrors M_1/M_2 or M_3 (Fig. 5b). The dependence indicates the presence of the longitudinal modes of the self-starting laser. The ratio of the mode bandwidth and the intermode interval is only $1.5/(2\cdot\pi)$, which corresponds to the small Q-factor of the cavity with the nonlinear mirror.

Therefore, the numerical calculations showed, in accordance with our previous analytical estimations and numerical calculations,[6-9] that the reciprocal cavity of the self-starting Nd:YAG laser oscillator can be completed by moving resonant RIGs, which are induced in the inverted laser crystal by generating beams. Above the threshold for amplifier gain, the self-consistent increase of the moving RIGs, induced initially by the noise of spontaneous emission and growth of these noised optical waves, gives rise to the generation of self-Q-switched pulses.

3. EXPERIMENTS

We have studied different experimental schemes of the self-starting laser oscillators comprising one, two or three tNd:YAG laser amplifiers.[3, 6-11] It has been demonstrated that these schemes have a number of new properties, such as generation in the presence of strong intracavity aberrations and high quality of generated beam with average power up to 60W. So, these schemes seem to be attractive for the creation of lasers with the output power more than 100W. The key question for application of the self-starting lasers is whether they are able to generate beams with extremely high average power and good beam quality, much better than in the lasers with ordinary mirrors.

3.1. Experimental schemes

The laser oscillator based on two flash-lamp pumped Nd:YAG amplifiers with a loop cavity completed by dynamic holographic gratings of population was investigated experimentally (Fig. 6). In the amplifiers we used Nd:YAG crystals (6 mm in diameter and 100 mm in length) produced by the Czochralski method. The concentration of Nd^{3+} ions in these crystals was about 1 % (the concentration of other Mo and Fe mixture ions was less than 5×10^{-3} % and 10^{-3} %, respectively). The logarithmic gain of each amplifier was varied up to 4.8 at pump pulse energy of as great as 80 J per each amplifier. The pump pulse duration was varied from 0.2 ms to 0.45 ms. The repetition rate was 30 Hz. The possibility of Q-switch operation was investigated using saturation absorbers $LiF:F_2^-$ and $Cr^{4+}:YAG$.

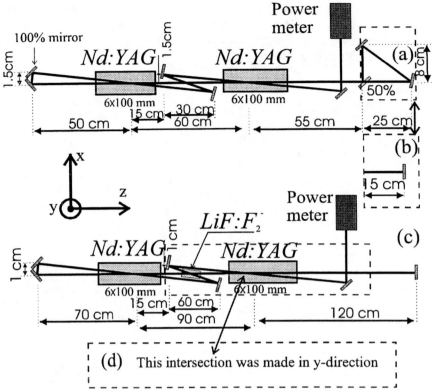

Fig. 6. Schematic of the laser oscillator based on two flash-lamp pumped Nd:YAG amplifiers with a loop cavity.

The laser generation was observed in the scheme shown in Fig. 6a. The dependencies of the output beam average power on the pump pulse energy at different pulse durations were investigated (Fig. 7). A increase in the beam power was observed when the pulse duration increased from 0.2 ms to 0.3 ms. This result can be explained by that the amplitude of the RIG caused by the population of the high lying level $^2F(2)_{5/2}$ depends on the simultaneous presence of the flash-lamp pumping and the population of the metastable level $^4F_{3/2}$ (a product of U_p and N_{ij} in the right-hand side of equation 1). The amplitude of the RIG increases at increasing pump pulse duration.

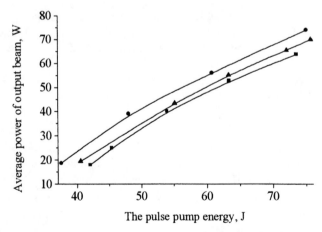

Fig. 7. Dependencies of the output beam average power on pump pulse energy of each amplifier at different durations of flash-lamp pump pulses: 0.3 ms (circles), 0.4 ms (triangles) and 0.2 ms (squares).

A further increase in the pump pulse duration (more than 300…400 μs) led to a decrease in the average power of the output beam. This can be caused by a decrease in the amplification and the pump power, which took place when the pump duration increased at a constant pump-pulse energy. This decrease had two results: first, movement toward generation threshold and, second, a decrease in the amplitude of the RIG (the amplitude of the RIG caused by population of the higher-lying level $^2F2_{5/2}$ strongly depends on the amplification and pump power). It was determined that the optimal pump pulse duration for the scheme with two amplifier scheme was about 0.3 ms.

It should be noted that the same average power of the generated beam is realized when a Sagnac interferometer, which provides the spatio-angular selection of the generating beam, was replaced by an ordinary mirror positioned at a definite distance from the amplifier (Fig. 6b). In this case, the quality of the output beam was approximately the same as in the case of Fig. 6a. For quantitative measurements of the quality and comparison, we used a standard test method.

3.2. Measurements of the quality of generated beams

We used a method for testing laser beam parameters (such as beam widths, divergence angle, and beam propagation factor) in accordance with the standards of the International Organization for Standardization[13].

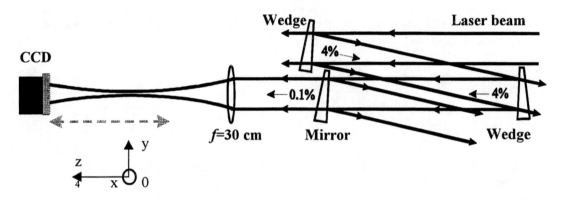

Fig. 8 Schematic of an experimental setup used for measurements of laser beam parameters (beam widths, divergence angle, beam propagation factor, and diffraction limit factor)

The setup shown in Fig. 8 was used for precise measurements of the beam quality. A part of the beam was coupled by two wedge plates and a mirror placed on the wedge plate with transmittance of ~ 0.1% and then was focused by a lens to obtain a beam waist suitable for the measurement. The focal length (30 cm) was chosen such as to minimize the lens aberration and, therefore, the influence on beam quality. The resulting beam waists were in the range from 50 μm to 300 μm. The transverse profile of the beam at different positions along the z axis was recorded by a CCD camera through a glass filter with varied transmittance (less than 0.01%). Using the CCD images (two-dimensional intensity distributions $F(x,y)$) the average distributions along the x and y directions ($F_x(x)$ and $F_y(y)$) were calculated using the following procedure:

$$F_x(x) = \int\limits_{-\infty}^{+\infty} F(x,y^{'})dy^{'}, \quad F_y(y) = \int\limits_{-\infty}^{+\infty} F(x^{'},y)dx^{'}. \tag{8}$$

Thus, the distributions $F_x(x)$ and $F_y(y)$ are analogous to those usually obtained by the method of moving slit and can be used for the determination of laser beam parameters according to the rules of the International Organization for Standardization.[13] The widths of the distributions $F_x(x)$ and $F_y(y)$ at the level of 13.5% are assumed to be the waist diameters along the x and y directions. The square of the determined beam diameter was approximated using a parabolic fit, depending on the position of z:

$$d^2 = A + B \cdot z + C \cdot z^2. \tag{9}$$

$$\theta = \sqrt{C}, \qquad d_0 = \sqrt{A - \frac{B^2}{4C}}. \tag{10}$$

In the first approximation for the beam propagation factor K_s we get

$$K_s = \frac{4\lambda}{\pi \cdot \theta \cdot d_0}. \tag{11}$$

Since the definition of the beam radius uses the second momentum of the intensity distribution of the transversal beam profile, the calculated values of d differ from the real beam radius. Therefore, we have to correct K_s using the following formula:[13]

$$K = \left[0.95\left(\frac{1}{\sqrt{K}} - 1 \right) + 1 \right]^{-2}. \tag{12}$$

Knowing this beam propagation factor, the diffraction limit factor M^2 is given by:

$$M^2 = \frac{1}{K}. \tag{13}$$

The transverse profiles of the beams generated in the schemes are shown in Table 1. It is seen that, indeed, the generated beams had near-Gaussian quality. Using the parameters of this fit, the beam divergence θ, the beam waist diameter d_0, the propagation factor K, and the diffraction limit factor M^2 were determined for each scheme (see Table 1).

Table 1.

Scheme	Fig. 6a		Fig. 6b		Fig. 6c		Fig. 6d		Fig.6c+LiF	
direction	x	y	x	y	x	y	x	y	x	y
d_0, μm	151	242	133	231	250	240	105	146	155	168
θ, $(\times 10^{-2})$	2.3	4.1	2.3	3.5	0.9	1.1	1.8	2.3	1.1	1.6
K	0.4	0.15	0.46	0.18	0.67	0.53	0.71	0.43	0.8	0.53
M^2	2.5	6.8	2.2	5.6	1.5	1.9	1.4	2.3	1.25	1.9
Average power, P, W	75		75		60		60		57	
profile of the beam in focal plane										

Thus the exact determination of the beam parameters showed high quality ($M^2 \approx 1.5$) of the generated beams with high average power (60 W).

3.3. Experiments with a three-amplifier scheme

The possibility of increasing the output beam power by adding a third Nd:YAG amplifier to the scheme was investigated (Fig. 8). In the first scheme, all amplifiers were included in the cavity, the population grating completing the cavity was induced in the output amplifier by four-wave mixing. In the second scheme, the last Nd:YAG amplifier was not included in the nonlinear cavity and served as an ordinary power amplifier.

It was observed that the average power of the generated beams increases up to 105-120 W in the both schemes (Fig. 9).

The precise determination of the beam parameters showed high quality ($M^2 \approx 1.4$ for scheme Fig. 8a and $M^2 \approx 1.8$ for scheme Fig. 8b) of the generated beams with high average power (Fig. 9b). The better beam quality in the scheme shown in Fig. 8a rather than in Fig. 8b demonstrated adaptive properties of the nonlinear holographic mirror to intracavity distortions.

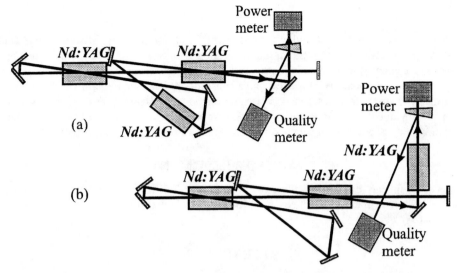

Fig. 8. Two variants of the experimental setup with three Nd:YAG amplifiers.

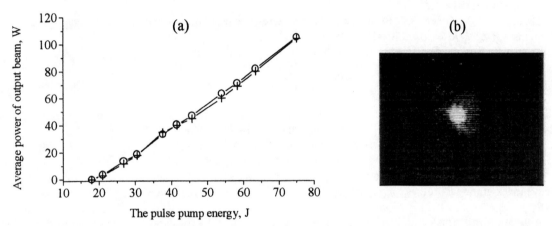

Fig. 9. (a) - Energy of the generated beam pulse as a function of the pump pulse energy of each amplifier in the scheme with three Nd:YAG amplifiers (circles – scheme Fig. 8a, crosses – scheme Fig. 8b). The repetition rate is 30 Hz. (b) - The CCD image of transverse structure of the generated beam in the scheme with three amplifiers (Fig. 8a) in the focal plane of the lens.

The temporal parameters of the generated pulses were as follows: pulse duration τ_p was 400 – 500 ns, and time period of pulses in the train T_p was 2 – 4 μs (Fig. 10).

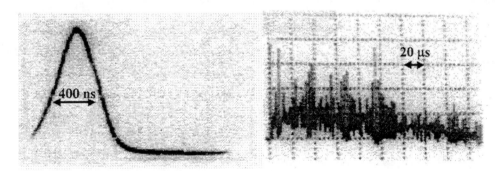

Fig. 10. Oscillograms of the generated pulses in the scheme with three amplifiers on time scales 200 ns and 20 μs.

4. CONCLUSION

We have studied experimentally and numerically a self-starting laser oscillator with cavity completed by population gratings in Nd:YAG laser crystals. It was shown that this generation is caused by self-consistent formation of the refractive index and gain gratings accompanying the population gratings. It was found that the Nd:YAG laser with the cavity formed by dynamic gratings possesses self-adaptive properties and a capability to generate radiation with high beam quality in the pulse-repetitive regime. The single transverse mode operation of the self-starting laser (comprising three amplifiers and having loop geometry) with an output average power up to 120 W was demonstrated. Similar laser architectures appear to be attractive for creation of more powerful generators with an output power of sub-kilowatt level.

ACKNOWLEDGMENT

This work was supported in part by INTAS (through the grant I 97-2112), EOARD/Air Force Research Laboratory through grant No. SPC99-4028 (under contract No. F61775-99-WE028) and the NATO "Science For Peace" foundation (through grant No. 974143).

REFERENCES

1. I.M. Bel'dyugin, V.A. Berenberg, A.E. Vasil'ev, I.V. Mochalov, V.M. Petnikova, G.T. Petrovskii, M.A. Kharchenko, and V.V. Shuvalov, "Solid-state lasers with self-pumped PC mirrors in the active medium", *Sov. J. Quantum Electr.* **16**, pp. 1142-1145 (1989).

2. M.J. Damzen, R.P.M. Green, K.S. Syed, "Self-adaptive solid-state oscillator formed by dynamic gain-gratings holograms, ", *Opt. Lett.* **20**, pp. 1704-1706 (1995).

3. O.L. Antipov, A.S. Kuzhelev, V.A. Vorob'yov, A.P. Zinov'ev, "Millisecond pulse repetitive Nd:YAG-laser with self-adaptive cavity formed by population gratings," *Optics Comm.* **152**, pp. 313-318 (1998).

4. A. Minassian, G.J. Crofts, M.J. Damzen, "Self-starting Ti:sapphire holographic laser oscillator," *Opt. Lett.* **22**, pp. 697-699 (1997).

5. P. Sillard, A. Brignon, and J.-P. Huignard, "Gain-grating analysis of a self-starting self-pumped phase-conjugate Nd:YAG loop resonator," *IEEE J. Quant. Electron.* **34**, pp. 465-472 (1998).

6. O.L. Antipov, S.I. Belyaev, A.S. Kuzhelev, A.P. Zinov'ev, "Nd:YAG laser with cavity formed by population inversion gratings", *SPIE Proceeding* (Edited by P. Galarneau and A.V. Kudryashov) **3267**, pp. 181-190 (1998).

7. O.L. Antipov, A.S. Kuzhelev, A.P. Zinov'ev, "High average-power solid-state lasers with cavity formed by self-induced refractive index gratings," in Laser resonators II, A.V. Kudryashov and P. Galarneau, eds., *Proc. SPIE* **3611**, 147-156 (1999).

8. O.L. Antipov, A.S. Kuzhelev, A.P. Zinoviev, et all., "Single-mode Nd:YAG laser with cavity formed by population gratings," in Nonlinear and Coherent Optics, V.E. Sherstobitov, eds., *Proc. SPIE* **3684**, pp. 59-63 (1998).

9. O.L. Antipov, A.S. Kuzhelev, D.V. Chausov, "Formation of the cavity in a self-starting high-average power Nd:YAG laser oscillator," Optics Express **5**, No. 12, pp. 286-292 (1999).

10. O.L. Antipov, S.I. Belyaev, A.S. Kuzhelev, D.V. Chausov, "Resonant two-wave mixing of optical beams by refractive index and gain gratings in inverted Nd:YAG," *J. Opt. Soc. America* B **15**, pp. 2276-2281 (1998).

11. O.L. Antipov, A.S. Kuzhelev, D.V. Chausov, and A.P. Zinov'ev, "Dynamics of refractive index changes in a Nd:YAG laser crystal under Nd^{3+}-ions excitation," *J. Opt. Soc. America* B **16**, pp. 1072-1079 (1999).

12. A.N. Oraevsky, "Quantum fluctuations and formation of coherency in laser," J. Opt. Soc. of Am. 5, pp. 933-945 (1988).

13. "Optics and optical instruments – Test methods for laser beam parameters: beam width, divergence angle and beam propagation factor" (ISO/DIS 11146:1995).

Intracavity resonator design to generate high-quality harmonized beams in diode-pumped solid-state lasers

Shuichi Fujikawa[*], Yoko Inoue, Susumu Konno, Tetsuo Kojima, and Koji Yasui

Advanced Technology R&D Center, Mitsubishi Electric Corporation,

8-1-1 Tsukaguchi-honmachi, Amagasaki, Hyogo, Japan

ABSTRACT

We have described resonator designs for stable and high-brightness beam generation by intracavity frequency doubling. The combination of a novel pump configuration and the bifocusing compensation resonator design provides efficient and stable harmonized beam generations with good beam quality.

Keywords: frequency conversion, laser resonators, neodimium:YAG laser, thermal lensing.

1. INTRODUCTION

Nonlinear frequency conversion has been driving forces to cultivate various applications of solid-state lasers such as material processing[1], sensing[2], entertainment[3,4] and pumping source for tunable lasers. From practical point of view, intracavity harmonic generation is an attractive scheme because of the high conversion efficiency and of the simple design of the laser system. However, there are several drawbacks associated with intracavity harmonic generation. Resonator instability easily stimulates the power fluctuation induced by the interaction between the gain medium and the nonlinear process. For high-power and high-quality harmonic beam generations, the stability of the resonator becomes more serious. The beam absorption leads to a nonuniform temperature distribution in the nonlinear crystal which distorts the wave front of the laser beam. The strong thermal lensing of the laser medium narrows the stable zone of the resonator. Consequently, the resonator configuration of intracavity frequency conversion must be carefully designed considering the optical property both for the nonlinear crystal and for the gain medium. In this paper, we describe resonator designs for stable and high-brightness beam generation by intracavity frequency doubled Nd:YAG lasers both for CW and Q-switched operations.

2. LASER HEAD DESIGN

We have proposed a simple and high-efficient diode-side-pumping configuration CIDER (close-coupled internal diffusive exciting reflector)[5,6]. A schematic cross section of a pump unit of the CIDER is shown in Fig. 1. The pump unit contains four 25 W diode arrays with 808-nm wavelength. Pump heads used for harmonic generations consisted of two or four pump units connected in series along the rod axis. A 4-mm diameter Nd:YAG rod was placed in a glass flow tube enclosed by a cylindrical diffusive reflector. Pump beams from diode arrays were coupled into the diffusive reflector through thin glass plates by total internal reflection. The transfer efficiency through the thin glass plate was more than 97 % with free adjustment.

Figure 2 shows the fluorescence intensity profile at 1064-nm, observed from the rod axis with a CCD camera. The pump distribution has nearly top-hat profile. Uniform pump distribution is required for stable high-quality beam generation by suppressing thermal lensing variation within the rod cross section[7].

---

[*] Correspondence: Email: fujikawa@lap.crl.melco.co.jp; Telephone: +81-6-6497-7110; Fax: +81-6-6497-7288

In *Laser Resonators III*, Alexis V. Kudryashov, Alan H. Paxton, Editors,
Proceedings of SPIE Vol. 3930 (2000) • 0277-786X/00/$15.00

115

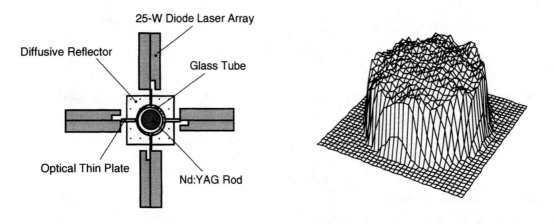

Fig.1 Schematic cross section the CIDER pump unit.   Fig.2 Fluorescence intensity profile at 1064 nm.

3. CW GREEN BEAM GENERATION

High-power CW all-solid-state green lasers have been demanded for various applications, such as a medical equipment, as a pumping source of Ti:Sapphire lasers, and as a light source for laser displays. We have demonstrated high-power CW green beam generation by intracavity frequency doubling of a diode-side-pumped Nd:YAG laser[8]. Figure 3 shows a schematic drawing of a CW green laser system. The laser system has two laser heads, each consisting of two pump units. A 90°-polarization rotator is placed between two uniformly pumped Nd:YAG rods for polarization-dependent bifocusing compensation[7]. The Z-shaped resonator has four mirrors consisting of two end mirrors with concave curvature and two flat folding mirrors. A type-II phase-matched KTiOPO$_4$(KTP) crystal was placed between the end mirror and the folding mirror (harmonic separator). The 532-nm green beam was extracted from the harmonic separator in single direction. A quarter-wave plate at 1064 nm is placed between the harmonic separator and the laser head to avoid "green problem"[9].

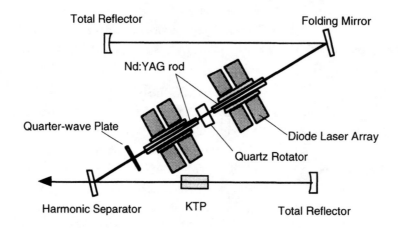

Fig. 3 Schematic drawing of the green laser system.

We examined two KTP crystals that have different absorption losses measured at 1064 nm. The absorption loss of KTP-A (10-mm long) was 0.03 % and that of KTP-B (15-mm long) was 0.3 %. Figure 4 shows the green output power as a function of the total diode output power. A stable green output power of 27 W was obtained with KTP-A at a total diode output power of 331 W. The beam quality was measured to be $M^2 = 8$. On the other hand, the green output power with

KTP-B was unstable although the highest output power was nearly the same (26 W) as that for KTP-A. Figure 5 shows the time traces of green output power both for KTP-A (a) and for KTP-B (b). A periodic spike was observed in the time trace for KTP-B whereas small noises existed in the time trace for KTP-A.

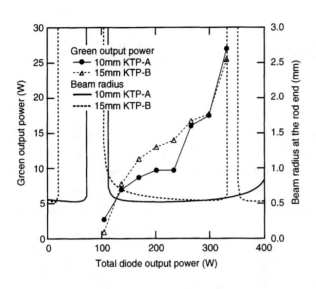

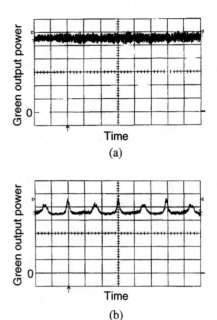

Fig. 4 Green output power and the calculated stable zone as a function of the total diode output power.

Fig. 5 Time traces of green output power at the total diode output power of 331W by use of (a) 10-mm KTP-A and (b) 15-mm KTP-B. Horizontal scale is 20μs/div.

We calculated the stable zone of the resonator considering the thermal lensing of the KTP crystals. The focal length of KTP-A was estimated to be about 1650 mm and that of KTP-B was estimated to be about 200 mm at the total diode output power of 331 W[8]. The calculated stable zones are also shown in Fig. 4. It was found that the periodic spike of the green output power with KTP-B appeared when the operating point approached the unstable zone of the resonator. To confirm the fact that the green power fluctuation is related to the unstable zone of the resonator, we modified the resonator configuration for KTP-B so that the laser should be operated in the stable zone by shortening the resonator length to account for thermal lensing of KTP-B. The time trace of the green output power with the modified resonator is shown in Fig. 6. The periodic spike in the time trace disappeared and the green output power was effectively stabilized. A stable green output power of 24 W with the beam quality of $M^2 = 11$ was obtained at a total diode output power of 331 W. These results indicate that the thermal lensing of the nonlinear crystal should not be negligible for intracavity frequency doubling resonator design.

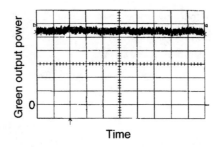

Fig. 6 Time trace of green output power with 15-mm KTP-B using the modified resonator at the total diode output power of 331 W. Horizontal scale is 20μs/div.

4. CW RED BEAM GENERATION

All-solid-state red lasers are attractive light sources for Cr: LiSAF pumping or laser projection We have demonstrated red beam generation by intracavity frequency doubling of a Nd:YAG laser operating at 1.319 μm[10]. Figure 7 shows a schematic drawing of the CW red laser system. To suppress laser oscillation at 1.06 μm, the dielectric coating of resonator mirrors specified the reflectivity to be higher than 99 % at 1.3 μm and lower than 30 % at 1.06 μm. A 10-mm length KTP crystal cut for type-II second harmonic generation at 1.319 μm was placed between the end mirror and the harmonic separator. Both 90°-polarization rotator and quarter-wave plate were also specific to the wavelength of 1.319 μm.

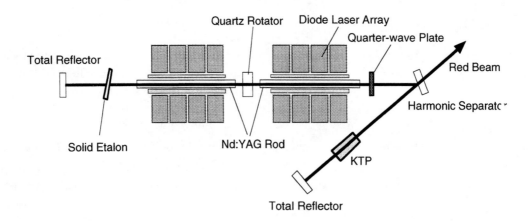

Fig. 7 Schematic drawing of the red laser system.

It should be noted that the laser oscillation at 1.3 μm of Nd:YAG enhances the thermal lensing of the laser medium in contrast to the 1.06-μm transition that reduces the thermal loading on the laser medium, because of the increase of quantum defect[11]. The resonator configuration was designed taking into account the thermal lensing both for the laser medium and for the nonlinear crystal.

Figure 8 (a) shows the measured time trace of the fundamental and the red output power without wavelength selective element. Neither output power was stable. The magnitude of fluctuation was 26.8 % for the fundamental output and 36.3 % for the red output. The spectrum of the red beam is also shown in Fig. 8 (b). The red beam was composed of two different wavelengths, 659.5 nm and 664 nm. The stable laser oscillations at two lines, 1.319 μm (R2-X1 transition) and 1.338 μm (R2-X3 transition), were simultaneously observed when we removed the KTP crystal. The 659.5-nm and the 664-nm radiation were produced by the frequency doubling of 1.319 μm and the sum frequency mixing of 1.319 μm and 1.338 μm respectively. The power fluctuation was considered to be caused by the gain competition between 1.319 μm and 1.338 μm through the nonlinear process.

Then we introduced a solid etalon to select a single transition at 1.319 μm. The non-coated fused silica etalon was inserted between the end mirror and the laser rod as shown in Fig. 7. The finesse and the free spectral range of the etalon were 0.6 and 2.07 THz respectively. Fig. 9 (a) and (b) show the time trace and the spectrum of the laser output with the etalon. Red radiation at a single wavelength of 659.5 nm was obtained by suppressing the 1.338-μm laser oscillation. Both the fundamental and the red output were well stabilized to the fluctuation of 15.3 % and 17.1 % respectively by inserting the solid etalon. Consequently the maximum red output power of 6.1 W was achieved with the beam quality of $M^2 = 5.6$ at the total diode output power of 486 W whereas the maximum output power was 4.5 W without the wavelength selective element.

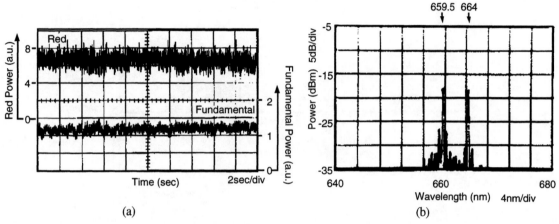

(a)   (b)

Fig. 8 (a) Time trace and (b) spectrum of red laser output obtained without wavelength selective element. Fundamental power measured simultaneously with the red output is also shown in (a). The total diode output power is 330 W.

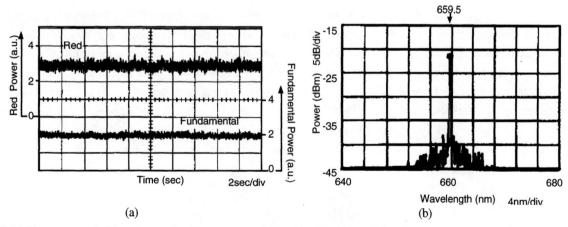

(a)   (b)

Fig. 9 (a) Time trace and (b) spectrum of red laser output obtained by inserting solid etalon in the resonator. Fundamental power measured simultaneously with the red output is also shown in (a). The total diode output power is 330 W.

5. Q-SWITCHED GREEN BEAM GENERATION

Intracavity frequency doubling of Q-switched Nd:YAG laser is an useful scheme for high-power green beam generation with more than 100-W output power. Conventional high-power green lasers, however have relative low beam quality: $M^2 > 20$ that limits the applications[12, 13]. We have developed high-brightness green laser with the output power of more than 100 W while maintaining good beam quality based on the combination of the CIDER configuration and the bifocusing compensation resonator design[14].

Figure 10 shows the schematic drawing of the high-brightness green laser system. A 700-mm long L-shaped resonator has two end mirrors with convex curvature and a flat harmonic separator. Two laser heads and a acousto-optical Q-switch were placed in one arm of the resonator. The Q-switch was operated at the 10-kHz repetition rate. A type-II phase-matched 15-mm long LBO crystal was placed in the second arm of the resonator. The green beam was extracted from the harmonic separator in single direction. The resonator was designed to provide stable operation from the threshold diode output power to the maximum-rated diode output power considering the optical property both for the non-linear crystal and for the acousto-optical Q-switch

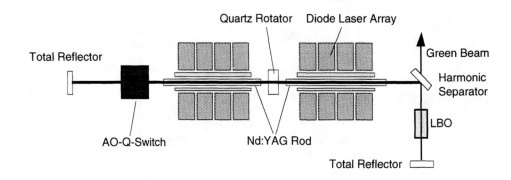

Fig. 10 Schematic drawing of the Q-switched green laser system.

Figure 11 shows the comparison between the Q-switched fundamental and the green output power as a function of the total diode output power. The fundamental operation was carried out by replacing one end mirror by a 15 % transmittance output coupler and removing the LBO crystal from the resonator. The maximum green output power of 138 W was generated with 800-W total diode output power and 1750-W electric input power, corresponding to the optical-to-optical efficiency of 17.3 % and the electric-to-optical efficiency of 7.9 %. The pulse width of the green beam was 70 ns at 138-W green output power. The ratio of the green output to the Q-switched output was approximately 90 %. The beam quality of the green beam at 138-W output was measured to be $M^2 = 11$. The corresponding brightness is calculated to be 434 MW/cm^{2}sr which is to our knowledge, the highest value reported for intracavity frequency doubled > 100-W green lasers. For more than 200 hours of continuous operation, the green output power remained stable at 100-W level with the power fluctuation of less than 3 % and the pulse-to-pulse instability of less than 5 %. No degradation of the LBO crystal was observed

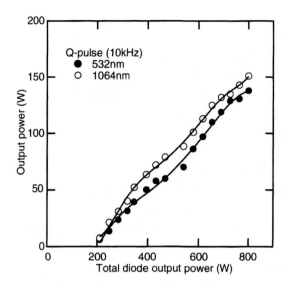

Fig. 11 Q-switched fundamental and green output power as a function of the total diode output power.

This green laser system has been utilized as a pumping source for ultrashort-pulse Ti:Sapphire laser systems and more than 7 W of 30-fs pulse output power has been proved at high repetition rate of 10 kHz[15]. An ultra-violet generation was also demonstrated using the green laser system. The green beam was externally doubled by using a 15-mm long CLBO crystal. The maximum 266-nm output power of 20.5 W was obtained at the green power of 106 W[16]. This value is the highest 266-nm UV power in solid-state lasers.

6. CONCLUSION

We have described resonator designs for stable and high-brightness beam generation by intracavity frequency doubling. The combination of the CIDER pump configuration and the bifocusing compensation resonator design taking into account the optical property both for the laser medium and for the nonlinear crystal provided efficient and stable harmonized beam generations with good beam quality.

ACKNOWLEDGEMENT

The authors thank K. Yoshizawa and M. Tanaka for supporting and encouraging their research.

REFERENCES

1. J. Golden, "Green laser score good marks in semiconductor material processing," Laser Focus World, pp. 75-88, June 1992.
2. G. Yang, P. R. Herzfeld, R. I. Billmers, V. M. Contarion, M. Rankin, E. Dressler and A. Laux, "Temporal response of extremely narrow band optical filters at 532 nm and their application in lidar systems," in in Digest of Conference on Laser and Electro-Optics, Paper CTuT5, 1996.
3. I. Dryer, "Laser color display and entertainment applications," pp. 61-70, Laser Focus World, September 1991.
4. W. E. Glenn, "Solid-state light sources for color projection," OSA TOPS on Advanced Solid-State Lasers, vol. 10, pp. 38-45, 1997.

5. T. Kojima and K. Yasui, "Efficient diode side-pumping configuration of a Nd:YAG rod laser with a diffusive cavity," Appl. Opt., vol. 36, pp. 4981-4984, 1997.

6. S. Fujikawa, T. Kojima and K. Yasui,"High-power and high-efficiency operation of a CW-diode-side-pumped Nd:YAG rod laser," IEEE JSTQE., vol.3, pp.40-44, 1997.

7. K. Yasui, "Efficient and stable operation of a high-brightness cw 500-W Nd: YAG rod laser," Appl. Opt., vol.35, pp. 2566-2569, 1996.

8. T. Kojima, S. Fujikawa, and K. Yasui, "Stabilization of a high-power diode-side-pumped intracavity-frequency-doubled CW Nd:YAG laser by compensating for thermal lensing of a KTP crystal and Nd:YAG rods," IEEE J. Quantum Electron., vol.35, pp. 377-380, 1999.

9. T. Bear, "Large-amplitude fluctuations due to longitudinal mode coupling in diode-pumped intracavity-doubled Nd:YAG lasers. " J. Opt. Soc. Amer. B, vol. 3, pp.1175-1180, 1986.

10. Y. Inoue, S. Konno, T. Kojima and S. Fujikawa, "High-power red beam generation by frequency-doubling of a Nd:YAG laser," IEEE J. Quantum Electron., vol.35, pp. 1737-1740, 1999.

11. Y. Inoue and S. Fujikawa, "Diode-pumped Nd:YAG laser oscillating at 1.319 mm producing 122 W for CW operation," IEEE J. Quantum Electron., submitted for publication.

12. B. J. Le Garrec et. al., "High-average-power diode-array-pumped frequency-doubled YAG laser," Opt. Lett., vol. 21, pp. 1990-1992, 1996.

13. J. J. Chang, E. P. Dragon, C. A. Ebbers, I. L. Bass, and C. W. Cochran, "An efficient diode- pimped Nd:YAG laser with 451 W of CW IR and 182 W of pulsed green output," OSA TOPS on Advanced Solid-State Lasers, vol. 19, pp. 300-304, 1998.

14. S. Konno, T. Kojima, S. Fujikawa and K. Yasui, "High-brightness 138-W green laser based on an intracavity-frequency-doubled diode-side-pumped Q-switched Nd:YAG laser," Opt. Lett., vol. 25, pp. 1-3, 2000.

15. Y. Nabekawa, T. Togashi, T. Sekikawa, S. Watanabe, S. Konno, T. Kojima, S. Fujikawa and K. Yasui, "All-solid-state 10-kHz high-peak-power Ti:Sapphire laser system," in proceedings of ultrafast optics 1999.

16. T. Kojima, S. Konno, S. Fujikawa, K. Yasui, K. Yoshizawa, Y. Mori, T. Sasaki, M. Tanaka and Y. Okada, "20-W, 10-kHz UV beam generation by an all-solid-state laser," OSA TOPS on Advanced Solid-State Lasers, vol. 26, pp. 93-95, 1999.

Invited Paper

Self-adapting thermal lens to compensate for the thermally induced lens in solid-state lasers

Thomas Graf[*], Rudolf Weber, Eduard Wyss, and Heinz P. Weber

University of Berne, Institute of Applied Physics, Sidlerstrasse 5, CH-3012 Berne, Switzerland

ABSTRACT

The thermally induced lens is a critical issue in high-power diode-pumped solid-state lasers. A self-adaptive scheme to balance the thermal lenses in laser resonators is presented. The requirements for the thermo-optical self-adaptive element and its influence on the resonator are discussed. With an appropriate compensating element and the correct resonator design, constant beam parameters are expected to be achieved over a pump range of several kilowatts.

Keywords: thermal lens, adaptive optics, laser resonators, resonator stability

1. INTRODUCTION

The power-dependent thermally induced lens is a major problem that has to be considered in the development of high-power solid-state lasers. The thermal lens is caused by inhomogeneous temperature distributions inside optical materials. The heating of the material results from absorption of either the pump power (gain medium) or the laser power itself (other optical elements). In general, three different effects contribute to the thermal lens: Firstly, the temperature dependence of the refractive index, in this paper referred to as dn/dT-part. Secondly, the axial expansion of the optical material leading to curved end faces, referred to as end-effect. Thirdly, the local strain of the material due to thermally induced stress leading to birefringence, which is not considered in this article. While birefringence reduction schemes were demonstrated successfully,[1, 2] compensation of the phase front distortions resulting from the temperature dependence of the refractive index and the expansion of the material are addressed in alternative active medium designs [3] but have not been solved yet for lasers with rods as gain elements. Moreover, it would be desirable to maintain constant beam parameters over a wide range of pump powers. This can be reached by adapting the resonator parameters to the respective pump power level. Although active mirrors [4] or resonator length adjustments [5] are possible means, they require sophisticated mechanical set-ups and/or electronic control. In addition, they usually do not allow to compensate for the aberrations.

Another possibility to compensate for the thermally induced lens is to take advantage of the effect itself by using a heated optical element. In the case of longitudinally pumped lasers, the absorption of the pump light in an end mirror or a lens was proposed to heat the compensating element.[6] For high-power lasers, transversal pumping is preferred due to its simpler scalability and since it is less expensive. In this case a compensating element, which is placed inside the resonator and absorbs part of the laser radiation, is more promising since there is no simple way to use part of the pump power to heat the compensating element unless a separate pumping unit is used (Figure 1). Fortunately, optical glasses with a strong negative

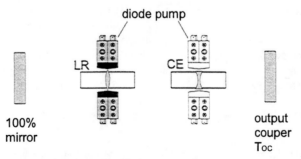

Figure 1. A laser resonator with an intra-cavity compensation scheme. LR: laser rod, CE: compensating element. The diode-pump for the CE is optional and is only needed if the CE does not absorb the laser radiation.

---

[*] Correspondence: Email: thomas.graf@iap.unibe.ch, WWW: http://www.cx.unibe.ch/iap/

In *Laser Resonators III*, Alexis V. Kudryashov, Alan H. Paxton, Editors,
Proceedings of SPIE Vol. 3930 (2000) ● 0277-786X/00/$15.00

123

dn/dT, with an absolute value comparable to that of Nd:YAG, such as the phosphate laser glasses Schott LG-760 or Hoya LHG-8,[6, 7] are readily commercially available. A compensating element, which is heated by the intra cavity power, shows many advantages. On the one hand, no additional pump source is required. On the other hand, the thermal lens is directly correlated to the output power and aberrations will be inherently and passively compensated. In addition, if the compensating element is doped with the same active laser ion as the gain medium, the amount of absorption (from the lower laser level) can be temperature controlled.

2. THERMAL CONSIDERATIONS

A good insight into the basic thermal behavior of heated optical materials is achieved with a simplifying analytical approach that describes the thermally induced lens as presented in the references 7-9. With this, the values of the relevant quantities involved can be estimated in a first approach. In this first step, we assume homogeneously heated rods with closely spaced thin lenses for the different contributions (dn/dT, end-effect) in both the active medium and the compensating element. Under these conditions the dioptric power of the total thermal lens in a heated optical element is given by the sum of the dioptric powers of the different contributions with $D_{therm} = D_{dn/dT} + D_{end}$, where $D_{dn/dT}$ is the contribution due to the temperature dependence of the refractive index, and D_{end} results from the bending of the end surfaces of the rod.

Homogeneous heating of a rod produces a temperature difference between the center the circumference which is independent of the cooling given by[7]

$$\Delta T = \frac{1}{4 \cdot \pi \cdot k} \cdot \frac{P_{heat}}{\ell_{heat}} \tag{1}$$

where k is the heat conductivity, ℓ_{heat} the length over which the element is heated, and P_{heat} is the total heating power. Together with the coefficient for the temperature dependence of the refractive index dn/dT and the expansion coefficient a_{exp} this temperature gradient is responsible for the thermal lens. With the above approximations the two contributions to the dioptric power D of the thermal lens in a heated rod of radius R_{Rod} are given by[7]

$$D_{dn/dT} = \Delta T \cdot \frac{2 \cdot dn/dT \cdot \ell_{heat}}{R_{Rod}^2} = \frac{dn/dT}{2 \cdot \pi \cdot k} \cdot \frac{P_{heat}}{R_{Rod}^2} \tag{2}$$

$$D_{end} = \Delta T \cdot \frac{4 \cdot a_{exp} \cdot (n_0 - 1)}{R_{Rod}} = \frac{a_{exp} \cdot (n_0 - 1)}{\pi \cdot k} \cdot \frac{P_{heat}}{\ell_{heat} \cdot R_{Rod}} \tag{3}$$

where n_0 is the undisturbed refractive index.

Apart from material and geometrical parameters, equations (2) and (3) depend on the heat source P_{heat}. In the case of the laser rod, the heating power is given by the fraction of the pump power P_{pump} which is absorbed and converted to heat:

$$P_{heat,LaserRod} = \eta_{transf} \cdot \eta_{abs} \cdot \eta_{heat} \cdot P_{pump} = \eta_h \cdot P_{pump}. \tag{4}$$

For Nd:YAG, the heat conversion factor η_{heat} is about 35% under lasing conditions [10] and the transfer η_{transf} and absorption η_{abs} efficiencies are typically about 80%.

If an external pump heats the compensating element, the heating is described by the same formula as the heating of the laser rod. In the scheme proposed in this paper, the heating of the compensating element is assumed to result from weak absorption of the intra-cavity circulating power. Therefore, the heated volume is given by the mode size at the location of the compensating element. For the following analytical approach, the multimode-beam in the compensating element is assumed to have about the same size as the rods (laser and compensating element). Furthermore, heating takes place over the whole length of the compensating element given by

$$P_{heat,compEl} = 2 \cdot \alpha_{abs} \cdot \ell_{compEl} \cdot P_{cir} = 2 \cdot \alpha_{abs} \cdot \ell_{compEl} \cdot \frac{P_{out}}{T_{OC}} \tag{5}$$

where α_{abs} is the absorption coefficient at the laser wavelength in the compensating element, ℓ_{compEl} is its length, P_{cir} is the intra cavity circulating power, P_{out} is the output power and T_{OC} is the transmission of the output coupler. The output power is related to the pump power, the laser threshold P_{th}, and the slope efficiency η_{slope} by

$$P_{out} = \eta_{slope} \cdot (P_{pump} - P_{th}) \tag{6}$$

With this and the eqs. (1-3) , the dioptric powers of the thermally induced lenses in the laser rod (subscript *LaserRod*) and the compensating element (subscript *compEl*) are found to be

$$D_{LaserRod} = \frac{dn/dT_{LaserRod}}{2 \cdot \pi \cdot k_{LaserRod}} \cdot \frac{\eta_h \cdot P_{pump}}{R^2_{LaserRod}}$$
(7)

$$D_{compEl} = \left(\frac{dn/dT_{compEl}}{2 \cdot R_{compEl}} + \frac{a_{exp} \cdot (n_0 - 1)}{\ell_{compEl}} \right) \cdot \frac{L_{compEl} \cdot \eta_{slope} \cdot (P_{pump} - P_{th})}{\pi \cdot k_{compEl} \cdot R_{compEl} \cdot T_{OC}}$$
(8)

where $L_{compEl} = 2 \cdot \alpha_{abs} \cdot \ell_{compEl}$ is the loss introduced by the compensating element. These are the two lenses which are produced inside the cavity, and which should compensate each other. We note that the end effect in the laser rod is neglected. This is allowed when the unpumped length at the ends of the rod is longer than the rod diameter, which is usually fulfilled for transversal pumping. The end effect in the compensating element can also be avoided if a composite rod with non-absorbing (e.g. undoped) ends is used.

The aim of the compensation is that the sum of the dioptric powers of the two lenses vanishes, i.e. $D_{LaserRod} + D_{compEl} = 0$. This condition can be combined with the equations (7) and (8) and solved for L_{compEl} in order to find the required absorption for the compensating element. For this estimation the laser is assumed to operate well above threshold, i.e. the threshold power is set to $P_{th} = 0$. Furthermore η_{slope} is set to be independent of L_{compEl}, which is allowed if this additional loss is small. With a length of the compensating element of a few centimetres, the end effect is about ten times weaker than the dn/dT-part and can be neglected. Finally, with additional realistic assumptions for the quantities involved ($dn/dT_{LaserRod} \approx -dn/dT_{compEl}$, $\eta_{heat} \approx \eta_{slope}$ for Nd:YAG, and $R_{compEl} \approx R_{LaserRod}$) the required absorption loss inside the compensating element is found to be

$$L_{compEl} \approx T_{OC} \frac{k_{compEl}}{k_{LaserRod}}$$
(9)

As the heat conductivity of technical glass is typically ten times smaller than that of Nd:YAG, and typical output coupler transmissions are about 10%, one gets a required overall absorption in the compensating element of 1%. This number is very encouraging and shows that the principle of passive intra-cavity compensation with optical materials with a negative dn/dT is possible without significant influence on the laser performance.

3. RESONATOR CONSIDERATIONS

Obviously the performance of a thermally compensated resonator significantly depends on where the compensating element is placed. Ideally, for an optimised compensation the negative thermal lens should be placed exactly at the same location as the positive thermal lens. Since this is not possible physically, one has to superimpose the two lenses optically such that the beam parameter (q-parameter) at the location of the positive lens is the same as at the location of the negative lens. This is achieved with an optical system that has a total ray transfer matrix given by the identity matrix (with positive or negative sign). For heated rods this optical system has to be inserted between the principal planes of the two elements. Equivalently, the thermal lenses can be treated as thin lenses sandwiched between two unperturbed rod pieces of length $L_{Rod}/2$.[11] For the sake of simplicity only the thin thermal lenses without the rods are shown in figure 2. The dashed line in the graph of figure 2 shows the fundamental-mode radius at the location of the thermal lens in the laser rod as a function of the pump power for

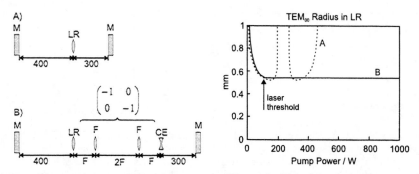

Figure 2. Influence of the compensating element (CE) on the mode radius in the laser rod (LR). M: mirror, F: focal length of lenses. Distances in mm.

a typical, transversally pumped Nd:YAG laser with the uncompensated cavity labelled A. In the cavity B the compensating element is optically superimposed to the laser rod with two lenses F. As soon as the laser threshold is reached (at 100 W) the lens in the compensating element sets in (absorbing about 1% of the circulating power) and the mode radius does not change any more with increasing power. If this is combined with a telescope to get a large fundamental-mode radius in the laser rod, this set-up can lead to a diffraction limited beam over the whole pump power range.

4. THE COMPENSATING ELEMENT MADE OF LG-760

As discussed in section 2 the required material properties for the generation of the compensating thermal lens are a negative dn/dT value (to compensate for laser materials with positive thermal lenses), low heat conductivity and a weak double-pass absorption of order 1%. The first two requirements are met by a whole variety of commercially available glass types. However, the absorption coefficient at the laser wavelength (1064 nm in our case) is much too small in most of the glass types coming into question. New glass types will have to be developed for a fully optimised implementation of the proposed self-adaptive thermo-optical compensation scheme.

As an alternative we are currently investigating the use of laser glasses such as the Nd-doped LG-760.[7] In these glasses, the absorption is due to the Nd-ion transitions between the lower and the upper laser level. Since, in this case, the absorption depends on the thermal Nd-ion population in the lower laser level, the absorption coefficient in the glass rod depends on the local temperature, as shown in figure 3 (left). As mentioned above, this has some interesting aspects. On the one hand, the absorption coefficient - and with it the slope of the dioptric power of the thermal lens versus the incident laser power – can be adjusted by controlling the temperature of the glass rod. This is very convenient since otherwise the dependence of the dioptric power of the thermal lens on the intra-cavity circulating laser power can only be adjusted by adapting the length of the glass rod. On the other hand, as soon as the absorption sets in, the glass rod is heated, leading to increased absorption and further increased heating and so on. This positive feedback mechanism may eventually lead to the fracture of the glass rod due to the thermally induced stresses caused by the elevated temperatures. This behaviour has been demonstrated both experimentally and with numerical simulations. However, as shown with the finite-difference numerical simulation in figure 3 (right), we have found that the uncontrolled increase of the temperature can be avoided by suitable cooling efficiency. If the LG-760 glass rod is not cooled (simple mount, surrounded by air) the above mentioned feed-back mechanism leads to a continuous temperature increase ending with the fracture of the rod (dashed line). With more efficient edge cooling (either directly water-cooled or wrapped in an indium foil and mounted in a water-cooled copper mount) we were able to demonstrate controlled thermal lenses that could be adjusted by tuning the temperature in preliminary extra-cavity experiments. In this case the temperature reaches a stable distribution after a few seconds (solid line). The experiments were carried out with the LG-760 rod in a Mach-Zehnder interferometer and a 100-W Nd:YAG laser beam, leading to comparatively weak thermal lenses that are difficult to measure with reasonable accuracy. Intra-cavity experiments which will lead to stronger lenses are currently being carried out.

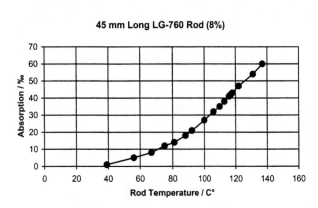
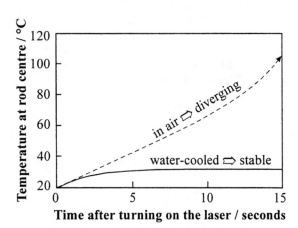

Figure 3. Left: Absorption in a LG-760, measured at 1053 nm. Due to the gain bandwidth of 19.5 nm (FWHM) the absorption at 1064 nm is approximately half of the absorption at 1053 nm. Right: Numerical simulations of the transient behaviour of the temperature at the centre of an 8% Nd-doped LG-760 rod with 4 mm of diameter after turning on the laser. The incident power is 1.2 kW at a wavelength of 1064 nm. 50% of the absorbed power was assumed to be converted to heat. The heat transfer coefficients were set to 0.025 Wcm^{-2}K^{-1} and 2 Wcm^{-2}K^{-1} for glass in air and glass in water, respectively.

5. THE COMPOSITE LASER ROD

As an alternative to the compensation scheme with the compensating thermal lens that is generated by weak absorption of the intra-cavity laser power, we are also investigating composite laser rods made of a transversally pumped Nd:YAG section sandwiched between two glass pieces as shown in figure 4 (left). In this situation the glass sections are heated by heat conduction through the YAG-glass transition. Therewith, the positive thermal lens in the Nd:YAG section is balanced with the negative thermal lens generated in the glass sections. The temperature distribution in the composite rod and the total thermally induced lens were simulated with numerical finite-element methods for a large variety of different situations. A typical temperature distribution in the composite rod, which is homogeneously pumped with 50 W of pump power in the Nd:YAG section, is shown in figure 4 (left). The rod diameter is 4 mm. The right-hand side of figure 4 shows the dependence of the total dioptric power of the composite rod with varying length of the two PK51 glass sections at 50 W and 100 W of pump power. With increasing glass thickness, the contribution of the negative lenses in the PK51 increases due to the longer path length with a negative gradient-index lens (radially increasing refractive index towards the rod edge). At the same time, the temperature in the Nd:YAG section is also slightly increased with increasing glass thickness because of the decreasing cooling through the glass endfaces which are in contact with air. This is a comparatively weak effect, which is believed to lead to the minimum of the dioptric power around a glass thickness of 4 mm shown in figure 4 (right). For glass pieces longer than about 6 mm, the total dioptric power does not change anymore, since the heating does not reach deeper into the glass section. With this example, the dioptric power of the rod can be reduced by more than a factor of 4. Further optimisations and experiments will be carried out.

For the experiments the different sections of the rod will be bonded together and are believed to resist pump powers of the order of several tens of Watts. Further power scaling may be achieved by manufacturing longer rods with several side-pumped Nd:YAG sections separated with compensating glass. The Fresnell losses at the YAG-glass transitions are smaller than 1% and may be further reduced by suitable coatings.

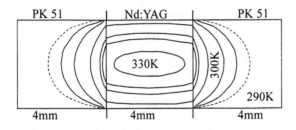
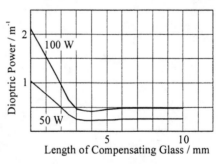

Figure 4. Left: Finite-elements simulation of the temperature distribution in the edge-cooled composite laser rod with a diameter of 4 mm. The Nd:YAG section is homogeneously pumped with a power of 50 W. The glass sections (PK51) are heated by heat conduction, generating negative thermal lenses to compensate for the positive lens in the Nd:YAG. Right: The total dioptric power of the composite rod with varying length of the PK51-glass sections at two different pump powers.

6. CONCLUSIONS AND OUTLOOK

In conclusion we have shown theoretically and with preliminary experiments that the thermo-optical compensation scheme to balance the positive thermal lens in a Nd:YAG laser rod with a negative thermal lens in a glass rod which is heated by weak absorption of the intra-cavity laser radiation is realistic and very promising. As an alternative we are currently also investigating novel composite laser rods, where the compensating thermal lenses are generated by heat conduction form the pumped Nd:YAG section(s) into the glass sections with negative thermal dispersion.

7. REFERENCES

1. Q. Lü, N. Kugler, H. Weber, S. Dong, N. Müller, U. Wittrock, "A novel approach for compensation of birefringence in cylindrical Nd:YAG rods", Optical and Quantum Electronics **28**, 57-69, 1996
2. N. Kugler, S. Dong, Q. Lü, H. Weber, "Investigation of the misalignment sensitivity of a birefringence-compensated two-rod Nd:YAG laser system", Applied Optics **36**, 9359-9366, 1997

3. A. Giesen, H. Hügel, A. Voss, K. Witting, U. Brauch, H. Opower, "Scaleable concept for diode-pumped high-power solid-state lasers", Appl. Phys. **B58**, 365-372, 1994

4. U. J. Greiner, H.H. Klingenberg, "Thermal lens correction of a diode-pumped Nd:YAG laser of high TEMoo power by an adjustable-curvature mirror", Optics Letters **19**, 1207-1209, 1994

5. D.C.Hanna, C.G. Sawyers, M.A.Yuratich, "Telescopic resonators for large-volume TEMoo-mode operation", Optical and Quantum Electronics **13**, 493-507, 1981

6. R. Koch, "Self-adaptive elements for compensation of thermal lensing effects in diode end-pumped solid-state lasers - proposals and preliminary experiments", Optics Communications **140**, 158-164, 1997

7. W. Koechner, *Solid-State Laser Engineering*, Springer Series in Optical Sciences

8. W. Koechner, "Absorbed pump power, thermal profile and stresses in a cw pumped Nd:YAG crystal", Applied Optics **9**, 1429-1434, 1970

9. R. Weber, B. Neuenschwander, H.P. Weber, "Thermal effects in solid-state laser materials", Optical Materials **11**, 245-254, 1999

10. T.Y. Fan, "Heat Generation in Nd:YAG and Yb:YAG", IEEE J. Quantum Electron. **29**, 1457-1459, 1993

11. Vittorio Magni, "Resonators for solid-state lasers with large-volume fundamental mode and high alignment stability", Applied Optics **25**, 107-117, 1986

SESSION 6

Laser Resonator Design

Orbital angular momentum, optical spanners and the rotational frequency shift

Miles J Padgett

Department of Physics and Astronomy, Kelvin Building, University of Glasgow, Scotland. G12 8QQ. UK.

ABSTRACT

Light possesses both spin and orbital angular momentum. Whereas the spin component arises from the spin of individual photons and is associated with circular polarization, the orbital component arises from the azimuthal component of the Poynting vector and is associated with helical wavefronts. This papers considers the generation of beams with orbital angular momentum, their interaction with mater, behaviour in nonlinear media and frequency shifts arising from their rotation. In each case the properties of beams containing orbital and/or spin angular momentum is compared and explained.

1. INTRODUCTION

It is well understood that circularly polarised light corresponds to the case when the intrinsic spin of every photon in the beam is aligned giving an angular momentum of $\sigma\hbar$ per photon, where $\sigma = \pm 1$ for left and right handed polarization respectively.

Light also possesses a linear momentum associated with each photon which if acting on a radius vector may give rise to an an additional angular momentum independent of the polarization state. Despite this, it was not until the early 1990's that it was established that Laguerre-Gaussian modes, characterized by an $\exp(il\phi)$ helical phase structure, possess an angular momentum $l\hbar$ per photon[1].

This paper outlines the origin of "orbital angular momentum" and methods for the generation of beams possessing such angular momentum. It also discusses the interaction of such beams with matter, nonlinear media and analogies with circularly polarised beams.

2. ORBITAL ANGULAR MOMENTUM AND LAGUERRE-GAUSSIAN MODES

The transverse profiles of coherent light beams emitted from laser cavities are frequently described in terms of the rectangular symmetric Hermite-Gaussian polynomials[2]. These are solutions to the paraxial form of Maxwell's equations and are the eigenmode solutions of the resonating cavity. The Hermite-Gaussian modes form a complete basis set, hence any arbitrary light beam can be formed by their linear superposition. They are characterized in terms of two mode indices m, n which state the order of the Hemite polynomial in the x and y directions respectively. The Hermite-Gaussian modes are denoted HG_{nm} and the mode order, N, is usefully defined as $N = n + m$.

However, the Hermite-Gaussian modes are not the only solutions to Maxwell's equations and other basis sets such as the Laguerre-Gaussian modes also exist. The Laguerre-Gaussian modes have circular symmetry with intensity profiles comprising concentric rings. They are also characterised in terms of two mode indices p the radial mode index which gives the number of rings and l, the azimuthal mode index giving the $\exp(il\phi)$ phase structure. The Laguerre-Gaussian modes are denoted LG^l_p and the mode order, N, is usefully defined as $N = 2p + |l|$. For $l \neq 0$, the $\exp(-il\phi)$ term produces l intertwined helical wavefronts and a phase singularity at the beam focus.

Figure 1 shows examples of the intensity distribution of Hermite-Gaussian and Laguerre-Gaussian modes of various mode order. Notice that the azimuthal indices of the Laguerre-Gaussian modes are impossible to determine from the intensity profile alone. To determine unambiguously the azimuthal mode index it is usual to interfere the Laguerre-Gaussian mode with a plane wave of the same frequency. The azimuthal phase variation gives rise to l spiral fringes in the interferogram. If the interfering beams are slightly misaligned then the spiral fringe pattern is superimposed on a straight line fringe pattern producing an l pronged discontinuity on the beam axis.

Email: m.padgett@physics.gla.ac.uk

In *Laser Resonators III*, Alexis V. Kudryashov, Alan H. Paxton, Editors,
Proceedings of SPIE Vol. 3930 (2000) • 0277-786X/00/$15.00

For a linearly polarised Laguerre-Gaussian modes, the helical wavefronts result in a Poynting vector of the form

$$\bar{S}(r,\phi,z)\alpha\frac{z_r}{z_r^2+z^2}\left(\frac{zr}{z_r^2+z^2}\hat{r}+\frac{l}{kr}\hat{\phi}+\hat{z}\right)$$

Note that in addition to the z-direction, there is an energy flow in the radial direction corresponding to the diffractive spreading of the beam and in the azimuthal direction. This "azimuthal energy flow" is the origin of the orbital angular momentum carried by these beams.

In 1992 Allen and co-workers established that beams of this type with an $exp(-il\phi)$ phase structure, have an angular momentum in the propagation direction of $l\hbar$ per photon. More general phase structures lead to arbitrary and non-integer values of orbital angular momentum[3]. This "orbital angular momentum" can be derived directly from Maxwell's equations and although expressed as a "per photon" value it is not exclusively a quantum property.

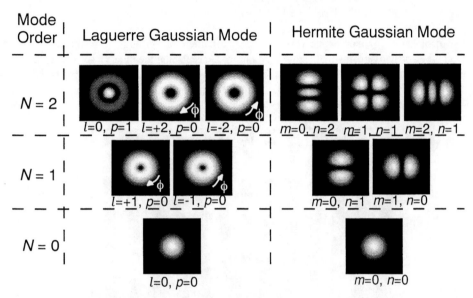

Figure 1. Intensity profiles of Hermite-Gaussian and Laguerre-Gaussian modes of low mode order.

As the beams can also be polarised this means that both spin and orbital angular momentum have well defined components in the direction of the beam. Consequently, the total angular momentum, J_z, of such a beam when circularly polarized is given by,

$$J_z = l + \sigma .$$

This was subsequently confirmed for a non-approximate solution of Maxwell's equations, except that there is now a very small correction term which is only zero for linearly polarised light[4].

The phase singularities that characterize Laguerre-Gaussian modes, are sometimes called optical vortices. Such beams have been studied previously by a number of groups[5,6,7,8], but not with reference to their angular momentum properties.

3. GENERATION OF LAGUERRE-GAUSSIAN MODES

A number of methods have been demonstrated for generating Laguerre-Gaussian modes including direct laser output[9,10], cylindrical lens mode converters[11], spiral phase plates[12,13] and computer-generated holograms[14], Of these, the computer generated holograms are perhaps the most versatile. Although the term "hologram" is correct, it is perhaps more informative to describe them as computer-generated diffraction patterns giving diffracted beams of the desired form. At its simplest, a computer-generated hologram is the calculated interference pattern that results when the desired beam intersects the beam with planar wavefronts emitted from a conventional laser. This pattern is transferred to high-resolution holographic film. For a Laguerre-Gaussian mode, the interference pattern with a plane wave is a diffraction grating with an l pronged dislocation on the beam axis. When the developed hologram is placed in the laser output, the first order diffracted beam has the desired form.

There are various levels of sophistication in hologram design. For a binary hologram the relative intensities of the two interfering beams are ignored and the transmission of the hologram is set to be zero for a phase difference between 0 and π or unity for a phase difference between π and 2π. For $l = 1$ the form of the hologram is shown in figure 2

For a binary hologram, the radial intensity profile of the diffracted beam depends on the profile of the illuminating beam. Consequently, the resulting beam is a superposition of different Laguerre-Gaussian modes with same azimuthal mode index, l, but differing radial mode indices, p. However, if the illuminating beam is itself a Gaussian mode then the resulting beam is a close approximation to a single-ringed, $p = 0$, Laguerre-Gaussian mode.

A further limitation of binary holograms is that very little of the incident power ends up in the first order diffracted spot. This last limitation can be partly overcome by blazing the grating, i.e. replacing the binary "square-wave" line structure with a grey-scale "saw-tooth". In addition, bleaching of the hologram transforms the variation in transmission into ones of refractive index. The resulting phase hologram has a substantially higher optical efficiency and conversions of over 50% can be obtained for the generation of single-ringed[15], $p = 0$, or multi-ringed[16], $p > 0$, Laguerre-Gaussian modes.

Figure 2. Binary hologram for generation for Laguerre-Gaussian modes with $l = 1$

When mode purity is of particular importance it is also possible to create more sophisticated hologram where the contrast of the pattern is varied as a function of radius such that the diffracted beam has the correct radial profile.

4. THE ROLE OF ORBITAL ANGULAR MOMENTUM IN NONLINEAR OPTICS

Sum-frequency mixing and parametric down-conversion are both examples of second order nonlinear processes where energy is exchanged between three optical fields within a nonlinear optical medium. Conservation of energy sets the relationship between the frequencies of the three fields to be

$$\omega_1 + \omega_2 = \omega_3.$$

When $\omega_1 = \omega_2$, the process is that of second harmonic generation, a technique widely used for converting the output of infrared lasers into the visible region of the spectrum, $\omega_3 = 2\omega_1$. The two fields are usually referred to as the "fundamental" and "second-harmonic" frequencies. In addition to the conservation of energy, requirement for the conservation of momentum means that the temperature and/or direction of propagation in the medium has to be adjusted such that the refractive indices, n_i, satisfy the equation

$$\frac{\omega_1}{n_1} + \frac{\omega_2}{n_2} = \frac{\omega_3}{n_3}, \text{ i.e. } k_1 + k_2 = k_3$$

When the three fields propagate co-linearly this "phase-matching" relationship is a scalar one. This is usually the case for efficient interactions of Hermite-Gaussian beams when co-linear wavevectors ensure a large interaction length within the nonlinear crystal. However, in general the three fields do not need to be co-linear and the relationship should be expressed in vectorial form,

$$\bar{k}_1 + \bar{k}_2 = \bar{k}_3.$$

4.1. Frequency doubling of single-ringed Laguerre-Gaussian modes

In an isotropic media the wavevector is co-linear to the Poynting vector. Although this is no longer quite true in a birefringent material (such as most nonlinear materials) both vectors have similar forms. Hence, for a Laguerre-Gaussian mode the wavevector has a spiral form with a well-defined azimuthal component. It seems reasonable to anticipate that this will significantly complicate the phase-matching of such beams. Perhaps surprising therefore is that the phase-matching conditions for Laguerre-Gaussian modes are the same as for their Hermite-Gaussian counterparts. This can only be the case if, even for the Laguerre-Gaussian modes, the wavevectors for the fundamental and second-harmonic fields remain co-linear[17].

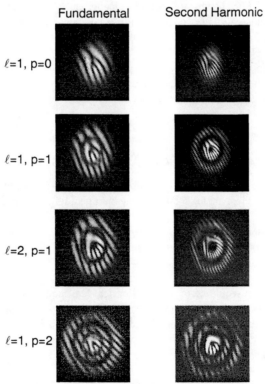

Figure 3. Interferograms of fundamental and second harmonic beams showing the doubling of the azimuthal mode index

The azimuthal component of the Poynting vector and hence the wavevector is l/kr [18]. Upon frequency doubling, k doubles and for the wavevectors of the fundamental and second harmonic fields to remain co-linear, l, the azimuthal mode index, must double also[19]. This deceptively simply result implies that the orbital angular momentum per photon is doubled and since upon frequency doubling the number of photons halves, that the orbital angular momentum in the light fields is conserved. This should be compared to the frequency doubling of circularly polarized light beams where the spin angular momentum is not conserved within the light fields.

Figure 3 shows interferograms of various Laguerre-Gaussian modes interfered with their own mirror image, both before and after frequency doubling. In each case, the azimuthal mode index can be deduced from the $2l$ fork dislocation on the beam axis.

4.2. Parametric down-conversion with Laguerre-Gaussian modes

In many ways degenerate parametric down-conversion is simply second harmonic generation in reverse, i.e. an incident field produces two fields of half the frequency. This input beam is usually referred to as the "pump" and the two down converted beams as the "signal" and "idler". It is tempting to think that a Laguerre-Gaussian pump beam with azimuthal mode index $2l$ would produce two down-converted beams with azimuthal mode indices l. However, experimental results showed that this was not the case[20].

Second-order non-linear interactions relate to the coupling of three fields. The relative phase of the fields is given by

$$\phi_3 - \phi_2 - \phi_1 = \pm \frac{\pi}{2}.$$

In second harmonic generation, two of the fields have a defined azimuthal phase structure resulting in a spatially coherent output beam also with a uniquely defined phase structure. In parametric down-conversion, only the pump field has a defined phase. The signal and idler fields, although having a unique relative phase relation, do not have a unique absolute phase. Consequently, the signal and idler beams are spatially incoherent with respect to the pump field. Since the orbital angular momentum arises from the phase structure of the beam it is concluded that the down converted beams do not carry a unique orbital angular momentum.

However, even though neither of the signal or idler beams has a well-defined orbital angular momentum their fixed relative phases does create a link between their orbital angular momentum states. Recent work by Zeilinger has shown that, on a photon level, the orbital angular momentum per photon, L_i is conserved, i.e.

$$L_1 + L_2 = L_3.$$

5. TRANSFER OF THE ORBITAL ANGULAR MOMENTUM FROM LIGHT TO MATER

Given that Laguerre-Gaussian modes had been predicted to carry an angular momentum, the obviously experiment was to use these beams to rotate an object in direct analogy to the Beth experiment for circular polarization.

As mentioned in the opening section of this paper, Beth used a quarter-wave plate, suspended from a quartz fibre, to transform circularly polarized light to a linear polarization. In this process the spin angular momentum of $\hbar$ per photon was transferred to the wave plate giving a measurable torque.

The direct analogue to this experiment would be to use a cylindrical lens mode converter to transform a beam with helical wavefronts and associated orbital angular momentum to a beam with planar wavefronts and no angular momentum. In the process, the angular momentum would be transferred to the lenses giving a measurable torque. Unfortunately, such an experiment is extremely challenging since slight misalignment of the beam results in its linear momentum exerting an additional torque about the suspension axis. Although an experiment of this kind has been attempted, to date, no confirmation of the exerted torque has been reported.

5.1. Optical tweezers

In 1987, after a number year of related research, Ashkin and co-workers used a tightly focused beam of laser light to trap a micron sized glass sphere in three dimensions: a technique called optical tweezers[21]. Optical tweezers rely on the force experienced by a dielectric material when placed in an electric field gradient. This force is easily demonstrated within the laboratory using an air-spaced, parallel-plate capacitor and a polystyrene slab. Reversing the polarity of the capacitor does not change the direction of the force. A tightly focussed laser beam produces an extremely large gradient in the electric field, resulting in a force on a small dielectric particle directed towards the beam focus. At optical frequencies, the requirement for a dielectric material is one that is highly transparent.

5.2. Optical Spanners

In 1994, it was suggested that orbital angular momentum could be transferred to a particle held in an optical tweezers[22]. Using a Laguerre-Gaussian mode within an optical tweezers enables rotation of the particle about the beam axis without any additional suspension mechanism. However, rather than trap a miniature mode converter, which would be difficult to make and impossible to align, the transfer of angular momentum would rely on partial absorption of the light by the trapped particle.

The first demonstration[23] of this angular momentum transfer was reported in 1995. The micron-sized particles were of a highly absorbing ceramic powder and consequently could not be trapped in the conventional way. The annular nature of a Laguerre-Gaussian mode means that the absorbing particle can in fact be laterally trapped in two-dimensions near the intensity null on the beam axis. Restraint in the third, or axial, dimension is provided by the microscope cover slip. When trapped with a Laguerre-Gaussian mode of azimuthal mode index $l = 3$, rotation of the particle was observed which was attributed to the transfer of orbital angular momentum from the beam. Although correct to within an order of magnitude the experimental uncertainty meant that it was impossible to quantify the orbital angular momentum in terms of $\hbar$.

Later the experiment was refined to use only partially absorbing Teflon particles, suspended in alcohol, where the gradient force was still sufficient to trap the particle in three dimensions and hence isolate it from the sample cell walls. Furthermore it was possible to compare the transfer of both spin and orbital angular momentum in the same beam. Power levels of a few 10's milliWatts gave rotation speeds of a few hertz. By switching the handedness of a circularly polarized Laguerre-Gaussian mode with $l = 1$, the total angular momentum could be changed from $\hbar - \hbar = 0$ to $\hbar + \hbar = 2\hbar$ per photon. The resulting "stop-start" rotation of the particle, shown in figure 4 confirmed both the magnitude of the orbital angular momentum and that both the orbital and spin angular momentum components of a light beam could be transferred in an equivalent fashion[24].

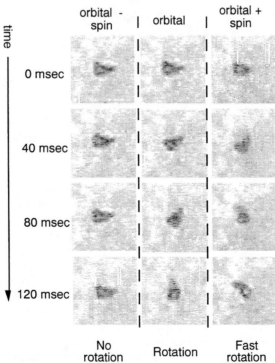

Figure 4. Successive frames of video showing the stop start behaviour of the trapped particle

5.3. An upper limit to the transfer of orbital angular momentum to a small particle.

One obvious question relating to the optical spanner is "how large are the particles it can rotate?". The momentum of inertia of a solid sphere and hence the torque required to accelerate it scales with the 5th power of its radius. Significant increases in laser power may result in damage either to the focussing objective or the particle itself, instead the angular momentum per photon must be maximized.

No mater how strong the focussing, the azimuthal momentum component cannot exceed the linear momentum in the direction of propagation $\hbar k$. This sets the absolute upper limit to the angular momentum, L_{max}, transferred to a particle to be the linear momentum multiplied by the radius of the particle, r, i.e.

$$L_{max} = \hbar k r \text{ per photon.}$$

However, in most experimentally realizable configurations the azimuthal component of the linear momentum will be much smaller and hence the angular momentum transferred to the particle will be significantly less than this limit.

Under more moderate focussing this geometrical argument can readily be extended. Assuming that the radius of the particle is much smaller than the radius of the focussing lens, R, the maximum orbital angular momentum is transferred to the particle is simply shown to be[25]

$$L_{max} = \frac{\hbar k R r}{\sqrt{f^2 + R^2}} \text{ per photon.}$$

6. THE ROTATIONAL DOPPLER SHIFT

The first-order translation Doppler shift is a phenomenon familiar to all, where the relative velocity, v, between an optical source and the observer gives rise to a frequency shift, $\Delta \omega$.

$$\Delta \omega = \frac{v}{c} \omega = \frac{p}{\hbar} v,$$

where p is the linear momentum per photon.

Much less well know is that there is a direct rotational equivalent, where frequency shift depends on the angular momentum per photon, L, and the rotational velocity, Ω,

$$\Delta \omega = \frac{L}{\hbar} \Omega.$$

It is important to realize that this effect it distinct from the rotational effect seen from object such as galaxies where the rotation gives rise to a linear component of velocity between the source and observer. The rotational Doppler shift is observed in the direction of the angular velocity vector where the normal "translational" Doppler shift is zero.

6.1. Rotational Doppler effect for circularly polarized light

For circularly polarized light, $L/\hbar = \sigma = \pm 1$. The rotation of a light source or detector exactly about the beam axis presents a number of experimental challenges since any translation motion may result in additional frequency shifts due to the linear Doppler shift. An elegant solution to this problem is to use a rotating half-wave plate inserted into the light beam.

It is well know that a half-wave plate rotates the linear polarization state of a light beam. If the half-wave plate is itself rotated with an angular velocity Ω then the polarization is rotated with an angular velocity of 2Ω in the opposite direction. For circularly polarized light, the situation is more complicated since the electric field vector is also rotating. If the wave plate is stationary the transmitted beam is still circularly polarized but the electric field vector is now rotating in the opposite direction, i.e. the handedness of circular polarization has been reversed. If the wave plate is now rotated with an angular velocity Ω, the handedness of the circular polarization is still reversed but its rotation frequency is shifted by an additional 2Ω. Consequently, a rotating half-wave plate can be considered as introducing a rotation to a polarized light beam. Note that when using circularly polarized light, it is necessary to introduce a second, stationary, half-wave plate so that the handedness of circularly polarization is returned to its initial state.

In 1980 Garetz identified this frequency shift as an annular Doppler shift[26] and sighted a number of related examples[27]. He analyzed this wave-plate induced frequency shift in terms of Jones polarization matrices, however, in later work it was also shown to be an example of a dynamically evolving Berry phase.

Geometrically, the effect for circularly polarized light is simple to understand and is the equivalent of saying that, if placed on a turntable, the second hand of a watch appears to rotate more quickly.

6.2. Rotational Doppler effect for helically phased light

A half-wave plate reverses the handedness of circularly polarized light. The equivalent optical component, a π mode converter, for light with helical wavefronts is any device which transforms x to $-x$. The simplest example of this is a Dove prism which inverts an image about its axis and hence if rotated with an angular velocity Ω rotates the image at 2Ω. In 1996 Nienhuis predicted that a rotating converter based on cylindrical lenses would induce a frequency shift to a Laguerre-Gaussian beam[28]. Although not initially recognized as a related effect the parallels between rotating waveplate and converter are striking. In both cases, the frequency shift can be expressed in terms of the beam rotation frequency multiplied by the angular momentum per photon, i.e.

$$\Delta\omega = \frac{\sigma\hbar}{\hbar}\Omega \ (Garetz) \qquad or \qquad \frac{l\hbar}{\hbar}\Omega \ (Nienhuis).$$

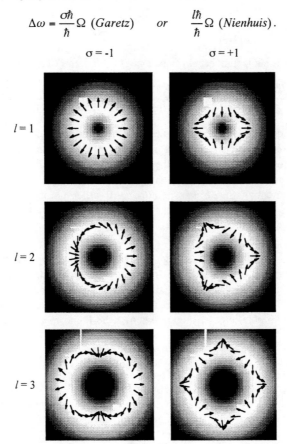

Figure 5. The transverse field structure of circularly polarized Laguerre-Gaussian modes

6.3. Rotational Doppler effect for circularly polarized Laguerre-Gaussian modes

Most recently, both these phenomena were combined in the same experiment. To minimize difficulties associated with the precise alignment and rotation of optical components, the experiment used a mm-wave source at $96GHz$. The corresponding wavelength is $\approx 3mm$ and optical component such as lenses and prisms can be made from PTFE. Another advantage is that in this region of the spectrum the frequency can be measured directly. The beam rotation was introduced using a Dove prism[29] combined with a half-wave plate, which rotated the phase structure and the polarization of the beam respectively. The results showed that rather than act independently, the spin and the orbital angular momentum act indistinguishably such that the observed frequency shift is given by[30]

$$\Delta\omega = \frac{(l+\sigma)\hbar}{\hbar}\Omega.$$

The origin of this effect can be understood by examining the transverse field distributions of circularly polarized Laguerre-Gaussian modes. Figure 5 shows clearly the $(l+\sigma)$ fold rotational symmetry of the modes, such that the rotation of the beam results in an additional $(l+\sigma)$ cycles in optical phase.

Since circularly polarized Laguerre-Gaussian modes form a complete basis set from which any arbitrary monochromatic beam can be described, this effect must be considered a general one and not a peculiarity of esoteric research.

7. REPRESENTATION OF BEAMS POSSESSING BOTH SPIN AND ORBITAL ANGULAR MOMENTUM

In the previous section, attention was drawn to the fact that wave plates and mode converters perform similar operations for light containing spin and orbital angular momentum respectively.

7.1 The Poincaré sphere for polarization states

The polarization state of a monochromatic light beam can be expressed in terms of the Stokes parameters.

$$p_1 = \frac{I(0°) - I((90°))}{I(0°) + I(90°)}, \quad p_2 = \frac{I(45°) - I((135°))}{I(45°) + I(135°)}, \quad p_3 = \frac{I(right) - I((left))}{I(right) + I(left)}$$

where $I(0°)$, $I(90°)$, $I(45°)$, and $I(135°)$ are the intensities of the light recorded through various orientations of linear polarizers, $I(right)$ and $I(left)$ are the intensities of the circularly polarized components in the beam. Any completely polarized light beam can be represented by a point on the sphere $p_1^2 + p_2^2 + p_3^2 = 1$. This sphere is called the Poincaré sphere. The north and south poles correspond to right and left hand circularly polarized light respectively and a corresponding spin angular momentum of $\pm\hbar$ per photon. All other points on the sphere can be considered as superpositions of these two states where the latitude of the point gives the relative weighting and the longitude the relative phase.

The change in polarization state after transmission through a wave plate can be represented by a geometrical transformation on the sphere. For example, starting with right-hand circularly polarized light, a quarter-wave plate produces a linear polarization at 45° to the axis of the wave plate. On the Poincaré sphere, this is represented by a move from the north-pole to a point on the equator. Similarly, a half-wave plate changes right-handed circular polarization to left-handed polarization. This is represented by a move from pole to pole. Figure 6 shows the Poincaré sphere and the geometry for the two wave plate transformations.

7.2 The Poincaré sphere equivalent for orbital angular momentum states

By analogy, there is an equivalent to the Poincaré sphere for the simplest case of light beams containing orbital angular momentum, i.e. of mode order $N = 1$. The north and south poles represent Laguerre-Gaussian LG^1_0 and LG^{-1}_0 modes respectively and a corresponding orbital angular momentum of $\pm\hbar$ per photon. All other points on the sphere are superpositions of these two modes[31]. For example the equator represents a $HG_{1,0}$, the orientation of which depends on the relative phase of the two Laguerre-Gaussian modes which corresponds to different longitudes. π and $\pi/2$ mode converters perform analogous transformations and half-wave and quarter-wave plates respectively. Figure 7 shows such a sphere, the modes it represents and two mode converter transformations.

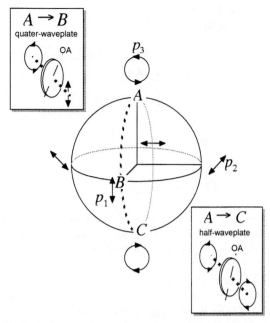

Figure 6. The Poincaré sphere and the representation of polarization states

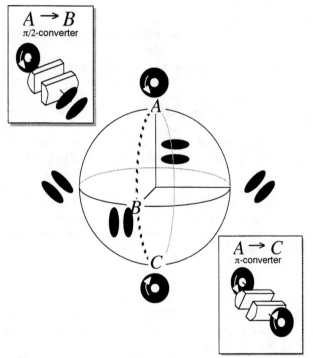

Figure 7. The Poincaré sphere equivalent and the representation of modes of order 1

7.4 Representation of mode of higher order

For the two-state systems of polarisation and Laguerre-Gaussian modes of order $N = 1$, the geometrical representation of the Poincaré sphere, or its equivalent, works well. However, a Poincaré sphere approach to modes of arbitrary order of N is not useful since additional spatial dimensions would be required. A matrix approach suffers no such limitation. In the case of polarisation, the Jones matrices are the equivalent representation to the Poincaré sphere and the same matrices can be used to represent modes of order $N = 1$. However, more interesting is that it is a simple matter to increase the dimensionality of these matrices to represent modes of arbitrary order.

Any mode of order N can be expanded as the sum of $(N + 1)$ Hermite-Gaussian modes of the same order[32]. It follows that any such mode may be represented by a column vector with $(N + 1)$ elements,

$$\begin{bmatrix} a_{N,0} \\ a_{N-1,0} \\ ... \\ ... \\ a_{1,N-1} \\ a_{0,N} \end{bmatrix}$$

where $a_{n,m}$ is the complex amplitude coefficient of the (n,m) Hermite-Gaussian mode. $\lfloor (N+1) \times (N+1) \rfloor$ – matrices modeling the operation of mode converters, filters and rotations can all be simply derived. The technique can also be readily extended to include polarization and mode structure simultaneously.

Such matrices are able to model the propagation of both spin and orbital angular momentum, through optical systems comprising both polarizing and mode transforming components. They have been used to model the $(l + \sigma)$ dependence of the rotational frequency shift and it seems likely that more applications will be found.

8. OPTICAL BOTTLE BEAMS

By definition, both the Hermite-Gaussian and Laguerre-Gaussian modes are structurally stable; i.e. ignoring radial scaling their form does not change upon propagation. However, for applications such as atom trapping or possibly optical tweezers there is an interest in producing beams which have an intensity zero at the beam focus surrounded in three dimensions by regions of higher light intensity. Recently single beam atom trapping has been reported in the dark region of a composite beam formed by inserting a small half-waveplate into the center of the beam[33]. The resulting beam can be analyzed in terms of two HG_{00} Gaussian modes with differing beam waists and hence differing Rayleigh ranges. Their relative phase changes with propagation such that they destructively interfere at the beam focus. The principle of interfering beams is a general one and rather than relying on the gradual variation of the Gouy phase due to the differing Rayleigh ranges a much more rapid phase change can be introduced by combining modes of different order. Such beams can be termed optical bottle beams[34]. Similar interference effects between Bessel beams with differing radial and hence longitudinal wavevectors have also been observed[35].

The intensity cross-section of the desired "bottle beam" at the beam focus differs from that either side and therefore, by definition, no structurally stable beam meets this requirement. However, a superposition of two or more structurally stable beams with different Gouy phases results in an intensity distribution that is not itself structurally stable. For a Laguerre-Gaussian mode, the Gouy phase is given by,

$$\psi(z) = (2p + l + 1) \arctan(z/z_r)$$

Consequently, any two Laguerre-Gaussian modes with the same wavenumber, k, but differing radial mode indices undergo a relative phase change in the vicinity of the beam focus. For example, the previous figure shows intensity profiles for two Laguerre-Gaussian modes with indices $l = 0$, $p = 0$ (the fundamental Gaussian mode) and $l = 0$, $p = 2$. When appropriately weighted, and phased, these two modes destructively interfere giving zero on-axis intensity at the beam focus, but due to their differing Gouy phases they constructively interfere either side. The Figure 8 shows the calculated through focus intensity distribution of their superposition.

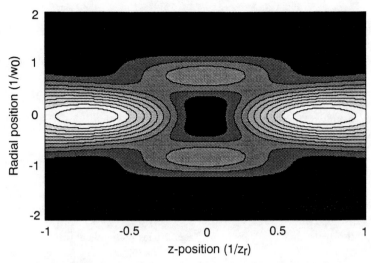

Figure 8. The calculated through focus intensity of a optical bottle beam

The required superposition of the two modes can readily be produced using a computer generated hologram to transform the HG_{00} output from a conventional laser into one with the desired form. Figure 9 shows the intensity cross section, at the focus and those immediately either side for the bottle beam formed between Laguerre-Gaussian modes with indices $l = 0$, $p = 0$ and $l = 0$, $p = 2$.

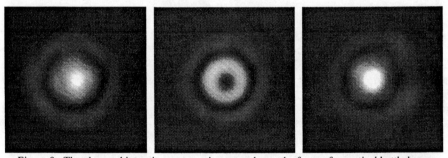

Figure 9. The observed intensity cross sections at and near the focus of an optical bottle beam

9. CONCLUSIONS

This paper has given a brief introduction to the nature of orbital angular momentum and how beams containing it can be generated. It has covered some of the recent research undertaken investigating the properties of such beams, including optical spanners, non linear interactions and the rotational Doppler shift. The field of orbital angular momentum is still young and undoubtedly numerous exciting experiments will be completed over the next few years by various groups.

ACKNOWLEDGEMENTS

The author would like to thank the Royal Society for their support, Les Allen his introduction to and co-worker throughout much of this field, Jochen Arlt, Johannes Courtial, Kishan Dholakia, Neil Simpson and Anna O'Neil, past and present members of the group.

REFERENCES

1. L Allen, M W Beijersbergen, R J C Spreeuw and J P Woerdman, "Orbital angular momentum of light and the transformation of Laguerre-Gaussian laser modes", Phys. Rev. A **45**, 8185-8189, 1992.
2. A E Siegman, *Lasers*, University Science books, Mill Valley, CA, 1986.

3.   J Courtial, K Dholakia, L Allen and M J Padgett, "Gaussian beams with very high orbital angular momentum", Opt. Commun. **144**, 210-213, 1997.

4.   S M Barnett and L Allen, "Orbital angular momentum and non-paraxial light beams", Opt.Commun. **110** 670-678 1994.

5.   J M Vaughan and D V Willetts, "Interference properties of a light beam having a helical wave surface", Opt. Commun. **20**, 263-267, 1979.

6.   C Tamm, "Frequency locking of 2 transverse optical modes of a laser", Phys. Rev. A **38**, 5960-5963, 1988.

7.   V Yu Bazhenov, M V Vasnetsov and M S Soskin, "Laser beams with screw dislocations in their wavefronts", JEPT Lett., Vol. **52**, 429-431, 1991.

8.   A G White, C P Smith, N R Heckenberg, H Rubinsztein-Dunlop, R McDuff, C O Weiss and C Tamm, "Interferometric measurements of phase singularities in the output of a visible laser", J. Mod. Opt. **38**, 2531-2534, 1991.

9.   C Tamm and C O Weiss, "Bistability and optical switching of spatial patterns in a laser" J. Opt. Soc Am. B **7**, 1034-1038, 1990.

10. M Harris, C A Hill, P R Tapster and J M Vaughan, "Laser modes with helical wave fronts", Phys. Rev. A, **49**, 3119-3122, 1994.

11. M W Beijersbergen, L Allen, H E L O van der Veen and J P Woerdman, "Astigmatic laser mode converters and transfer of orbital angular momentum", Opt. Commun. **96**, 123-132, 1993.

12. M W Beijersbergen, R P C Coerwinkel, M Kristensen and J P Woerdman, "Helical-wave-front laser-beams produced with a spiral phaseplate", Opt. Commun. **112**, 321-327, 1994.

13   G A Turnbull, D A Robertson, G M Smith, L Allen and M J Padgett, The generation of free-space Laguerre-Gaussian modes at millimetre-wave frequencies by use of a spiral phaseplate, Opt. Commun. **127** 183-188 (1996)

14. N R Heckenberg, R McDuff, C P Smith and A G White, "Generation of optical-phase singularities by computer-generated holograms", Opt. Lett, **17**, 221-223, 1992.

15. N R Heckenberg, R McDuff, C P Smith, H Rubinsztein-Dunlop and M J Wegener, "Laser beams with phase singularities", Opt. Quant. Elec. **24**, S951-S962, 1992.

16. J Arlt, K Dholakia, L Allen and M J Padgett, The production of multi-ringed Laguerre-Gaussian modes by computer generated holograms, J. Mod. Opt. **45**, 1231-1237, 1998.

17. K Dholakia, N B Simpson, L Allen and M J Padgett, "Second harmonic generation using Laguerre-Gaussian laser modes", Phys. Rev. A **54**, R3742-R3745, 1996.

18. M J Padgett and L Allen, "The Poynting vector in Laguerre-Gaussian laser modes", Opt. Commun. **121** 36-40, 1995.

19. J Courtial, K Dholakia, L Allen and M J Padgett, "Second harmonic generation and the conservation of orbital angular momentum with high-order Laguerre-Gaussian modes", Phys. Rev. A **56**, 4193-4196, 1997.

20. J Arlt, K Dholakia, L Allen and M J Padgett, "Parametric down conversion for light beams possessing orbital angular momentum", Phys. Rev. A **59**, 3950-3952, 1999.

21. A Ashkin, J M Dziedzic, J E Bjorkholm and S Chu, "Observation of a single-beam gradient force optical trap for dielectric particles", Opt. Lett. **11**, 288-290, 1986.

22. M J Padgett, and L Allen at the Rank Prize Mini-Symposium on Optical Tweezers, Auguest 1994; quoted by B Amos and P Gill, Optical Tweezers, Meas. Sci. Technol., **6**, 248, 1995.

23. H He, M E J Friese, N R Heckenberg and H Rubinsztein-Dunlop, "Direct observation of transfer of angular momentum to absorptive particles from a laser beam with a phase singularity", Phys. Rev. Lett. **75**, 826-829, 1995.

24. N B Simpson, K Dholakia, L Allen and M J Padgett, "The mechanical equivalence of the spin and orbital angular momentum of light: an optical spanner", Opt. Lett. **22**, 52-55, 1997.

25. J Courtial and M J Padgett, "Limit to the orbital angular momentum per unit power in a light beam that can be focussed onto a small particle" Accepted for publication in Opt. Commun. 1999.

26. B A Garetz and S Arnold, "Variable frequency shifting of circularly polarized laser radiation via a rotating half-wave retardation plate", Opt. Commun. **31**, 1-3, 1979.

27. B A Garetz "Angular doppler effect", J. Opt. Soc. Am. **71**, 609-611, 1981.

28. G Nienhuis, "Doppler effect induced by rotating lenses", Opt. Commun. **132**, 8-14, 1996.

29. M J Padgett and J P Lesso, "Dove Prisms and polarised light", J. Mod. Opt. **46**, 175-179, 1999.

30. J Courtial, D A Robertson, K Dholakia, L Allen and M J Padgett, "Rotational frequency shift of a light beam", Phys. Rev. Lett. **81**, 4828-4830, 1998.

31. M J Padgett and J Courtial, "A Poincaré-sphere equivalent for light beams containing orbital angular momentum", Opt. Lett. **24**, 430-432, 1999.

32. L.Allen, J.Courtial and M J Padgett, "A Matrix Formulation for the Propagation of Light Beams with Orbital and Spin Angular Momenta", Accepted for publication in Phys. Rev. E, 1999.

33. R Ozeri, Lev Khaykovich and N Davidson, "Long spin relaxation times in a single-beam blue-detuned optical trap", Phys. Rev. A **59**, R1750-1753, 1999.

34. J Arlt and M J Padgett, "Generation of a beam with a dark focus surrounded by regions of higher intensity: an optical bottle beam", Accepted for publication in Opt. Lett. 1999.

35. S Chavez-Cerda, E Tepichin, M A Meneses-Nava andn G Ramirez, "Experimental observations of interfering Bessel beams", Opt. Exp. **3** 524-529, 1998.

Fundamental- and High-Order-Mode Losses in Slab Waveguide Resonators for CO_2-Lasers

Thierry Teuma, Gerhard Schiffner, Vladimir Saetchnikov

Dept. Electrical Engineering, Chair AEE0, Ruhr-University

Universitaetsstrasse 150, 44780 Bochum/Germany

ABSTRACT

Several papers [1-3] contain calculated optical losses of CO_2 waveguide resonators considering attenuation and coupling losses due to interference between modes within the space between the end of the waveguide and the mirror. A strong dependence of these losses on guide dimensions, especially those dimensions affecting the Fresnel number of the resonator, was found. To measure these losses we used a method based on a scanning Fabry-Perot interferometer. By determining the bandwidth total losses are calculated.

The slab concept has the advantage that a very compact laser-design is possible and a significant reduction of the gas consumption can be achieved. The disadvantage is the strong dependence of optical losses and beam quality on the electrode gap, on the alignment of the electrodes, and on surface quality. All electrode parameters may change with temperature. For loss measurements the electrode gap must be changed in very small steps in the order of 5 μm or less. To realize such a fine adjustment mechanical systems have been developed which can further be used to misalign the electrodes, too. The results obtained confirm calculated data by Gerlach and Hill [1,2] for waveguides or those found by Saetchnikov [4] for slab structures. Our experimental data may help to further improve models used for computations.

Keywords: CO_2 laser, resonator, slab waveguide, optical loss, roughness, electrode gap, Fresnel number, periodical, statistical

1. INTRODUCTION

The impact of small variation of the Fresnel number on the optical losses of slab waveguide resonators was analysed in [1-3]. A high sensitivity was found out. Great fluctuation of optical losses by small variation of waveguide length were observed, with an great impact on the beam quality. In the papers quoted above, it was proved that mode and wavelength selectivity can be achieve with an appropriate choice of waveguide length. Variation of the electrode gap of a slab waveguide resulting in a similar change of the fresnel number, achieve similar results for the optical losses [5]. Experimental results, which will be reported in this paper, treat the dependence of optical losses on the electrode gap in a stable slab waveguide resonator. Considering the fact that in high power slab waveguide laser (> 1 kW), due to the high temperature of the plasma in the near of the electrodes, mechanical deformations can happen, causing a change of the electrode gap, these measurements should also help to locate parametric regions where the level of losses is less sensitive to small variations of the electrode gap. Additionally, different electrode material will be used and several surface roughness will be achieved through different kind of mechanical treatment. Before describing the experimental results, some theory of the method used for the measurements will be explained.

2. BACKGROUND AND THEORY

For measurements of optical losses of CO_2 slab waveguide, a method based on a scanning Fabry-Perot-Interferometer (FPI) shown in fig. 1 was used.

In *Laser Resonators III*, Alexis V. Kudryashov, Alan H. Paxton, Editors,
Proceedings of SPIE Vol. 3930 (2000) ● 0277-786X/00/$15.00

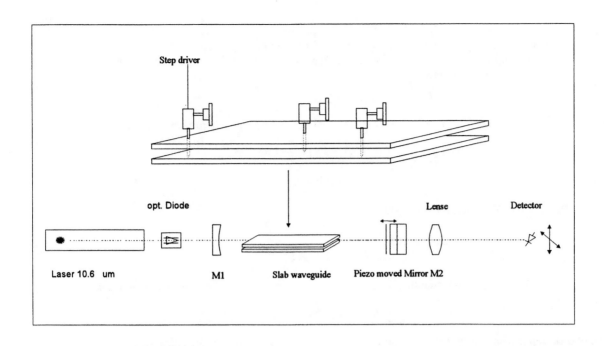

Fig. 1. Scanning Fabry-Perot-Interferometer for measurement of optical losses of a slab waveguide resonator

A laser beam is coupled inside the resonator through the mirror M_1. Moving the mirror M_2 forth and back parallel to the optical axis through three piezo translators, the total length of the resonator is varied periodically and the output beam intensity leaving the resonator, measured with a infrared photo voltaic detector (transforming the infrared signal in voltage) is described through the Airy function (1), for more details see [6]. The bandwidth of the interferometer is used to calculate the losses arising during a single trip (half round trip) of the resonator. The piezo translators move linearly and continuously in a range of zero to 60 μm, trigged by a triangular voltage. Through this periodical change of the resonator length, several resonator modes went in resonance, when their resonance frequency is reached [6]. The output signal of such a scanning interferometer is represent in fig. 2.

$$\frac{I_T}{I_0} = \frac{(1-R_1)(1-R_2)T_m}{\left(1 - T_m\sqrt{R_1 R_2}\right)^2 + 4 T_m \sqrt{R_1 R_2} \sin^2 \frac{\Theta}{2}} \tag{1}$$

$$\Theta = 2\pi \frac{L_0}{\lambda/2} \tag{2}$$

T_m is the transmission factor of the medium between the mirrors, I_T is the transmitted intensity from the resonator, I_0 is the input intensity coupled in the resonator, L_0 is the total length of the resonator

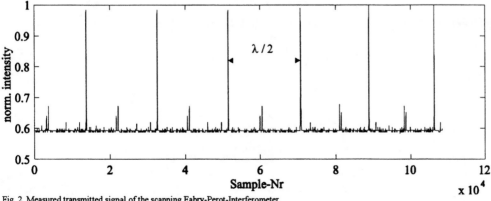

Fig. 2. Measured transmitted signal of the scanning Fabry-Perot-Interferometer

The expression for the transmitted intensity I_T depends on the input intensity. To remove this dependence, I_T should be scaled with the maximal intensity $I_{T,max}$ which happen by identical, loss free mirrors and a loss free medium with $T_m = 1$. In this case $I_{T,max} = I_0$. The input intensity is completely transmitted. Always when the interferometer length is an integer number of half wavelength of a resonator mode,

$$L_0 = i \cdot \lambda / 2, i = 1,2,3...$$

the resonance condition is fulfilled. The expression for the maximal transmitted intensity can be written as

$$\frac{I_{T,max}}{I_0} = \frac{(1-R_1)(1-R_2)T_m}{\left(1-T_m\sqrt{R_1 R_2}\right)^2} \tag{3}$$

Dividing (1) through (3), the following expression can be obtained

$$\frac{I_T}{I_{T,max}} = \frac{\left(1-T_m\sqrt{R_1 R_2}\right)^2}{\left(1-T_m\sqrt{R_1 R_2}\right)^2 + 4T_m\sqrt{R_1 R_2}\sin^2\frac{\Theta}{2}} \tag{4}$$

The reflectivity of the mirrors and the transmission factor of the resonator medium can be summarised in one parameter as

$$R_g = T_m\sqrt{R_1 R_2}$$

also called complete reflectivity factor of the resonator. This factor describes the part of the intensity, which stay inside the interferometer after a single trip. The loss factor, describing the change of the intensity inside the interferometer after a single trip, can be expressed as,

$$V = 1 - R_g$$

Considering the expression of R_g, (4) can be rewritten in the following form:

$$\frac{I_T}{I_{T,max}} = \frac{\left(1-R_g\right)^2}{\left(1-R_g\right)^2 + 4R_g\sin^2\frac{\Theta}{2}} \tag{5}$$

The peaks in fig. 2 show the positions where the resonance condition is fulfilled for several modes propagating inside the interferometer. The presence of several modes within the interferometer can be recognised through the different size of peak intensities observed in the output signal. The amount of the optical losses will be derived from the With of the peaks and the distance between them. Variation of the interferometer length through piezo translators leads in variation of the angle Θ in (5). Solving (5) in the angle Θ, the width of a peak $\Delta\Theta$ can be expressed as function of $I_t/I_{t,max}=a^{-1}$.

$$\Delta\Theta = 4\arcsin\left(\sqrt{\frac{(a-1)(1-R_g)^2}{4R_g}}\right) \tag{6}$$

Considering (2), the variation of the angle Θ in fig. 3a can be expressed as variation of the

146

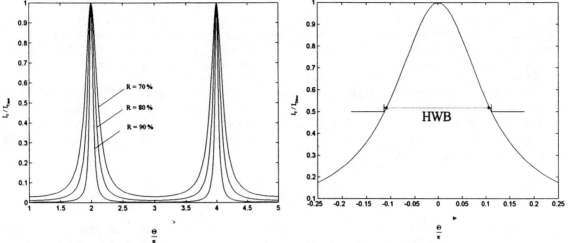

Fig. 3a. Transmitted intensity versus the angle Θ

Fig. 3b. Transmitted intensity versus the angle Θ, for a single peak

resonator length as follows,

$$\Delta L_o = \frac{\lambda}{\pi} \arcsin\left(\sqrt{\frac{(a-1)(1-R_g)^2}{4R_g}} \right) \qquad (7)$$

The full width at half maximum of the peak, *HWB*, at $I_t/I_{t,max} = 1/2$, is shown on fig. 3b. (7) can be written in this case (a=2) as,

$$\Delta L_{o,\frac{1}{2}} = HWB = \frac{\lambda}{\pi} \arcsin\left(\frac{(1-R_g)}{2\sqrt{R_g}} \right) \qquad (8)$$

Solving (8), following expression can be found for R_g in dependence on the peak width *HWB* [7].

$$R_G = 1 + 2\Theta^2 - 2\Theta\sqrt{1+\Theta^2}$$
$$\Theta = \sin\left(\frac{\pi \cdot HWB}{\lambda} \right) \qquad (9)$$

With the value of R_g, the optical losses during a single trip can be calculated using the following formula.

$$V = 1 - R_g$$

3. MEASUREMENT SETUP

The beam coming from laser is coupled inside the resonator. Depending on the present length of the resonator a part of this beam is coupled out after propagation through the resonator and focused through a lens on the photo voltaic detector D. The resonator, consisting on both mirror M_1, M_2 and between them the object, which optical losses should be measured, for example a slab waveguide is the most important part of the measurement setup. The mirror M_1 has a radius of curvature of 1.7 m and M_2 is flat. Both mirror have a reflectivity of 99% by a wavelength of 10.6 μm, a diameter of 25.4 mm and are both on ZnSe. The resonator is stable and hemispherical, the beam waist is situated on the position of the mirror M_2. The distance between slab waveguide electrodes and the mirrors are on both sides 10 mm. The materials used for electrode were copper or aluminium, with a active surface of 400 mm x 20 mm. Several surface profiles of electrodes were achieved through different kind of mechanical treatment. To measured the transmitted intensity, a HgCdTe photo voltaic detector was used. The detector had a response time of 0.2 ns. Because of the very small voltage level of the output signal of the detector, a amplifier was used to get a treatable signal. The laser used as beam source was a 10 W, DC-excited sealed off CO_2 laser. the gas was closed in a glass capillar with a radius of 8 mm. The output beam was linear polarised, with multimode beam pattern and a beam radius of 4 mm and a beam divergence of 3 mrad. The wavelength was tuneable on a range of about 2 μm between 9.2 and 11 μm. The task of the optical diode was to prevent the reflected beam coming from the mirror M_1 to get back inside the laser and cause a modulation of the output power.

Before starting the measurements the complete setup shown in fig. 1 was aligned, using a red HeNe laser. During the measurements the electrode gap was moved starting by a gap of 2a = 1.4 mm. The step chosen varied between 2 and 20 μm. The mechanical positioning error should not exceed 5% repeating the measurements. To achieve these severe requirements a mechanical setup was build, consisting on three step driver positioned on the top electrode like shown on fig. 1. With this device it was possible to realise step of 1 μm.

4. EXPERIMENT

The laser used provides a multimode gaussian beam. That means additionally to the ground mode, some higher order modes have to be consider. After been coupled inside the resonator these gaussian modes are transformed in resonator own modes consisting on waveguide and free space modes. The mode with the highest intensity (the greatest transmission peaks within the output signal) will be called in this paper principal mode (which should not automatically means ground mode) and the other modes with smaller intensities will be called higher order modes. During the measurements only the higher order mode with the next greatest transmission peaks will be taken into account. Note that the principal mode with the greatest peaks should automatically have the lowest losses.

The flatness of the electrode was realised through a special steel base plate with minimal deformation along the length. Both top and bottom electrode were screwed with plates before their mechanical treatment, to achieve define roughness.

Before starting with concrete measurements, tests were made to determine the optimal step by variation of the electrode gap. Measurements were made with the steps 2a = 20, 10, 5 and 2 μm using a copper slab waveguide with the dimensions 400 mm x 20 mm. The results are shown on fig. 4.

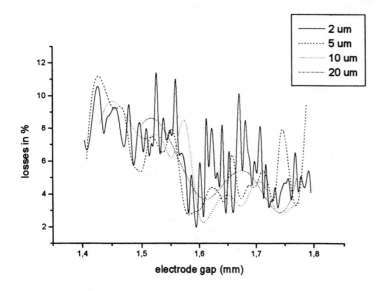

Fig. 4. Comparison of resonator losses using different steps for electrode gap

Comparing all 4 curves in pairs, the step of 5 μm was selected as the most appropriate one because showing more details then the curves with steps of 10 and 20 μm and having approximately the same content like the curve with the step of 2 μm but needing less than the half of the time for entire measurements

Roughness of the electrode surface of slab waveguide can influent strongly the quality of the beam of CO_2 laser. This can be analysed by comparing the losses of the slab waveguide by different treatments of the electrodes. After the first treatment of the copper electrode the surface had a statistical roughness structure like shown in fig. 5a, b.

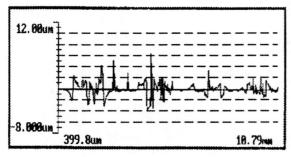 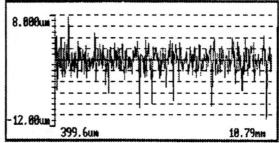

Fig. 5a. Statistical roughness profile in propagation direction   Fig. 5b. Statistical roughness profile in perpendicular direction

The mean value of the roughness at all was R_a = 0.24 μm and the mean value of the peak to peak (highness) roughness was R_t = 0.8 μm for the propagation direction and R_a = 2.4 μm, R_t = 9.5 μm for the direction perpendicular to the propagation direction. Treating the surface for the second time following values were obtained for the roughness, R_a = 0.82 μm, R_t = 3.5 μm for the propagation direction and R_a = 1.8 μm, R_t = 8.5 μm for the perpendicular direction. The results obtained after measurements of the losses by both surfaces are shown in fig. 6. It can be seen that the level of losses increases with the highness of roughness. But the qualitative behaviour is still very similar.

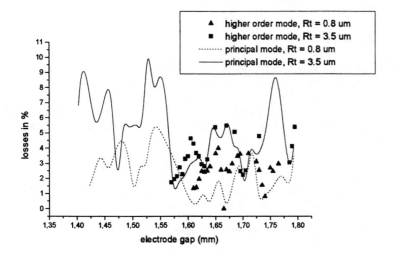

Fig. 6 Resonator loss versus electrode gap for two different highness of statistical roughness

The second kind of mechanical treatment of the electrode results in periodical roughness like shown on fig. 7.

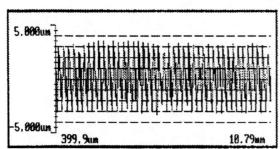

Fig. 7. Periodical roughness profile in propagation direction

The profile shows a sinusoidal structure with R_a = 1.3 µm and R_t = 5.5 µm for the propagation direction. The period amount P = 235 µm. To analyse the effect of periodical roughness several treatments were made to achieve different values of R_a and R_t. Fig. 8 shows the optical losses achieved with three different highness of roughness. It can be observed, that variations of the roughness highness, do not influence the level of losses in slab waveguides.

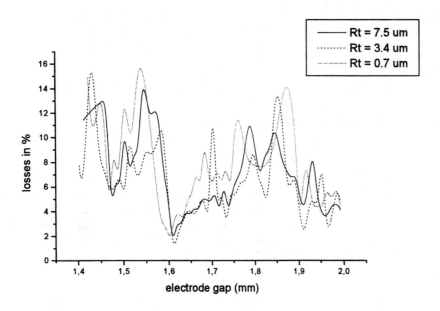

Fig. 8. Comparison of resonator losses by different highness of periodical roughness

Another important parameter by periodical roughness is the period P. Comparison of optical losses for surfaces with similar values for R_a and R_t but different period of roughness profiles P provides no more information, the qualitative and quantitative behaviour was nearly the same.

A comparison of optical losses by slab waveguide with similar parameter of roughness (R_a, R_t) by periodical and statistical roughness profiles provides very similar results like shown in fig. 9. The only difference is the complete spatial move the curve on the electrode gap axis, on about 50 µm.

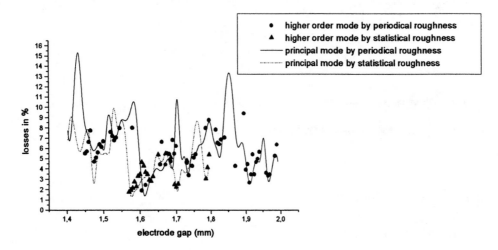

Fig. 9. Comparison of resonator losses by statistical and periodical roughness ($R_t \approx 3.5$ µm)

Additional to copper as electrode material, electrode with the same dimensions as mentioned above ware made on aluminium. Statistical roughness of the surface with the values R_a = 0.6 µm and R_t = 2.75 µm was realised. Comparison of the results obtained by the measurements of losses with aluminium and copper electrode by similar roughness are shown in fig. 10. The average level of losses is more or less similar, but the qualitative behaviour looks quite different. There is no absolute minimum by aluminium like by copper, and the losses decreases by aluminium continuously with higher electrode gaps.

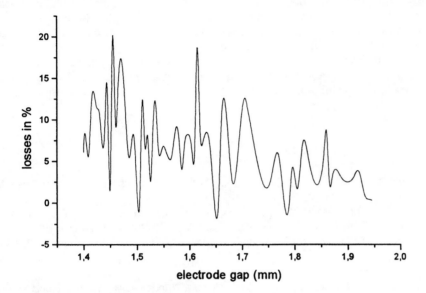

Fig. 10. Resonator losses by aluminium electrodes with statistical roughness (R_t ≈ 2.75 µm)

Finally Measurements were made with copper electrodes by different wavelengths (λ = 9.4, 10.2, 10.6 µm). The results are shown in fig. 11. Like by comparison of optical losses by statistical and periodical roughness, only an spatial move the curves on the electrode gap axis is observed.

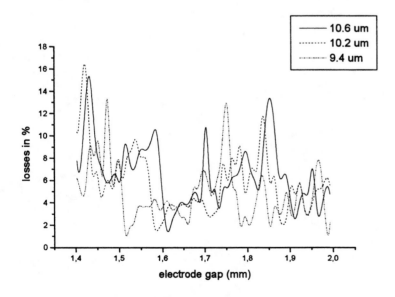

Fig. 11. Comparison of resonator losses by different wavelengths

5. CONCLUSIONS

Roughness of electrode surface influences the mode propagation in slab waveguide resonators in different ways. Variation of the highness by periodical roughness do not changes the dependence of optical losses on electrode gap. But like detailed explained in [3], due to reduction of higher order modes in the resonator, the loss maximums become higher and wider. Variation of the highness by statistical roughness modifies the dependence of optical losses on electrode gap. This effect and the reduction of higher order modes using periodical roughness will be analysed more distinctly making additional measurements using a much smaller electrode-mirror gap in order to reduce the influence of the coupling losses.

After first measurements using aluminium electrodes with statistical roughness were made, a stronger dependence of the optical losses on the higher order modes can be observed. This is characterised through the great number of narrow maximums, which are a indication for the great number of resonance's between the several existing modes [3]. Additional measurements with aluminium electrodes will help us to make a more distinct analysis of the influence of the highness by statistical and periodical roughness, and a better comparison with the behaviour of the optical losses using copper electrodes.

6. REFERENCES

[1]   J. Banerji, A.R. Davies, C.A. Hill, R.M. Jenkins, and J.R. Redding, "Effects of curved mirrors in waveguide resonators," Applied Optics, Vol. 34, No. 16, 3000-3008, 1995

[2]   Robert Gerlach, Dianyuan Wie, and Nabil M. Amer, "Coupling Efficiency of waveguide laser resonators formed by flat mirrors: Analisys and Experiment," IEEE J. of Quantum electronics, Vol. QE-20, No. 8, August 1984

[3]   Christopher A. Hill, "Transverse Modes of Plane-Mirror Waveguide Resonators," IEEE J. of Quantum Electronics, Vol. 24, No. 9, September 1988

[4]   V. Saetchnikov, private communication

[5]   Chistopher L. Petersen, Dietmar Eisel, Josef J. Brzezinski, and Herbert Gross, "Mode and wavelength Selectivity in Slab-Geometrie CO_2 Lasers," SPIE Vol. 2206 / 91

[6]   Gerhard Schiffner "Vorlesungsmanuskript: Elektrooptik I & II," Ruhr-Universität Bochum, Lehrstuhl für Allgemeine Elektrotechnik und Elektrooptik, 1995/1996

[7]   T. Goretzky "Dissertation: Untersuchungen eines Scanning-Fabry-Perot-Interferometer zur Bestimmung kleiner optischer Verluste," ," Ruhr-Universität Bochum, Lehrstuhl für Allgemeine Elektrotechnik und Elektrooptik, 1992

SESSION 7

Microresonators and Whispering-Gallery Modes

Invited Paper

Microsphere integration in active and passive photonics devices

Vladimir S. Ilchenko*, X. Steve Yao, Lute Maleki

Jet Propulsion Laboratory, California Institute of Technology,
MS 298-100, 4800 Oak Grove Dr., Pasadena, CA 91109-8099

ABSTRACT

As important step towards integration of microspheres in compact functional photonics devices, we demonstrate direct efficient coupling of light in and out of high-Q whispering-gallery (WG) modes in silica microspheres using angle-polished single mode fibers. Based on this principle, we present a 1-inch fiber-pigtailed microsphere module that can be used for fiber-optic applications, and a fiber-coupled erbium-doped microsphere laser at 1.55μm. In addition, we report preliminary data on the intensity modulation based on high-Q WG modes in a lithium niobate sphere. We also demonstrate a novel geometry WG-mode optical microcavity that combines Q~10^7, typical for microspheres, with few-nanometer mode spacing earlier available in lower quality factor Q~10^4 microfabricated planar rings.

Keywords: Microsphere, microcavity, photonics, microlaser, electrooptic modulator, erbium

1. INTRODUCTION

In the growing field of microcavity science and applications, microspheres [1-5] stand out as exceptional type of cavities with the dimensions compatible with microfabricated devices and the quality-factor normally obtained in large Fabry-Perot "supercavities". Despite numerous suggested applications – from compact filters, microlasers, diode laser stabilization, to microwave optoelectronic oscillators and sensors [6-12], - the roadpath towards miniaturized functional devices based on microspheres was blocked by the absence of compact and efficient couplers compatible with integrated and fiber optics. In this report, we describe in detail a new simple method for direct coupling from single-mode optical fibers to high-Q whispering-gallery (WG) modes in microspheres. In this method, the tip of the fiber is angle polished under the angle providing total internal reflection and synchronism of the evanescent wave in the truncated core area with the azimuthal propagation of WG modes in the microsphere. Based on this novel technique, we present a compact fiber-pigtailed optical filter module with a microsphere, and a fiber-coupled erbium-doped microsphere laser. By contrast to earlier microspehere laser demostrations, aimed at implementation of ultra-low threshold operation, our laser work focuses on obtaining significant output power, up to several microwatts, in the optical communication band 1.55μm.

The principle of angle-cut, or angle-folded dielectric waveguide can be utilized for obtaining similar phase-matched coupling of microsphere modes with planar integrated waveguides. Flexibility of this method allows efficient coupling between WG modes in low-refraction-index microspheres and guided modes in high-index (e.g. silicon) waveguides. This would pave the way to implementation of truly integrated narrow-line (sub-kHz) microsphere-stabilized semiconductor laser source and on-chip microwave optoelectronic oscillator.

As pointed out in many works, combination of strong field confinement and high quality-factor in whispering-gallery modes enables observation, at a low threshold, of various nonlinear optical effects. Beginning with extensive studies in liquid aerosol droplets (raman scattering, lasing in dye solution droplets), with further demonstrations of dispersive bistability in silica microspheres and low-threshold lasing in doped silica and ZBLAN spheres, nonlinear-optic studies with WG modes were focussed on isotropic, uniform materials that possessed the third-order nonlinear succeptibility. Meanwhile, novel applications will be possible if sphere is made of electro-optic, $\chi^{(2)}$ materials. Many of them are characterized by rather small optical losses so that quality-factors up to 10^7 can be expected if appropriate sphere fabrication and polishing technique is used for those crystalline materials. We present here observation of whispering-gallery modes in few-mm diameter lithium niobate spheres and demonstration, on their basis, of a prototype optical intensity modulator.

The final part of our report is devoted to the exploration of novel geometry for high-Q whispering-gallery mode microcavity. The goal is to reduce the cavity volume and the number of excited modes compared to microspheres and therefore create a miniature solid state cavity that would combine small size and very high Q with the type of spectrum one can expect in single-longitudinal-mode Fabry-Perot resonator. Applying an original new technique, we fabricated a dielectric

*Correspondence: E-mail: ilchenko@jpl.nasa.gov; Telephone: 818 354-8485; Fax: 818 393-6773

154

In *Laser Resonators III*, Alexis V. Kudryashov, Alan H. Paxton, Editors,
Proceedings of SPIE Vol. 3930 (2000) • 0277-786X/00/$15.00

structure of quasi-toroidal, or highly-oblate spheroidal geometry in the area of desired WG mode lozalization. As confirmed by experiment, such a cavity does provide a drastic reduction in the number of excited WG modes (two within free spectral range of 383GHz corresponding to the cavity diameter 160µm at the wavelength 1.55µm), while maintaining the Q ~ 10^7.

2. SINGLE-MODE FIBER COUPLER FOR MICROSPHERES

At present moment, in addition to the well-known prism coupler with frustrated total internal reflection [1-9], few attempts have been reported to directly couple a sphere to an optical fiber. However these fiber couplers had either limited efficiency due to residual phase mismatch (side-polished bent fiber coupler [13]), or still appreciable size including fragile

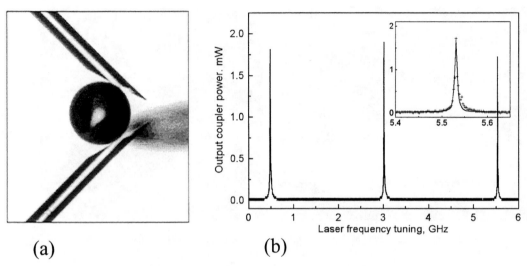

(a) (b)

Fig.1. Close-up view (a) and optical transmission characteristic (b) of a microsphere with two side-polished fiber couplers. Input power 7.5…8.3mW; maximum transmission at resonance ~23.5% (fiber-to-fiber loss 6.3dB); $Q_L > 3 \times 10^7$ at 1550nm; sphere diameter 405µm. Unloaded $Q_o \approx 1.2 \times 10^8$. Fiber: SMF-28, polished at ~76°.

core-to-cladding transformers (tapered fiber coupler [14]). Recently, we have demonstrated a new and simple method of direct fiber coupling to high-Q WG modes, which in essense is a hybrid of waveguide and prism coupler. A close-up of the experimental setup with microsphere and two couplers is shown in Fig.1a. The tip of a single-mode fiber is angle-polished under steep angle. Upon incidence on the angled surface, the light propagating inside the fiber core undergoes total internal reflection and escapes the fiber. With the sphere positioned in the range of the evanescent field from the core area, the configuration provides efficient energy exchange in resonance between the waveguide mode of single-mode fiber and the whispering-gallery mode in the sphere. The angle of the polish is chosen to secure the phase matching requirement: $\Phi = arcsin(n_{sphere}/n_{fiber})$. Here n_{fiber} stands for the effective refraction index to describe the guided wave in the fiber core truncation area, and n_{sphere} stands for the effective refraction index to describe azimuthal propagation of WG modes (considered as closed waves undergoing total internal reflection in the sphere), see plot in Fig. 2. Since the linear dimensions of the angle- cut core area match well the area of evanescent field overlap, the

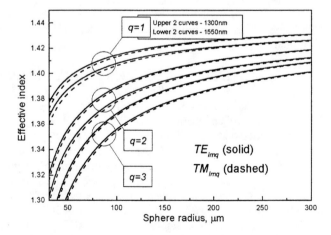

Fig.2. Effective index for WG mode azimuthal propagation (based on mode frequency approximation by C.C.Lam *et al* [15])

the new system is equivalent to a prism coupler with eliminated collimation/focusing optics. The effective refraction index to describe the azimuthal propagation of WG modes near the surface of the sphere can be calculated, for example, on the basis of asympthotic expressions [15] for WG mode frequencies ω_{lq}, where $n_{sphere} = 2cl / D\omega_{lq}$. Single coupler efficiency in our experiment was more than 60%, comparable with the best reported results for prism coupler (78%) and fiber taper (90%); total fiber-to-fiber transmission at resonance ~23% (insertion loss 6.3dB). The demonstrated simple "pigtailing" of the microspheres will lead to their wider use in fiber optics, enabling the realization of a whole class of new devices ranging from ultra-compact narrow band filters and spectrum analyzers and high-sensitivity modulators and sensors, to compact laser frequency stabilization schemes and opto-electronic microwave oscillators.

Fig.3 presents a prototype packaged microsphere-based fiber-optic filter module with in-line positioned angle polished input and output couplers, insertion loss ~12 dB and the loaded Q factor 3×10^7 at 1550nm.

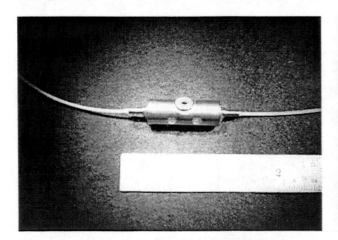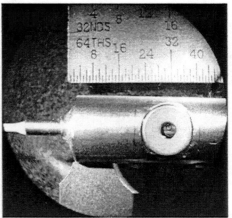

Fig.3. A prototype fiber pigtailed microsphere package. Insertion loss ~12dB;
loaded Q~3×10^7 at 1550nm

Apart from "discrete element" fiber optics applications, microspheres present further challenge for true integration with planar optics devices. Very high $Q \geq 10^8 ... 10^9$, uncommon for existing planar microcavities, would open way for sub-kHz-linewidth integrated laser sources [11] and single-chip microwave optoelectronic oscillators [12]. The ultimate goal is the replacement of laboratory hand preparation of spheres (fusion of silica preforms in ox-hydric flame or CO_2 laser beam) by appropriate microfabrication technology. It is worth to note here that such a task may not be extraordinary difficult, given the progress in planar silica waveguides and succesful demonstrations of thermal treatment techniques for formation of smooth curvilinear integrated optics elements. The other part of the integration is the development of appropriate waveguide coupler elements. This task can possibly be solved along three routes: a)precise tailoring of propagation constant in the waveguide to match that of the sphere [16]; b)truncation, or "reflection" of the waveguide from vertical cleave, by analogy with the above-described fiber coupler.

2. FIBER-COUPLED ERBIUM-DOPED MICROSPHERE LASER

Early reports on microsphere lasing demonstrated the stimulated emission in dye-doped polysterene spheres with relatively low-Q of ~10^4 and free-beam excitation [17]. Reports of high-Q microspheres acting as laser cavities were focused on the possibility to achieve extremely low threshold [7] of laser operation with Nd ions. Preliminary data on 1.55µm lasing in Er-doped ZBLAN spheres have been reported [18] with pumping and outcoupling of laser radiation obtained by means of bulky prism coupler. Our miniature design utilizes the earlier disclosed angle-polished single-mode fiber coupler for both delivering pump power into the sphere and the pickup of laser radiation. The schematic of experiment is presented in Fig.4, with the inset photograph showing the lasing sphere next to the fiber coupler. The circular lasing area is visualized by upconversion-pumped fluorescence at 525/545nm. The pumping radiation is provided by the multimode diode laser E-Tek LDPM7000 BBA10, stabilized by external fiber grating at the wavelength 977.6nm. Through polarization controller and wavelength division multiplexer, the pumping radiation reaches the angle-polished fiber coupler and excites the WG modes in the microspheres. The modes are excited in a travelling-wave regime, (counterclockwise direction in Fig.4). Quality-factor of microsphere modes at pumping wavelength was 1.5×10^5 in loaded regime (zero microsphere-coupler gap) and $(0.7\ldots1)\times10^6$ in the undercoupled regime. Relatively low undercoupled Q of the sphere, compared to projected $Q_P = 2 \times 10^7$ from the reported pump attenuation 2.8dB/m of the material, may be explained by scattering losses produced by residual optical inhomogeneities (refraction index variations) in the spheres. The spheres were produced from the 0.6mm rods of core material extracted from fiber preforms (INO 502 O1). Main part of the silica cladding was removed by machining, subsequently the remaining cladding was etched in hydrofluoric acid. Since the core of the preform (produced by CVD

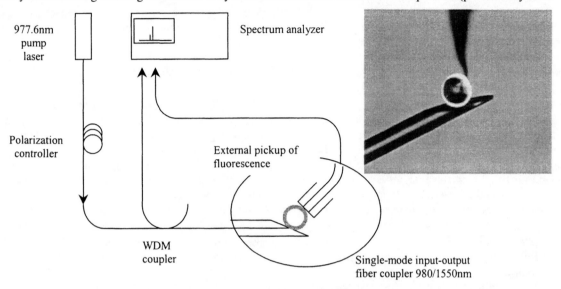

Fig. 4. Schematic of the laser experiment and a close-up of the microsphere with single-mode fiber coupler.

technique) contained alternating layers with varying concentration of components and refraction index, prior to formation of a sphere, the material was fused several times and mixed for homogenization.

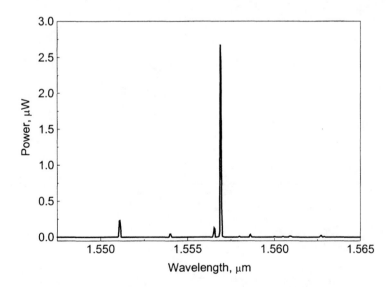

Fig.5 Typical oscillation spectrum of the fiber-coupled erbium-doped microsphere laser

After that, the sphere of diameter 50 to 150 micron was formed by microtorch fusion technique described elsewhere [4]. The laser radiation was outcoupled in the direction opposite to the pump (clockwise in Fig.4), split out by the WDM coupler and send to the optical spectrum analyzer (Advantest Q8383). Depending on the alignment of the coupler, the lasing could be obtained throughout the most of luminescence band of Er^{3+} ion - between 1530nm and 1560nm. Fig.5 presents a typical spectrum of oscillation of Er-doped sphere. Maximum obtained power in our preliminary experiment was ~3µW in the output fiber.

3. TUNABLE WHISPERING-GALLERY MODES IN LITHIUM NIOBATE SPHERE

Efficient recycling of light in WG modes of spherical cavities calls for variety of applications that may be suggested if a sphere is built of nonlinear material. In this case, responsivity of the device would increase compared to conventional single-path, or low-finesse interferometer configurations, because of the high-Q of WG mode resonances. As a first step towards nonlinear devices based on spherical cavities, we present here observation of WG modes in few-mm diameter lithium niobate spheres and demonstration, on their basis, of a prototype optical intensity modulator.

Lithium niobate, one of the most common nonlinear-optic material, is an electrooptic crystal with low losses $\alpha \leq 0.02$ cm^{-1}, allowing the quality-factor $Q=2\pi n/\alpha\lambda \sim 10^7$ of whispering-galery modes. By contrast to amorphous materials, high-quality spheres may not be obtained by fusion of crystalline materials. Spontaneously forming boundaries between misoriented crystalline grains, or blocks, create significant optical inhomogeneities through the bulk of the sphere and on the surface, thus increasing the scattering losses beyond the acceptable level. (Self organized spheres of cubic (non-birefringent) crystalline material were reported in [6]. Because of their sub-grain few-micron size, they supported WG modes of reasonable $Q\sim10^4$-10^5.) To obtain high-surface quality spheres of birefringent crystalline materials, one has to necessarily machine and polish them using conventional optical methods. On flat or low-curvature substrates, modern metods allow to obtain angstrom size residual inhomogeneities – compatible to the roughness of fire-polished silica and therefore allowing ultimate $Q>10^9$. These methods, however, have never been adapted for very small radii of curvature. In one of the early works [2], WG modes with quality-factor of about 1×10^8 were observed in a 3.8cm diameter mechanically for the Gravity Probe B experiment. In the meantime, with a given size of surface inhomogeneity, limitation of Q by scattering losses is in direct proportion to the sphere diameter.

In our preliminary experiments, we used a lithium niobate sphere of diameter $D = 4.8$mm, custom fabricated and polished out of commercially obtained LiNbO$_3$ specimen (from Casix Inc). Because evanescent wave coupling could be

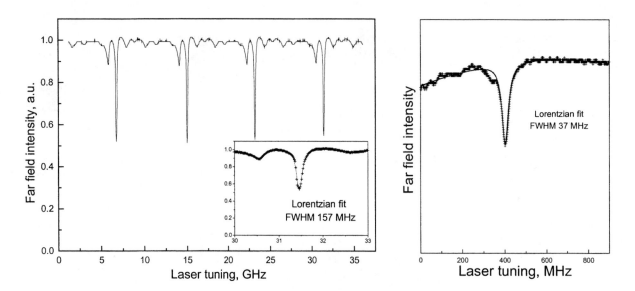

Fig.6. Whispering-gallery mode spectrum in a 4.8mm diameter LiNbO$_3$ sphere, at 1550nm: a)35GHz scan revealing the FSR=8.3GHz of TE mode sequence, excitation in normal plane to the crystal axis; b)closeup of an individual mode; resonance bandwidth 37MHz corresponding to the quality factor $Q > 5\times10^6$

provided by a prism of refraction index higher than that of the material ($n_0 = 2.220$; $n_e = 2.146$), we used a miniature diamond prism ($n = 2.4$). Spectral measurements were done using the E-Tek DFB laser at about 1550nm, frequency-scannable via current modulation. As seen from results presented in Fig.6, the employed prism coupling technique provided about 50% energy coupling efficiency in the loaded regime. The observed free spectral range FSR ~ 8.3GHz corresponded to the sequence of successive principal mode numbers l for TE-type WG modes excited in the plane perpendicular to the crystal axis. The estimated loaded quality-factor of the modes in Fig.6a is about 1.2×10^6, and the non-Lorentzian shape of the observed resonance dips suggested that they corresponded to clusters of slightly non-degenerate modes. Let us note that the character of the observed spectrum critically depended on the orientation of the sphere. With the excitation off the perpenducular plane to the crystal axis, the observed spectrum became dense, revealing a sequence of closely spaced modes with Q-factors up to 5×10^6 (see Fig.6b). Thise preliminary result confirms that inexpensive fabrication and polishing techniques allow to achieve the Q factor in the spheres of crystalline lithium niobate that is close to the limits defined by material attenuation.

With the far off symmetry plane orientation of the input beam, the observed frequency response in the far field represented a continuous sequence of fringes produced by partially overlapping modes (see Fig.7).

The electrooptical effect in the spheres of lithium niobate is in the simplest way manifested via the tuning of WG modes by applied electrical field. To demonstrate this, we applied RF electrodes near the poles of the sphere and observed the displacement of the mode spectrum in the frequency domain by the applied voltage U at the rate 0.5MHz /Volt. This electrical tunability is in satisfactory agreement with the crude estimate $|\Delta v/U| = \frac{1}{2}\, n^2 \gamma_{13}\, /D \approx 0.8$ MHz/V, where $v = 194$THz -- is the lightwave frequency; $\gamma_{13} = 10$pm/V -- is the electrooptical constant; and $D = 4.8$mm, diameter of the sphere, is used as electrode spacing.

Presented in Fig.7 is the demonstration of the low-frequency electrooptical intensity modulation by the lithium niobate sphere. During continous monitoring of WG mode spectrum in the sphere by means of frequency-tuned laser, a 100kHz rf voltage with $U_{eff} = 40$V (~125V peak-to-valley) was applied to electrodes, resulting in amplitude modulation maximal near the slopes of individual cavity resonances.

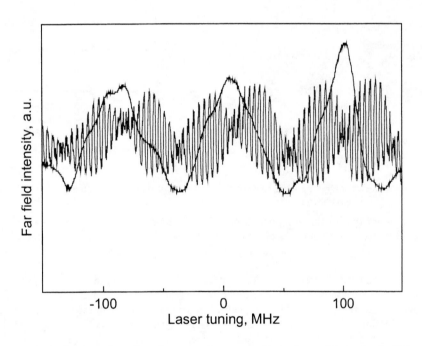

Fig.7. Demostration of low frequency electrooptical intensity modulation by the lithium niobate sphere: effect of radiofrequency (100kHz) voltage (U_{eff}=40V)

By definition, an electrooptical modulator, based on a cavity, may not be a wide-band device. It may however be suitable for a number of applications where optical carrier is fixed, and the cavity spectrum can be trimmed to have optical modes at the carrier frequency and the modulation sidebands. With development of appropriate fabrication techniques and reduction of the sphere size (at least 100 fold reduction is possible without compromising the optical Q), this inconvenience will be compensated by two serious advantages over existing modulators. First, the controlling voltage (analog of V_{π}) can be reduced into milliVolt domain. Second, tiny capacity of electrooptic microspheres will simplify application of microwave fields, compared to both plane-wave bulk electrooptical modulators and integrated Mach-Zender interferometer modulators.

4. HIGHLY OBLATE SPHEROIDAL MICROCAVITY:
FREE SPECTRAL RANGE OF FEW NANOMETERS AND FINESSE ~10^4

Despite their small dimensions, because of high symmetry microspheres exhibit relatively dense spectrum of high-Q whispering-gallery modes. Even with the optimized near-field coupler devices [13,14,19], the typical spectrum consists of *overlapping* groups of *TE (TM)* $_{lmq}$ modes spaced by "large" free spectral range (FSR). By "large" FSR we denote here, roughly, the interval in the sequence of *fundamental* resonances corresponding to successive number l of wavelengths packed along the "equatorial" circumference of the sphere. It is easy to estimate that with the silica sphere diameter between 150-400 micron, the "large" FSR should be 670...165GHz, or in the wavelength scale, 5.4...1.3nm, near the center wavelength 1550nm. Although degenerate in ideal spheres, modes with different *transverse* structure (mode index m defining the field extent in the "latitudinal" domain) are split in slightly eccentrical cavity. As a result, average mode spacing is between 1 and 10GHz, in typical microspheres of 150-400micron diameter and ~1% eccentricity obtained by microtorch (or CO_2-laser assisted) fusion of silica preforms.

For some of applications such as spectral analysis and laser stabilization, such a dense spectrum precludes simple technical solutions and calls for using intermediate filtering etc. To overcome this problem, the ultimate solution would be to create a single mode waveguide ring shaped by planar technology methods or somehow spliced out of a section of straight guide. The first option is implemented in planar ring cavities [20], but the quality-factor in this case is limited to $Q \leq 10^4...10^5$ by fundamental scattering losses at flat boundaries. Technically, direct splicing of small single mode fiber rings is not out of question, however because of bending losses the low-contrast core does not effectively confine radiation unless the radius is

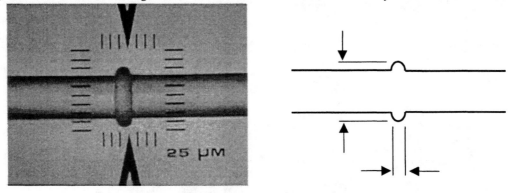

Fig.8. Microphotograph and the cross section of a novel geometry high-finesse
dielectric microcavity with whispering gallery modes

more than few millimeters. In our experiment, we approached the "single-mode" operation in high-Q WG mode microcavity by shaping the dielectric, in the area of WG mode localization, into geometry of highly eccentric ellipsoid of rotation. A pre-determined amount of low-loss silica glass, preshaped as a sphere, was heated and squeezed between flat cleaved tips of optical fiber.

As a result, the action of surface tension forces under axial compression resulted in desired geometry (see typical microcavity in Fig.8). One of the fiber "stems" was then cut, the whole structure installed in proper alignment next to a standard prism coupler, and the whispering-gallery mode spectrum observed using the frequency-tunable DFB semiconductor laser near the wavelength of 1550nm. The laser could be continuously frequency-scanned, by modulation the current, within the range of ~80GHz. To obtain high-resolution spectrum of WG modes over a wide range, we have combined 15 individual scans with ~60GHz frequency shifts between them. The frequency reference for "stitching" the spectral fragments was provided by simultaneous recording of the fringes of high-finesse (F~120) passively stabilized Fabry-Perot ethalon (FSR=30GHz). Additional frequency marks were obtained a system of sidebands resulting from 3.75GHz amplitude modulation of the laser signal before sending it into the Fabry-Perot interferometer. Total drift of the FP fringes was less than 400MHz over the total measurement time ~15min. After recording the individual scans, the combined spectrum was complilated on computer. Results are presented in Fig.9. As seen in Fig.9, only two whispering-gallery modes of selected polarization are excited within the free spectral range of the cavity FSR = 383GHz, or 3.06nm in the wavelength domain. The transmission of "parasitic" modes is at least 6dB smaller than that of principal ones. With individual mode bandwidth 23MHz, the finesse F = 1.7×10^4 is therefore demonstrated.

Presented here are preliminary experimental data on the novel, highly oblate spheroidal dielectric cavity with WG modes. They demonstrate that, because of the increased curvature of the cavity in transverse direction with respect to azimuthal circulation of light, the multiple "transverse" modes are effectively decoupled from the input/output device. Rigorous analysis of the transmission characteristic of the new cavity requires 1)electrodynamic analysis (solution of the Helmholtz equation in spheroid) and 2)calculation, on the basis of obtained eigenfunctions for the WG modes, of the input/output coefficients with evanescent wave coupler devices. Conventional solution for boundary problem for a spheroid is non-existent. Apart from numerical methods, the mode structure and frequencies may possibly be obtained using the eikonal methods previously applied to microspheres [21]. With the field configuration known, calculation of evanescent

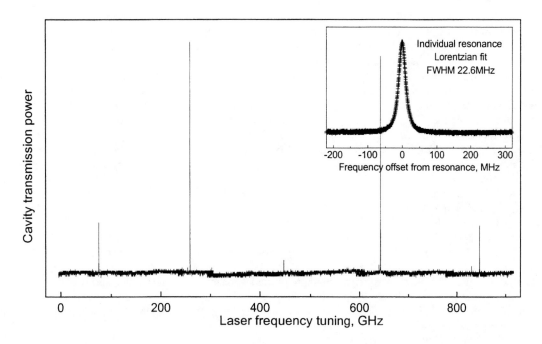

Fig.9. Spectrum of whispering-gallery modes in spheroidal dielectric microcavity (D = 160μm; d = 35μm). Free spectral range 383GHz (3.06nm) near central wavelength 1550nm. Individual resonance bandwidth 23MHz (loaded Q = 8.5×10^6). Finesse F = 1.7×10^4

coupling coefficients can be done within the guidelines described in [22].

Strong reduction of the number of effectively excited modes increases the applicability of the new microcavity compared to microspheres. In simple diode laser frequency locking schemes [11], robust single mode operation should be possible because WG mode spacing is now compatible with the gain bandwidth of the laser diode proper. In spectral analysis applications, the as yet unprecedented resolution becomes available with a truly miniature device. For an optoelectronic oscillator, the new cavity provides convenient reference for sub-millimeter-wave to teraHertz frequency sidebands, provided that appropriate detector and modulator are developed to complement such a very-high-frequency OEO.

ACKNOWLEDGEMENT

The research described in this paper was carried out by the Jet Propulsion Laboratory, California Institute of Technology, under a contract with the National Aeronautics and Space Administration.

REFERENCES

1. V.B.Braginsky, M.L.Gorodetsky, V.S.Ilchenko, "Quality-factor and nonlinear properties of optical whispering-gallery modes", *Phys.Lett.* **A137**, pp.393-6, 1989.
2. S.Schiller, R.Byer, "High-resolution spectroscopy of whispering-gallery modes in large dielectric spheres", *Opt.Lett.*, 16, pp.1138-40, 1991.

3. L.Collot, V.Lefèvre-Seguin, M.Brune, J.-M.Raimond, S.Haroche, "Very high-Q whispering-gallery mode resonances observed on fused silica microspheres", *Europhys.Lett.* **23(5)**, pp.327-33, 1993.

4. H.Mabuchi, H.J.Kimble, "Atom galleries for whispering atoms: binding atoms in stable orbits around a dielectric cavity", *Opt.Lett.*, **19**, pp.749-51, 1994.

5. D.W.Vernooy, A.Furusawa, N.P.Georgiades, V.S.Ilchenko, H.J.Kimble, "Cavity QED with high-Q whispering-gallery modes", *Phys.Rev.*, **A57**, pp.R2293-6, 1998.

6. M.Nagai, F.Hoshino, S.Yamamoto, R.Shimano, M.Kuwata-Gonokami, "Spherical cavity-mode laser with self-organized CuCl microspheres", *Opt.Lett.*, **22**, pp.1630-32, 1997.

7. V.Sandoghdar, F.Treussart, J.Hare, V.Lefèvre-Seguin, J.-M.Raimond, S.Haroche,, "A very low threshold whispering-gallery mode microsphere laser", Phys.Rev., **B54**, pp.R1777-80, 1996.

8. F.Treussart, V.S.Ilchenko, J.-F.Roch, P.Domokos, J.Hare, V.Lefèvre-Seguin, J.-M.Raimond, S.Haroche, "Whispering-gallery mode microlaser at liquid-helium temperature", *J.Lumin.*, **76**, pp.670-73, 1998.

9. M.L.Gorodetsky, S.P.Vyatchanin, V.S.Ilchenko, "Coupling and tunability of optical whispering-gallery modes: A basis for coordinate meter", *Opt.Commun.*, **107**, pp.41-8, 1994.

10. R.K.Chang, A.J.Campillo, eds., *Optical processes in microcavities*, World Scientific, Singapore, 1996.

11. V.V.Vassiliev, V. L.Velichansky, V.S.Ilchenko, M.L.Gorodetsky, L.Hollberg and A.V.Yarovitsky, "Narrow-line-width diode laser with a high-Q microsphere resonator", *Opt. Commun.* **158**, pp.305-12, 1998.

12. V.S.Ilchenko, X.S.Yao, L.Maleki, "High-Q microsphere cavity for laser stabilization and optoelectronic microwave oscillator", *Proc. SPIE*, **3611**, pp.190-198, 1999.

13. N.Dubreuil, J.C.Knight, D.Leventhal, V.Sandoghdar, J.Hare, and V.Lefèvre-Seguin, J.-M.Raimond, S.Haroche, "Eroded monomode optical fiber for whispering-gallery mode excitation in fused-silica microspheres", *Opt.Lett.*, **20**, pp.1515-8, 1995.
A.Serpengüzel, S.Arnold, G.Griffel, "Excitation of resonances in microspheres on an optical fiber", *Opt.Lett.*, **20**, pp.654-6, 1995.

14. J.C.Knight, G.Cheung, F.Jacques, T.A.Birks, "Phase-matched excitation of whispering gallery mode resonances using a fiber taper", *Opt.Lett.*, **22**, pp.1129-31, 1997.
Laine JP, Little BE, Haus HA, "Etch-eroded fiber coupler for whispering-gallery-mode excitation in high-Q silica microspheres", IEEE Photonic Tech. Lett. **11**, pp.1429-30, 1999.

15. C.C.Lam, P.T.Leung, K.Young, "Explicit asymphotic formulas for the positions, widths, and stregths of resonances in Mie scattering", *J.Opt.Soc.America,* **B9**, pp.1585, 1992.

16. B.E.Little, J.-P.Laine, D.R.Lim, H.A.Haus, L.C.Kimerling, S.T.Chu, "Pedestal antiresonant reflecting waveguides for robust coupling to microsphere resonators and for microphotonic circuits", *Opt.Lett.* **25**, pp.73-5, 2000.

17. M.Kuwata-Gonokami, K.Takeda, H.Yasuda, K.Ema "Laser emission from dye-doped polystyrene microsphere", *J.Appl.Phys.*, **31**, pp.L99-L101, 1992.

18. F.Lissilour, P.Feron, P.Dupriez, G.M.Stephan, "Whispering-gallery mode Er-ZBLAN microlasers at 1.56μm", *Proc.SPIE*, **3611**, pp.199-205, 1999.

19. V.S.Ilchenko, X.S.Yao, L.Maleki, "Pigtailing the high-Q microsphere cavity: a simple fiber coupler for optical whispering-gallery modes", *Opt.Lett.,* **24**, pp.723-5, 1999.

20. S.Suzuki, K.Shuto, Y.Hibino, "Integrated optic ring resonators with two stacked layers of silica waveguides on Si", *IEEE Photon. Technol. Lett.*, **4**, pp.1256-8, 1992.
B.E.Little, S.T.Chu, H.A.Haus, J.Foresi, J.-P.Laine, "Microring resonator channel dropping filters", *J. Lightwave Technol.,* **15**, pp.998-1005, 1997.

21. J. Hare, V.Lefevre-Seguin, 1997 (Private communication, Laboratoire Kastler Brossel, ENS, 24 rue Lhomond, F-75231 Paris Cedex 05, France).

22. M.L.Gorodetsky, V.S.Ilchenko, "Optical microsphere resonators: optimal coupling to high-Q whispering-gallery modes", *J. Opt. Soc. America B* **16**, pp.147-54, 1999.

Threshold minimization and directional laser emission from GaN microdisks

Nathan B. Rex[a], Richard K. Chang[*a], Louis J. Guido[b]

[a]Department of Applied Physics, Yale University, New Haven, CT 06520-8284
[b]Department of Materials Science and Engineering , Virginia Polytechnic Institute and State University, Blacksburg, VA 24061

ABSTRACT

2-μm thick GaN microdisks with the following shapes: circular; square; and quadrupolar with various deformation amplitudes have been investigated by optical pumping. For circular microdisks, the minimum laser threshold was found when the pump is a ring-shaped 355 nm laser beam. An imaging technique is used to photograph (with a CCD camera) the sidewall of the microdisk at various angles in the horizontal plane. From the imaging results as a function of observation angle, it is possible to extract information about the laser output location along the sidewall and its far-field angular pattern. Image results for the quadrupoles (with optical pumping of the top face) suggest that the directionality of laser emission is associated with chaotic-orbits that emerge just outside the highest curvature edges. With focused pumping of the same quadrupole structures at the middle of the top face, only Fabry-Perot modes involving the two lowest curvature interfaces are observed.

Keywords: GaN laser, microdisk, whispering-gallery modes, axicon, quadrupole, directional emission

1. INTRODUCTION

Recently, room temperature, electrically pumped laser action has been demonstrated in vertical cavity surface emitting microdisks, consisting of InGaN multiple quantum wells as the active region sandwiched between two nitride-based distributed Bragg reflectors[1]. Laser action has also been demonstrated in as-grown GaN hexagonal pyramid microcavities which have facets that can provide sufficient feedback for lasing [2]. Because of its material properties[3], GaN can benefit from total internal reflection at the sidewall of circular microdisks that provides high-Q feedback associated with whispering-gallery modes (WGMs), and thus leads to lower laser thresholds[4]. Our previous work has demonstrated laser action in GaN microdisks (2-μm epitaxial layer on sapphire substrate), when optically pumped on the top face with 355 nm and 266 nm radiation[5]. Current research with microdisk lasers has focused on two areas: (1) laser threshold minimization; and (2) directional laser emission.

There are several reports on methods to decrease the laser threshold in GaN and other semiconductor microdisks. Lower laser thresholds were achieved by etching small holes from the interior of GaAs microdisks and thereby reducing the amplified spontaneous emission (ASE) from competing nonlasing WGMs[6]. Lower laser thresholds have also been shown in micron and sub-micron scale microdisks where the small sizes reduce the excited-state population loss associated with ASE from the central region of the structure into which WGMs can not enter[7,8]. Other groups have studied electrode placement and current blocking as methods of reducing the laser threshold in microdisks[9]. Still others have improved the microdisk edge (sidewall) quality in order to reduce scattering losses[10].

Circular microdisks, however, have the inherent disadvantage of non-directional emission. Various groups have investigated methods of using the low laser thresholds of microdisks while achieving directional emission. Strategies have involved perforations and protrusions[11], waveguides[12], and cavity deformations[13]. Stadium-shaped microdisk lasers have shown increased directionality and feedback provided by regular-orbit "bow-tie" modes[13].

In this article, we report methods of minimizing the laser threshold in optically pumped circular GaN microdisks and of achieving directional emission from quadrupolar-deformed GaN microdisks. By selectively pumping only areas just within the circumference of the disk, we are able to measure lower stimulated emission (SE) thresholds and excite only high-Q WGMs. The optimized optical pumping geometry has broad implications for the eventual fabrication and integration of electrically pumped GaN microdisk lasers. In addition, we report directional emission from polygonal and quadrupolar shaped GaN microcavities that has the same laser threshold as the uniform emission from circular GaN microdisks. In the

In *Laser Resonators III*, Alexis V. Kudryashov, Alan H. Paxton, Editors,
Proceedings of SPIE Vol. 3930 (2000) ● 0277-786X/00/$15.00

163

case of the quadrupole, this directional emission is neither due to "bow-tie" modes nor any other type of regular-orbits, but rather is owing to chaotic orbits of the type that are known to co-exist with WGMs in the quadrupole microcavity.

2. EXPERIMENTAL PROCEDURE

This experiment uses GaN microstructures etched from 2 μm thick MOCVD-grown epitaxial layers. Various sizes of circular microdisks were fabricated with standard photolithography and reactive-ion etching techniques. The conventional reactive-ion etching techniques caused sidewall roughness on the sub-micron scale. The deformed cavity microstructures are mathematically defined quadrupole deformations, $r(\phi) = r_0[1+\varepsilon \, Cos(2\phi)]$, with r_0=100 μm and ε ranging from 0.12 to 0.20 in steps of 0.02. These quadrupole structures as well as the polygonal microcavities were formed by chemically assisted ion beam etched (CAIBE), which produced microdisks with remarkably smooth sidewalls.

All microstructures are optically pumped by the third harmonic (λ_p=355 nm) of a Q-switched Nd:YAG laser, with excitation direction perpendicular to the plane of the GaN microstructures. The principal data taken for the experiment is the spectrum emitted from the GaN microstructures and the image of the microdisk sidewall emitting the laser radiation. For the spectral measurements, a quartz lens, positioned at ~ 85° from the surface normal, collects the emitted light and images a small portion of the microdisk onto the slit of a 0.5 m focal-length spectrometer. A gated intensified charge-coupled device (ICCD) camera records the spectra. For the images, a regular camera lens, with the aperture set to accept 5° of emission and located ~85° from the surface normal, images the microstructures onto a second gated ICCD camera. In order to restrict the image data to the stimulated emission region of the GaN spectrum, a band pass filter (370-385 nm) is located in front of the ICCD camera. The imaging technique allows measurement of the angular distribution of the far-field intensity as well as the determination of the emission location from the GaN sidewall.

For circular microdisks, three different optical pumping geometries are used for the experiment: first, broad-beam or flood pumping, where the entire top face is uniformly pumped; second, focused-beam pumping, where the pumping beam is localized to a circular spot less than the GaN top face; and third, ring-shaped pumping just within the microdisk edge where the highest-Q WGMs are localized. For optically pumping the quadrupolar-shaped GaN disks, either broad-beam pumping or focused-beam pumping is used.

A ring-shaped beam, with a specific outer diameter and thickness can be achieved by using a conical lens referred to as an axicon[14] in conjunction with a spherical lens. Figure 1 (a) shows the experimental setup necessary to produce a ring excitation using an axicon and spherical lens combination. This configuration is highly flexible. The thickness, t, of the ring is adjustable by moving the GaN microdisk with respect to the focal length, f, of the converging lens, with the minimum thickness achieved when the disk is at the focal length. The diameter, d, of the ring is also adjustable by changing the distance, D, of the axicon from the microdisk.

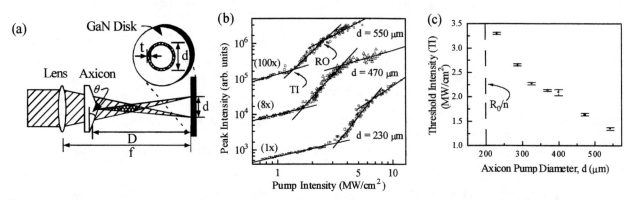

Figure 1. (a) Experimental setup for axicon pumping of a 550 μm GaN disk. The lens-axicon combination produces a hollow ring of laser light; the axicon pump thickness (t) and diameter (d) are adjustable. (b) Plot of the peak intensity emitted in the region of 365-380 nm versus pump intensity for three axicon-pump diameters and a constant t = 70μm. For d = 550 and d =470 μm the Threshold Intensity (TI) is much lower than for d = 230 μm. The SE peak saturates and rolls over (RO) at lower input pump intensity for d=550 μm pump than for smaller d's. (c) Plot of TI versus d [not all spectra are shown in (b)]. The threshold decreases for increasing d.

3. RESULTS AND DISCUSSION FOR CIRCULAR MICRODISKS

Figure 1 (b) is a log-log plot of the peak intensity for the stimulated emission (SE, in the region 365-380 nm) versus the pump intensity for a 550 μm diameter GaN microdisk with pump ring diameters of 550 μm, 470 μm, and 230 μm. We kept the pump-ring thickness (t = 70 μm) fixed and and varied the ring diameter, d. All three curves in Fig. 1(b) exhibit the same behavior as a function of increasing pump intensity: first a sublinear growth, followed by a superlinear growth, and finally back to sublinear growth. These curves are consistent with the standard transitions that are expected as the spontaneous emission becomes SE, and finally exceeds the laser threshold[15] while the pump intensity is increased. The threshold intensity from photo-luminescence to SE is designated in Fig. 1(b) as TI and the roll-over intensity from SE to saturation regime is designated in Fig. 1(b) as RO. Technically, TI is not the laser threshold, which is defined to be the midpoint of the superlinear regime[15].

Noteworthy trends of Fig. 1(b) from the three curves with different d values are the following: (1) the largest ring-shaped pumping (d = 550 μm) reaches the superlinear-growth regime at lowest pump intensities in comparison with smaller ring-shaped pumping diameters (d = 470 μm and d = 230 μm); (2) the region of superlinear growth decreases as d is increased, a result that is expected, based on the fact that the higher-Q modes are localized nearer to the microdisk edge; and (3) lasing with higher-Q WGMs causes the roll-over point to be reached at a lower pump intensity.

Figure 1 (c) is a plot of the threshold intensity (TI from photoluminescence to SE) versus the ring-pump diameter d. The TI is taken from the intersection of the best fit of the sublinear and superlinear regions, similar to the three curves shown in Fig. 1 (b). The SE threshold decreases as the pump diameter d is increased toward the GaN microdisk diameter [see Fig. 1(c)]. The lasing threshold, which is specified at the midpoint of the superlinear regime also decreases as d is increased to the microdisk diameter R_0 [16]. Lasing threshold could not be reached when d is less than R_0/n, where n is the index of refraction (n= 2.65 for GaN). In the ray model, R_0/n is the radius of the caustic circle where no WGM rays can enter and, hence, the photoluminescence and SE created within R_0/n does not couple to WGMs[17].

4. RESULTS AND DISCUSSION FOR GaN SQUARE AND QUADRUPOLAR MICRODISKS

(a) GaN Square

We investigated first the emission image of a GaN square, as an introduction to the imaging technique and to the types of information that can obtained, e.g., from where on the sidewall the SE and laser emission are emerging and what their angular profile in the far field is. The flood-pumping configuration is used for all the GaN square measurements. Some of the WGMs of a square are shown in Fig. 2(a), consisting of rays which form an inscribed square (rotated by 45° with respect to the GaN square sample) as well as rotated rectangles. The inscribed square and rectangles all have the same path length and thereby should have the same resonance wavelength, forming degenerate WGMs. The internal angle of incidence of these rays are all greater than the critical angle for total internal reflection $\theta_c = \sin^{-1}(1/n)$ and, hence, the refracted rays emerge as evanescent waves along the surface of the GaN square. Because the faces of the GaN square are not infinite, the evanescent wave converts to a propagating wave, once it reaches the slightly rounded corners of the square.

An image of one sidewall of the GaN square (either $\phi = 0°$ or 90°) will consist of two bright spots at the edge of the square. Figure 2(b) shows in gray scale the image intensity as a function of edge position for different observation angles (ϕ), defined in the standard way relative to the horizontal x-axis. In Fig. 2(b) for the initial angle $\phi = 90°$, the camera is looking at top edge of the GaN square. The crystal is then rotated counter-clockwise until the initial right edge is now on the top $\phi = 0°$. In Fig. 2(c) the same data is plotted in a 3-d format, where the intensity is shown at different edge positions as ϕ is rotated from $\phi = 90°$ to $\phi = 0°$. The appearance of the double peak which starts around $\phi = 20°$ and extends to $\phi = 60°$ is conjectured to be caused by the rounded corners of CAIBE fabricated GaN square.

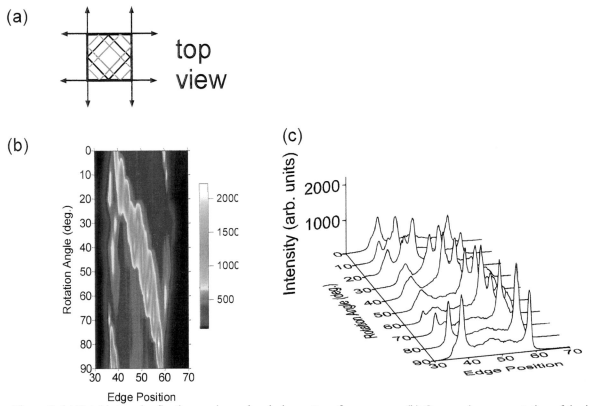

Figure 2. (a) Total internal reflection modes and emission pattern for a square. (b) Gray-scale representation of the image data of a GaN square. (c) Same data as (b) but in a 3-dimensional plot. At angles 0° and 90° the camera is perpendicular to the sidewall of the square and shows brightest emission from the edges of the face of the square.

(b) GaN Quadrupoles

Figure 3(a) shows the images of the sidewall of a $\varepsilon = 0.20$ GaN quadrupole that is **flood pumped** with 355 nm radiation. The 3-d plots show the intensity versus edge position of the image profiles for $\phi = 90°$ to $0°$. At $\phi = 90°$, the emission is from the highest curvature edges of the quadrupole, arising from diffractive-escaped rays which are tangent to these two edges. The edge-position length, for the image taken at $\phi = 90°$, corresponds to the major axis of the quadrupole. At $\phi = 0°$, the weak emission is from the lowest curvature portion of the quadrupole, where diffractive escape is the lowest. The edge-position length at $\phi = 0°$ corresponds to the minor axis of the quadrupole. The diffractive escape of internal rays with incident angles greater than θ_c is much larger at the highest curvature region than at the lowest curvature region of the quadrupole.

The far-field intensity at any observation angle (polar plot) is shown on the right column of Fig. 3(a). The far-field intensity at a specific ϕ can be extracted from the image data shown on the left column of Fig. 3(a). The far-field intensity at ϕ is simply the integral of the intensity along the entire image length, i.e., the area under each intensity versus edge-position curve. The emission-polar plot clearly shows that some amount of laser-emission directionality can be achieved with a quadrupolar-shaped GaN microdisk.

The source of the directional maximum, in terms of location along the sidewall and of the emission-cone angle, can be deduced by a more detailed analysis of the image data shown on the left column of Fig. 3(a). In principle, the image data recorded at several camera-aperature angles can provide information on the emission-cone angle. The maximum emission emerges from a region on the sidewall that is slightly away from the highest curvature edge. Furthermore, the emergence angle is not tangent to the GaN-air interface. We tentatively conclude that the internal rays that have escaped had incident angles less than θ_c, in order to give rise to the maximum emission around $\phi = 50°$. Hence, these rays escaped by refraction, in accordance with Snell's law of refraction. In the chaos-theory framework, these rays belong to chaotic orbits.

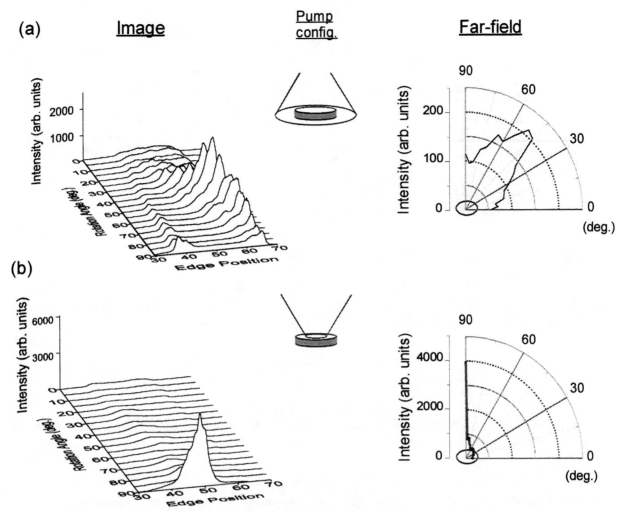

Figure 3. (a) Emission from a quadrupole of ε=0.20 with uniform pumping. Maximum emission is at 50°. Image profiles show where along the sidewall the laser radiation is emerging from the quadrupole. Maximum emission at 50° is non tangential, indicative of refractive escape. (b) Emission from a quadrupole of ε=0.20 with focused beam pumping. The beam is in the center of the quadrupole, preferentially pumping Fabry-Perot type modes.

Figure 3(b) shows the dramatic change of the SE and laser images that occurs when the GaN quadrupole (with ε = 0.20) is pumped by a **focused beam** (diameter approximately 50 µm) which is incident at the center of the top face. From the image data shown on the left column of Fig. 3(b), one peak is observed, only at ϕ = 90°. Furthermore, radiation emerges from the middle portion of the sidewall, when viewed perpendicular to the major axis direction (at ϕ = 90°). We conclude that this highly directional emission is associated with the Fabry-Perot modes established because of reflections between the two lowest curvature interfaces which form a stable concave mirror-like cavity. These Fabry-Perot modes have spatial overlap with the focused-pump beam when it is focused at the center of the GaN quadrupole. Because these Fabry-Perot modes are not observed with using the flood-pump geometry (see Fig. 3(a)), we conclude that the WGMs and the chaotic-orbit modes have a lower laser threshold than the Fabry-Perot modes.

The dependence of the far-field profile of GaN quadrupoles (defined by $r(\phi) = r_0[1+\varepsilon \cos(2\phi)]$, with r_0=100 µm) as ε is varied, from 0.12 to 0.20 in increments of 0.02, is shown in Fig. 4. The flood-pumping configuration is used for all these measurements. The far-field curves were extracted by integrating the image data as a function of edge position at each observation angle, i.e., the integrated intensity of the total light impinging on the ICCD camera at each ϕ, that is varied in 5-degree increments. These emission patterns indicate that the emitted light becomes more directional as the quadrupole-

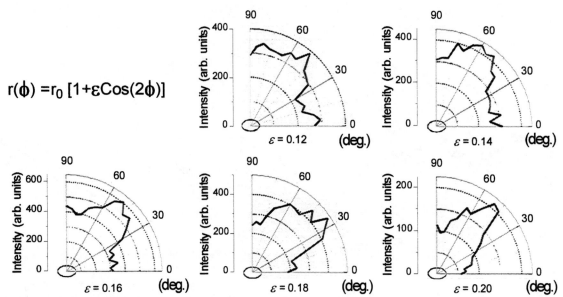

$$r(\phi) = r_0 [1 + \varepsilon \cos(2\phi)]$$

Figure 4. Far-field intensity angular patterns for quadrupoles, $r(\phi) = r_0[1 + \varepsilon \cos(2\phi)]$, of varying deformation parameter ε. Emission becomes more directional as ε increases.

deformation parameter, ε, increases. This agrees with theoretical predictions from surface-of-section (SOS) plots for ray trajectories. As fewer WGMs are supported with larger ε, the emission should be more directional.

ACKNOWLEDGEMENTS

We would like to acknowledge NSF for financial support (Grant PHY-9612200). We express our appreciation to Dr. David Bour and Dr. Michael Kneissl at Xerox PARC for their work with sample preparation.

REFERENCES

1. T. Someya, R. Werner, A. Forchel, M. Catalano, R. Cingolani, and Y. Arakawa, "Room temperature lasing at blue wavelengths in Gallium Nitride microcavities," Science **285** (1999) pp. 1905-1906.
2. H. Jiang, J. Lin, K. Zeng, and W. Yang, "Optical resonance modes in GaN pyramid microcavities," Appl. Phys. Lett **75** (1999) pp. 763-765.
3. A. Hangleiter, G. Frankowsky, V. Härle, and F. Scholz, "Optical gain in the nitrides: are there differences to other III-V semiconductors?" Mat. Sci. & Eng. **B43** (1997) pp. 201-206.
4. S. McCall, A. Levi, R. Slusher, S. Pearton, and R. Logan, "Whispering-gallery mode microdisk lasers," Appl. Phys. Lett. **60** (1992) pp. 289-290.
5. S. Chang, N. Rex, R. Chang, G. Chong, and L Guido, "Stimulated emission and lasing in whispering-gallery modes of GaN microdisk cavities," Appl. Phys. Lett. **75** (1999) pp. 166-168.
6. S. Backes, J. Cleaver, A. Heberle, J. Baumberg, and K. Köhler, "Threshold reduction in pierced microdisk lasers," Appl. Phys. Lett. **74** (1999) pp. 176-178.
7. A. Levi, S. McCall, S. Pearton, and R. Logan, "Room temperature operation of submicrometre raidus disk laser," Electron. Lett. **29** (1993) pp. 1666-1667.
8. T. Baba, M. Fujita, A. Sakai, M. Kihara, and R. Watanabe, "Lasing characteristics of GaInAsP-InP strained quantum-well microdisk injection lasers with diameter of 2-10 μm," IEEE Photonics Technol. Lett. **9** (1997) pp. 878-880.
9. S. Thiyagarajan, D. Cohen, A. Levi, S. Ryu, R. Li, and P. Dapkus, "Continuous room-temperature operation of microdisk laser diodes," Electron. Lett. **35** (1999) pp. 1252-1254.
10. S. Ando, N. Kobayashi, and H. Ando, "Hexagonal-facet laser with optical waveguides grown by selective area metalorganic chemical vapor deposition," Jpn. J. Appl. Phys. **34** Part 2 (1995) pp. L4-L6.
11. S. Backes, J. Cleaver, A. Heberle, and K. Köhler, "Microdisk laser structures for mode control and directional emission," J. Vac. Sci. & Technol. **B16** (1998) pp. 3817-3820.

12. D. Rafizadeh, J. Zhang, S. Hagness, A. Taflove, K. Stair, S. Ho, and R. Tiberio, "Waveguide-coupled AlGaAs/GaAs microcavity ring and disk resonators with high finesse and 21.6-nm free spectral range," Opics Lett. **22** (1997) pp. 1244-1246.

13. C. Gmachl, F. Capasso, E. Narimanov, J. Nöckel, A. Stone, J. Faist, D. Sivco, and A. Cho, "High-power directional emission from microlasers with chaotic resonators," Science **280** (1998) pp. 1556-1563.

14. I. Manek, Y. Ovchinnikov, and R. Grimm, "Generation of a hollow laser beam for atom trapping using an axicon," Optics Comm. **147** (1998) pp. 67-70.

15. A. Siegman, Lasers (University Science Books, Sausalito, CA, 1986), pp. 491-557.

16. Because of material damage above 10 MW/cm^2 at λ_p=355 nm, it is not possible to achieve sufficient lasing sublinear dependence for small pump diameters in order to find the midpoint of the superlinear regime.

17. S. Hill and R. Benner, Optical Effects Associated with Small Particles, P. Barber and R. Chang, ed. (World Scientific, Singapore, 1988), pp. 3-61.

* Correspondence: Email: richard.chang@yale.edu; Telephone: 203-432-4272; Fax: 203-432-4274

Invited Paper

Tight-Binding Coupling of the microsphere resonators

Makoto Kuwata-Gonokami

Department of Applied Physics, Faculty of Engineering, University of Tokyo
and
Cooperative Excitation Project of ERATO (JST),
7-3-1 Hongo, Bunkyo-ku, Tokyo 113-8656, Japan

ABSTRACT

We discuss the optical whispering gallery modes in single and coupled spheres in the small size range. We study the emission from a dye doped polystyrene spheres with diameters ranging from 2 to 5 μm under quasi-steady state optical excitation. By studying the spectral characteristics of the emission of the individual 5 μm sphere, we observe the transition from spontaneous emission to lasing. Although the dye molecule linewidth is broader than the free spectral range of whispering-gallery modes for such a sphere, more than 30% of the spontaneous emission is emitted into the lasing mode. By monitoring the frequencies of fluorescence peaks of individual spheres, we sort out two spheres with appropriate size matching and make them in contact. We observe coherent resonant coupling of optical whispering-gallery modes in fluorescence from two-sphere system (bisphere). By taking into account harmonic coupling of the whispering-gallery modes, the obtained features of the normal mode splitting are well explained by the tight binding photon picture.

Keywords: microsphere, tight binding, normal mode splitting

1. INTRODUCTION

Manipulation of light path in micrometer length scale has recently attracted a considerable attention both from fundamental and application point of view. Conventionally, the manipulation is based on the photonic crystal concept.[1-3] In photonic crystals, which have weak periodic modulation of the refractive index, propagation of the light wave can be descried in terms of the is wave packet propagation in a weak potential. Correspondingly, such an approach can be referred as nearly free photon approach analogous to the nearly free electron approach in band theory. Alternatively, the micro-manipulation of light can be achieved by exploring the possibility to confine the light in a small unit of the wavelength size. Light propagates through the system of such units due to the coupling between the nearest neighbours. Such tight binding photon approach[4] to the micromanipulation of light allows one to guide the optical waves by connecting the units, which may be referred to as photonic atoms, in the arbitrary shaped microstructures (photonic molecules).

The microspheres are the most natural choice of the unit to be employed in the tight binding photon device. It is known that a dielectric sphere acts as a unique optical microcavity which has very long photon storage time within a small mode volume.[5-8] In particular, Q-factors of the order of 10^{10} have been observed for whispering gallery modes (WGM's) in quartz spheres with diameter of several tens of micrometers,[9-13] and the mode structure of pair of these large spheres in contact has been studied.[14] However, in order to explore the feasibility of micro-manipulation of light, one has to confirm the existence of the coherent coupling between spheres of the size of a few times of optical wavelength. Lorenz-Mie theory predicts long photon lifetime even for small spheres, giving , for example, nearly 30 picoseconds for a 4 μm sphere with a refractive index of 1.59. This has allowed one to propose such relatively small spheres to be employed as photonic atoms[15] for tight binding scheme. However, the coherent coupling between two adjacent microspheres of such size range was not achieved for a long time. The coherent coupling results in the splitting of the corresponding WGM's and is manifestation of the well known phenomena of the normal mode splitting (NMS) in coupled harmonic oscillators. Correspondingly, the direct evidence of the coherent coupling is the observation of the NMS in the pair of two identical sphere in contact. However, although some attempts have been made,[16] NMS has has been observed very recently[17] mainly because of the difficulty in the precise size control of the spheres.

In this paper we present our results on the WGM effects in dielectric spheres as photonic atoms and coherent coupling between two spheres of the wavelength size. The structure of the paper is the following. In the next section

170

In *Laser Resonators III*, Alexis V. Kudryashov, Alan H. Paxton, Editors,
Proceedings of SPIE Vol. 3930 (2000) • 0277-786X/00/$15.00

we will give a brief introduction in the properties of the WGM and discuss the spontaneous emission and lasing from the 5 μm spheres. The results on the emission from two strongly coupled spheres (bisphere) of size from 2 to 5 μm are presented in the last section.

2. WGM EFFECTS IN LASING FROM THE 5 μM SPHERES

The WGM is a resonance of light wave trapped inside dielectric spheres (or disks) by total internal reflection. For a given sphere, the resonance occurs at specific value of the size parameter, $x = 2\pi r/\lambda$, where λ is the light wavelength in the vacuum and r is the sphere radius. Each WGM can be specified by the mode number n (indicates the number of light wavelengths around the circumference and indicates the order of spherical Bessel and Hankel functions, which gives the radial field distribution in the Lorentz-Mie theory[18]), the order number s (corresponds to the number of maxima in the radial dependence of the internal electric field) and the azimuthal mode number m (gives the orientation of the WGM's orbital plane). Since the frequency of WGM does not depend on m, each resonance has $(2n + 1)$-fold degeneracy.

Microsphere is a good candidate to study the cavity effects on lasing. Spherical geometry has two advantages. First, the Mie theory allows us to perform rigorous quasi-analytical analyses of the cavity QED effects.[19] Second, high-Q spherical cavities can be readily self-assembled by the strong surface tension. Efficient lasing has been observed with dye solution droplets[20] and dye doped polymer microsphere.[7,8] Systematic study of fluorescence and lasing spectra have been reported for dye droplet.[21] However, since in these experiments the sphere size was too large and, correspondingly, the mode structure was very complicated, the quantitative analysis of the lasing characteristics was not possible. In order to reduce the number of high-Q WGM, which gives rise to the multimode lasing, it has been proposed[8] to use the microspheres with enhanced surface concentration of the dye molecules. However, in order to reduce the threshold and to achieve a considerable lasing efficiency it is necessary to reduce the sphere size down to several microns. In this section we consider the WGM effects in stimulated emission, which takes place when the 5 μm dye doped polymer spheres are excited by the pulsed light. We compare the results with theory and show clear evidence of cavity QED effects on the micro spherical cavity laser with a broad band gain medium.

The properties of the micro-laser emission is determined by the Einstein A-coefficient, which is given by the vacuum field distribution. In vacuum the Einstein coefficient, which gives the probability of the spontaneous emission is given by[19]

$$A = \frac{2\pi}{\hbar^2} \int_0^\infty \left\langle |\mu E_{vac}|^2 \right\rangle \rho(\omega) L(\omega) d\omega \tag{1}$$

where $\rho(\omega) = \omega^2/\pi^2 c^3$ is the density of state for the free space, $L(\omega)$ is the material gain, E_{vac} is the vacuum electric field which corresponds to the vacuum Rabi frequency $g = |\mu E_{vac}|/\hbar$. In the bulk medium the Einstein coefficient can be reduced to $A_{bulk} = n_0 f A$, where f is the local field factor. In the spherical cavity, which size is comparable with the wave length, the mode structure of the vacuum field is dramatically altered and becomes a position-dependent:

$$A_{sphere}(r) = \frac{2\pi f}{\hbar^2} \int_0^\infty \left\langle |\mu E_{vac,sphere}(\omega, r)|^2 \right\rangle \rho(\omega) L(\omega) d\omega \tag{2}$$

In order to characterize the change of the properties of the medium we can introduce the enhancement factor $\eta(r) = A_{sphere}(r)/A_{bulk}$.

The dynamics of the microsphere emission can be described in terms of the following balance equations[22] for the number of photons in the effective mode volume S and number of excited atoms N:

$$\frac{dN}{dt} + [(1 + \beta S)]A_{sphere}N = P$$
$$\frac{dS}{dt} + \gamma S = \beta A_{sphere}(S + 1)N \tag{3}$$

S is the number of photons ion the effective mode volume, β is the ration of the spontaneous emission rate into the mode of interest to the total spontaneous emission rate, γ is the mode decay rate. Here we have introduced the averaged Einstein coefficient, which can be related with the bulk one by the phenomenological enhancement factor F as follows: $A_{sphere} = F A_{bulk}$.

In order to obtain F we calculate enhancement factor of the spontaneous emission rate of dye molecule as a function of position inside the sphere and frequency (see Fig.1a). Spontaneous emission is strongly enhanced at the resonance frequencies, however it is suppressed in between the resonances. We find that such alteration occurs periodically with a frequency interval, which is referred to as the free spectral range (FSR) and gives the frequency difference between the neighbouring modes of the same order number. For example, for a TE modes with $s = 1$ of polystyrene (refractive index of 1.58) spheres of 5 μm diameter, FSR is about 400 cm^{-1} at the wavelength approximately 580 nm (see insert in Fig.1). It has been shown[7] that the FSR is inversely proportional to the sphere diameter, however the homogeneous width of typical dye molecules are broader than the FSR for the 5 μm sphere. In order to obtain the overall enhancement factor for the sphere, i.e. to evaluate F from the balance equations (3), we perform the frequency averaging by integrating the enhancement factor FSR. Fig.1b shows the frequency averaged enhancement factor for the TE mode with $s = 1$ and $n = 38$.

The steady-state solution of (3) is given by

$$N = \frac{\gamma}{\beta A_{sphere}} \frac{S}{S+1} \tag{4}$$

$$S = \frac{P}{\gamma} - \frac{1-\beta}{\gamma} N A_{sphere} \tag{5}$$

For $S \gg 1$ (lasing regime) this reduces to

$$N_{las} = \frac{\gamma}{\beta A_{sphere}} \tag{6}$$

$$S_{las} = \frac{P}{\gamma} - \frac{1-\beta}{\beta} \tag{7}$$

giving the saturation of the excited atoms density. In the fluorescent regime ($S \ll 1$) the number of excited atoms and number of photons are proportional to the pump rate: $S_{fluo} = \beta P/\gamma$ and $N_{fluo} = P/A_{sphere}$. One may observe from the above equations that the number of photons is linear function of the pump rate in both regimes, however the slope efficiency is different. The lasing threshold corresponds to $S_{fluo} = S_{las}$ and $N_{fluo} = N_{las}$ giving $P_{th} = \gamma/\beta$. Threshold can be readily observed as the step-like behaviour of the $S(P)$ in the semi-logarithmic scale.[22]

We perform measurements of the lasing with the microspheres of the with the polystyrene spheres of 5 μm and 10 μm diameter, which are stained in a solution of 10^{-2} M Pyrromethene 580 [1] in methylene dichloride. Net amount of the dye concentration is determined by the linewidth analyses of the emission from 10 μm samples. The dye concentration is about 2×10^{-3} mol/l. Polystyrene sphere are placed on a glass plate under the microscope. We excite one sphere with a optical fiber probe and collect the emission from the sphere through the same fiber probe. With this method we can easily detect the emission from individual sphere. Multimode silica glass optical fiber with core diameter of 400 μm is made into fiber probe by pulling in a frame. In this experiment, the head of the probe has core diameter of about 20 μm. We use frequency doubled CW Q-switched YLF laser as a optical pumping source. Wave length is 527nm, pulse duration is 100 nsec and repetition rate is 1kHz. The pulse duration is much longer than the spontaneous emission lifetime of the dye molecule, thus we regard the emission as quasi steady state condition.

We measure emission from the spheres of 5 μm in diameter at various excitation level. The input and output characteristics of a 5 μm sphere show typical single mode laser behaviour [14]. In this case, the separation energy of adjacent modes is bigger than homogeneous width of the dye molecule and, therefore, the mode competition does not occur. The WGM structure for the 5 μm polystirene sphere is presented on Fig.2. The emission spectra below and above the lasing threshold for WGM with $n=38$ and $s =1$ are presented on Fig.2(a) and Fig.2(b), respectively. The observed threshold energy is only 10 pJ for 100 nsec excitation pulse duration. From the amplitude of the jump of the output intensity on Fig.3 we obtain the $\beta \simeq 0.3$ for 5 μm microcavity.

We calculate the vacuum field amplitude distribution of the whispering gallery mode inside the sphere as a function of position and frequency. Close to the surface region, the vacuum field amplitude becomes large at the frequency of WGM resonance of $s = 1$. On the other hand, vacuum field amplitude is suppressed between the resonance modes. As a result, the spontaneous emission rate of a molecule with broad homogeneous width does not change. Even in such case, however, the emitted photons are predominantly coupled into a high-Q WGM giving . This is the reason why the obtained spontaneous emission coupling coefficient for 5 μm sphere is much larger than that for 20 μm sphere[8] with enhances surface concentration of the dye molecules.

3. TIGHT-BINDING MODES OF THE RESONANT BISPHERES

In this section, we discuss the experimental observation of the normal mode coupling in the system of two polymer spheres in contact (bisphere) under extreme size control. Specifically, we study bispheres formed from the spheres of diameters ranging from 2 to $5\,\mu$m. The sufficiently narrow line width and wide separation of WGM's in this size range allows us to avoid intricate band mixing. We use monodispersive polystyrene spheres (refractive index 1.59) which are soaked with a solution of dye (Nile Red, concentration is about 10^{-2} mol/l, fluorescence FWHM is about 70nm). Dye doped spheres are sown on a glass plate under a microscope and can be manipulated with an optical fiber probe which is made by pulling the multi-mode optical fiber in a frame. The diameter of the fiber probe is about $3\,\mu$m. By using microscope objective lens, individual sphere is excited with the second harmonics of a CW Q-switch Nd-YLF laser ($\lambda = 527$ nm). The dye fluorescence is collected with the same fiber probe and sent to a spectrometer with a liquid nitrogen cooled CCD detector. The sketch of the experimental setup is presented on Fig.4. We use both 50 cm and 1 m focal length spectrometers with spectral resolution of 0.06 nm and 0.03 nm respectively.

Figure 5 shows an example of fluorescence spectrum from the individual sphere of about $4.1\,\mu$m in diameter. Sharp lines, which originate from the strong modification of the vacuum field inside the sphere,[6–8,23–25] appear at the resonances frequencies of WGM's with TE and TM polarizations.

For spheres of diameter from 2 to $5\,\mu$m we can identify the mode indices n and s of the high-Q WGM by comparing the positions of the lines in the fluorescence spectra with the predictions of the Lorenz-Mie theory. The width of the resonances $\lambda_{n.s}$ gives the Q-factor for WGM of a given n and order number s, $Q_{n.s} = \lambda_{n.s}/\Delta\lambda_{n.s}$. The obtained Q-factor is determined by $Q^{-1} = Q_{abs}^{-1} + Q_s^{-1} + Q_r^{-1}$.[8] Here, $Q_{abs}^{-1} = \alpha\lambda/2\pi n_0$ describes the absorption loss, α is the absorption coefficient, n_0 is the refraction index; Q_s^{-1} describes the losses due to light scattering by surface roughness and Rayleigh scattering inside the sphere; Q_r^{-1} corresponds to the intrinsic diffraction leakage. Spheres of $2\,\mu$m in diameter are found to have a Q-factor of the order of 10^2. This is comparable with the radiative leakage rate which follows from the Mie theory. Spheres of diameters more than $4\,\mu$m are found to have Q-factors of the order of 10^3. This is two orders less than the radiative leakage rate, and, therefore, the Q-factor is mainly determined by the light absorption by dye molecules and light scattering by imperfect surface. Note, that non-ideal spherical shape of spheres lifts the degeneracy with respect to the azimuthal mode indices, giving rise to the broadening and asymmetry of the fluorescence lines. The upper limit of the eccentricity of the spheres used in this experiment can be estimated from the width of narrowest line. For example, the narrowest line width of the $5\,\mu$m spheres is 0.4nm which ensure the eccentricity less than $0.08\,\%$.

Once we label the fluorescence lines with the mode indices n and s, we can judge the relative size of the sphere within an error of $1/Q$ which is about 0.05% for $5\,\mu$m sphere. In order to make the bisphere, we choose two spheres of the desired size by comparing the frequencies of the specific WGM resonance. By handling the optical fiber probe attached to the three axis mechanical stage, we pick up one sphere and attach it to the other.

The measurement configurations are presented on the Fig.6(a). In parallel configuration (A) the fiber probe is set parallel to the axis of the bisphere, while in the perpendicular configuration (B) the fiber probe collects fluorescence in the direction perpendicular to the axis of the bisphere. Fig.6(b) and (c) show the fluorescence spectra of the bisphere (A,B) and individual spheres (C) in the vicinity of the TE mode of $n = 30$ and $s = 1$ and TM mode of $n = 29$ and $s = 1$ resonances, respectively. For both the polarizations, we observe new peaks due to the inter-sphere coupling, in the parallel configuration (A) but not in the perpendicular configuration (B). To explain this, one should recall that each WGM of a given n and s has $(2n + 1)$-fold degeneracy associated with the azimuthal mode number m. Hence $2n + 1$ orientations of the WGM orbital plane are permitted and, correspondingly, $(2n + 1)^2$ combinations of the inter-sphere couplings are generally possible. However, the inter-sphere coupling is maximum for the pair of modes whose orbitals include the contact point and lay in the same plane. Therefore, the fluorescence of the coupled modes is maximum in the direction parallel to the bisphere axis. As a result, the signal from the coupled inter-sphere modes is more pronounced in the parallel configuration than in the perpendicular configuration. As we can see in Fig.6(b) and (c), the uncoupled modes fluorescence also contributes to the signal. By reducing the aperture of the fiber or keeping the probe away from the bisphere, this signal diminishes.

The dependence of the mode coupling on the orbital plane orientation breaks the degeneracy with respect to the azimuthal indices. In particular, bisphere modes originated from the coupling of the WGM with the different combinations of azimuthal numbers contribute differently to the signal. Correspondingly, the line shape of the signal represents the energy distribution among the coupled modes. The asymmetric line shape of the coupled modes on Fig.6(b),(c) indicates that stronger the coupling between WGM, more the energy in the resulting bisphere mode.

In order to study the dependence of the WGM coupling on the sphere size(size detuning), we fix the diameter of one sphere and change the diameter of the second sphere in the scale of 0.3 %. Figure 7 shows the detuning dependence of the bisphere fluorescence in the vicinity of TE29,1 mode for various pairs with size detuning. Open and closed circles show the uncoupled original resonance, and open and closed triangles show coupled modes. We clearly see the anti-crossing behaviour which is the signature of NMS.

Figure 8 shows the dependence of bisphere resonance on detuning for TE30,1 and TM29,1 modes. Positions of modes before contact are shown with dashed lines. Both of TE30,1 and TM29,1 modes show anti-crossing with some asymmetry. Such an asymmetry could be attributed to the influence of the broad second order ($s = 2$) WGM's, and their positions are shown in the figure with thick shaded lines.

The observed features of inter-sphere coupling can be explained by the resonance enhancement of the internal field in bispheres. By expanding the internal and external fields of each sphere in terms of the vector spherical harmonics, we solve numerically the coupled linear equation by the matrix inversion method.[26] The energy of the internal field is obtained from the expansion coefficients as a function of the incident plane wave parameters and sphere radii. Excellent numerical convergence is achieved with the summation of n up to 49. When the incident wave is almost parallel to the bisphere axis, the spectrum shows prominent peaks asymmetrically located on both sides of TM29,1 and TE30,1 modes. In between these prominent peaks there appear a series of less intense peaks converging to the original frequencies of WGMs. These substructures make the line shape asymmetric as we shown in Fig.6(b) and (c). Open circles of Fig.8 show the positions of the prominent peaks. One may observe an excellent agreement between calculation and experiments from the Fig.8. Note, that both the theory and experiment shows asymmetry in the mode splitting.

In order to explore the feasibility of the tight binding scheme for the micro-manipulation of light it would be desirable to introduce a mode overlap parameter which could describe the intersphere coupling. It is clear that these parameters will be determined by the convolution of the modes' electric fields. Therefore, one can expect the strong coupling for the modes with the matching frequencies whose orbitals lay nearly in the same plane and are close to the contact point. We will refer coupling between the modes with the same and different mode numbers as "intramode" and "intermode" coupling, respectively. By introducing the phenomenological coupling parameters of the dimension of frequency, which accounts for the inter- and intramode coupling for the relevant modes, we can describe the bisphere fluorescence in standard terms of the interacting harmonic systems. In order to reproduce the experimental results we take into account the intramode coupling between the WGMs with $s = 1$, intermode coupling of the WGMs with $s = 1$ and $s = 2$ and neglect the coupling between TE and TM modes. The best fit values of the coupling parameters are shown in Table 1(TE mode) and Table 2 (TM mode) and the corresponding detuning dependence are shown in Fig.8 as solid lines.

One can see from Fig.8 that this model well explains the overall features of the normal modes of bispheres and the asymmetric behaviour of the upper mode branch of the TM29,1 due to the strong mixing with the TM25,2 mode. The best fit values of the coupling parameters are almost constant for all the modes involved in the narrow region of mode number variation from $n = 25$ to 30.

We perform four series of measurements using spheres with diameters of 2, 3, 4.1 and 5 μm and estimate the coupling parameters. The obtained coupling parameters normalised to the resonance frequencies are 7×10^{-3} for 2 μm ($n = 14$), 5×10^{-3} for 3 μm ($n = 21$), 4×10^{-3} for 4.1 μm ($n = 29$) and 3×10^{-3} for 5 μm ($n = 36$). This indicates that the coupling parameters decreases with increasing sphere diameter or mode number. The results of the measurements are presented on Fig.9 on the "size parameter - Q-factor" plane. To explain such a behaviour, one can recall that the smaller the spheres, more the electric field outside the spheres and hence the coupling parameter increases. One could remind here that the mode number represents the order of the Hankel function which gives the field of WGM outside the sphere. In the experiment two spheres are in contact and hence we obtain the maximum available coupling for a given size. The coupling can be reduced by increasing the distance between spheres as it has been demonstrated for bigger spheres.[14] We would like to emphasise that the obtained coupling parameter are much larger than the line widths, indicating the dominance of coherent resonant coupling. This indicates that the coherent resonant coupling is dominating.

4. CONCLUSION

We demonstrate the efficient three-dimensional confinement of the light using the dielectric spheres of small size range. By studying the spectral characteristics of the emission of the individual 5 μm sphere, we observe the transition from spontaneous emission to lasing. Although the dye molecule linewidth is broader than the free spectral range of whispering-gallery modes for such a sphere, more than 30% of the spontaneous emission is emitted into the lasing mode. Our results show that the precise control of the sphere size on the sub-micron scale is the is necessary to achieve the coherent coupling of WGM's in bispheres. The anti-crossing behaviour of the mode coupling, which are predicted by the wave optics calculations, can be reproduced in the experiment.[17] Importantly, that overall features of obtained bisphere normal modes are well explained with the coupled harmonic oscillator model. For the spheres of the size of several microns, the coupling parameters are found to be much larger than the line width and smaller than the mean mode separation of the WGM. This ensures the feasibility of the tight binding manipulation of light waves in a structure composed of contacting spheres, which can be used in the development of new optoelectronic devices.

ACKNOWLEDGMENTS

The author would like to acknowledge Dr K. Takeda for sample preparation and discussion, and Mr. T. Mukaiyama, S. Ozawa for the experiment. We wish also to thank Dr. M. Morinaga, Dr Y. Miyamoto, Prof H. Miyazaki and Dr Yu. P. Svirko for theoretical discussions. This work was partially supported by grant-in-aid for Scientific Research in the Priority Area "Near-Field Nano Optics".

REFERENCES

1. A. Mekis, J. C. Chen, I. Kurland, S. H. Fan, P. R. Villeneuve, J. D. Joannopoulos, "High transmission through sharp bends in photonic crystal waveguides", Phys. Rev. Lett **77,** pp. 3787-3789, 1996.

2. J. D. Joannopoulos, P. R. Villeneuve and S. Fan, "Photonic crystals: Putting a new twist on light", Nature **386,** pp. 143-149, 1997.

3. J. D. Joannopoulos, P. R. Villeneuve, and S. Fan, "Photonic crystals", Solid State Commun., **102,** pp. 165-173, 1997.

4. E. Lidorikis, M. M. Sigalas, E. N. Economou and C. M. Soukoulis, "Tight-binding parametrization for photonic band gap materials", Phys. Rev. Lett.**81,** pp. 1405-1408, 1998.

5. P. R. Conwell, P. W. Barber, and C. K. Rushforth, "Resonant spectra odf dielectric spheres", J. Opt. Soc Amer. B **1,** pp. 62-67, 1984.

6. R. E. Benner, P. W. Barber, J. F. Owen and R. K. Chang, "Observation of structure resonances in the fluorescence-spectra from microspheres", Phys. Rev. Lett. **44,** pp. 475-478, 1980.

7. M. Kuwata-Gonokami, K. Takeda, H. Yasuda, and K. Ema, "Laser-emission from dye-doped polystyrene microsphere", Jpn. J. Appl. Phys. **31,** pp. L99-L101, 1992.

8. M. Kuwata-Gonokami, and K. Takeda, "Polymer whispering gallery mode lasers", Optical Materials **9,** pp. 12-17, 1998.

9. V. B. Braginsky, M. L. Gorodetsky, and V. S. Ilchenko, "Quality-factor and nonlinear properties of optical whispering-gallery modes", Phys. Lett. **A 137,** pp. 393-397, 1989.

10. L. Collot, V. Lefevreseguin, M. Brune, J. Raimond, S. Haroche, "Very high-q whispering-gallery mode resonances observed on fused-silica microspheres", Europhys. Lett. **23,** pp. 327-334, 1993.

11. V. Sandoghdar, F. Treussart, J. Hare, V. LefevreSeguin, J. M. Raimond, S. Haroche, "Very low threshold whispering-gallery-mode microsphere laser", Phys. Rev. A **54,** pp. R1777-R1780 (1996).

12. M. L. Gorodetsky, A. A. Savchenkov, and V. S. Ilchenko, "Ultimate Q of optical microsphere resonators", Opt. Lett. **21,** pp. 453-455, 1996.

13. D. W. Vernooy, V. S. Ilchenko, H. Mabuchi, E. W. Streed, H. J. Kimble, "High-Q measurements of fused-silica microspheres in the near infrared", Opt. Lett. **23,** pp. 247-249, 1998.

14. V. S. Ilchenko, M. L. Gorodetsky, and S. P. Vyatchanin, "Coupling and tunability of optical whispering-gallery modes - a basis for coordinate meter", Opt. Communs.**107,** pp. 41-48, 1994.

15. S. Arnold, J. Comunale, W. B. Whitten, J. M. Ramsey, K. A. Fuller, "Room-temperature microparticle-based persistent hole-burning spectroscopy", J. Opt. Soc. Amer. B **9,** pp. 819-824, 1992.

16. S. Arnold, A. Ghaemi, P. Hendrie, and K. A. Fuller, Opt. Lett.,"Morphological resonances detected from a cluster of 2 microspheres", **19,** pp. 156-158, 1994.

17. T. Mukaiyama, M. Takeda, H. Miyazaki, Y. Jimba and M. Kuwata-Gonokami, "Tight-binding photonic molecule modes of resonant bispheres", Phys. Rev. Lett. **82,** pp. 4623-4626, 1999.

18. M. Born and E. Wolf. *Principles of Optics*, Pergamon Press, Oxford, 1975.

19. M. O. Scully and M. S. Zubairy. *Quantum Optics*, Cambridge University Press, Cambridge, 1997.

20. H-M. Tzeng, K. F. Wall, M. B. Long, and R. K. Chang, "Laser-emission from individual droplets at wavelengths corresponding to morphology-dependent resonances",Opt. Lett. **9,** pp.499-501, 1984.

21. H.-B. Lin, J. D. Eversole, and A. J. Campillo, "Cavity-mode identification of fluorescence and lasing in dye-doped microdroplets", J. Opt. Soc. Am B. **9,** pp. 43-50, 1992.

22. K. Ujihara, M. Osige and M. Takagu, "Rate-equation analysis of a pulsed microcavity laser", J. Journ. Appl. Phys. **32,** pp. L1808-L1810, 1993.

23. A. J. Campillo, J. D. Eversole, and H.-B. Lin, "Cavity quantum electrodynamic enhancement of stimulated-emission in microdroplets", Phys. Rev. Lett. **67,** pp. 437-440, 1991.

24. H.-B. Lin, and A. J. Campillo, "CW nonlinear optics in droplet microcavities displaying enhanced gain", Phys. Rev. Lett. **73,** pp. 2440-2443, 1994.

25. S. Arnold, "Cavity-enhanced fluorescence decay rates from microdroplets", J. Chem. Phys. **106,** pp. 8280–8282, 1997.

26. M. Inoue, K. Ohtaka, "Enhanced raman-scattering by two-dimensional array of polarizable spheres", J. Phys. Soc. Jpn. **52,** pp. 1457-1468, 1983.

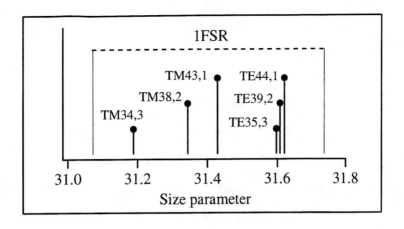

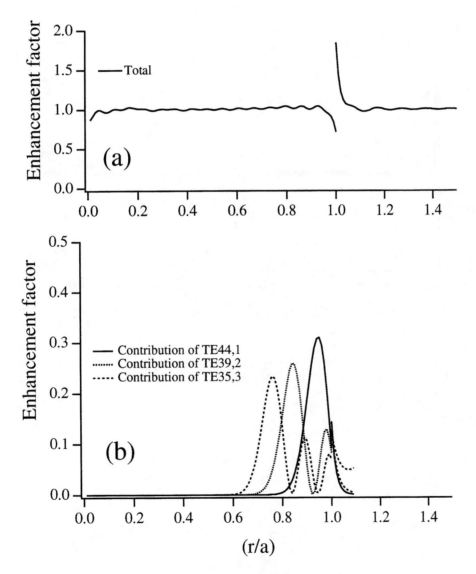

Figure 1. (a)The dependence of the enhancement factor on the position of the dye molecule within the 5 μm polystyrene sphere for several modes;(b)net enhancement factor after averaging over the FSR. Insert: FSR for the 5 μm polystyrene sphere.

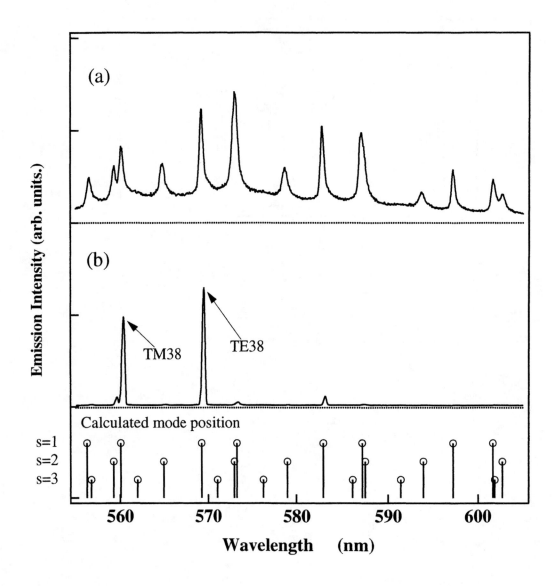

Figure 2. The emission spectrum of the of the dye doped 5 μm polystyrene sphere below (a) and above (b) laser threshold for WGM modes with $n=38$ and $s=1$. The calculated WGM positions for $s=1$, $s=2$ and $s=3$ are shown at the bottom of the figure.

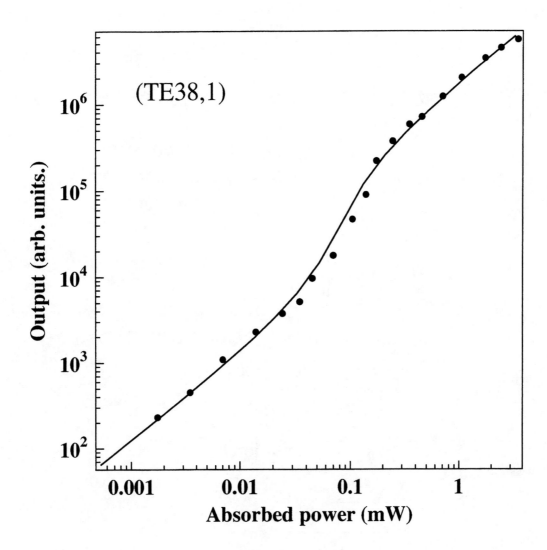

Figure 3. Dependence of the lasing output for TE38,s=1 mode of the 5 μ m microsphere on the absorbed pump power.

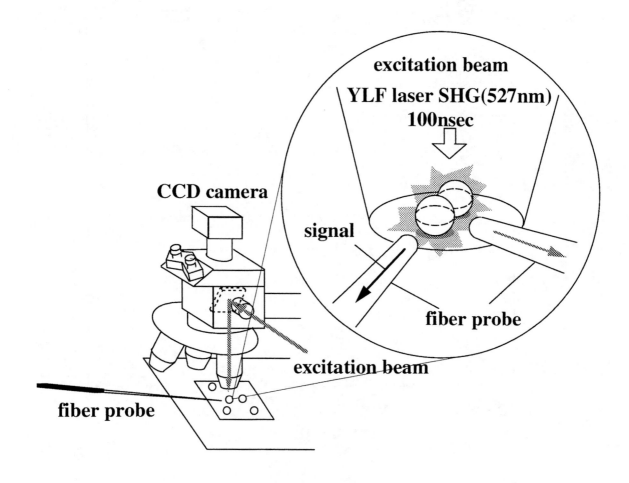

Figure 4. The experimental setup for observation fluorescence from the dye doped microspheres.

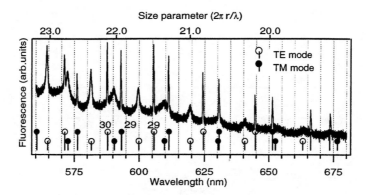

Figure 5. Typical fluorescence spectrum from 4.1 μm sphere. The mode assignment is shown at the bottom of the figure.

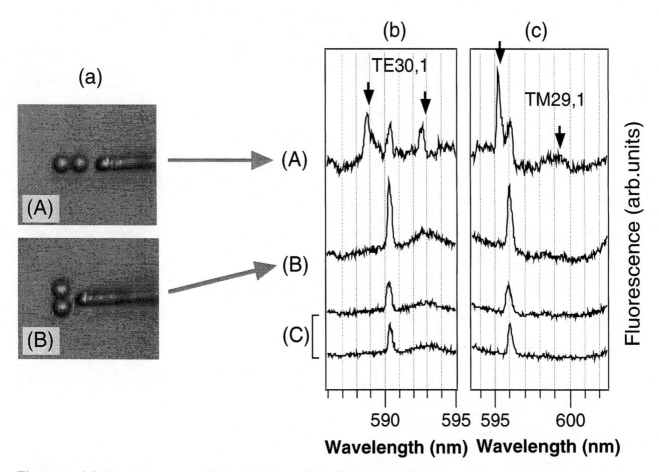

Figure 6. (a):Microscope images of the bisphere and the fiber probe. Diameter of the probe is about 5 to 10 μm. We detect the emission in two different geometries; one is parallel configuration (A), the other is perpendicular configuration(B). (b),(c):Spectra of resonant bisphere of TE30,1 mode and TM29,1 mode in parallel configuration (A) and perpendicular configuration (B). Spectra (C) show the fluorescence of individual spheres before contact. Two spheres have almost the same diameter. The arrows indicate the coupled modes.

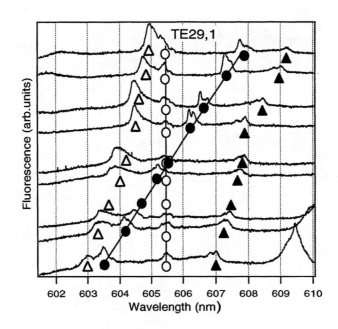

Figure 7. Emission spectra of nine bisphere with slightly different sized spheres. Fiber probe is set parallel to the axis of the bisphere. Open and closed circles indicate the uncoupled modes. Open and closed squares show the coupled modes of bisphere.

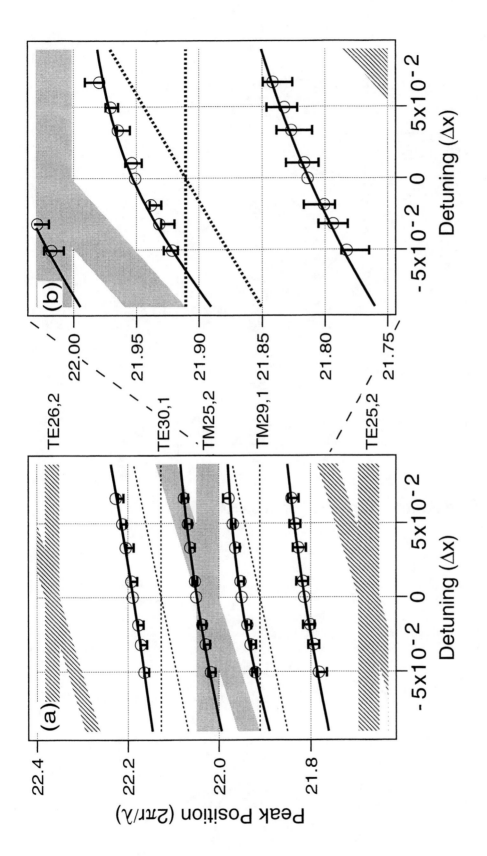

Figure 8. Resonance frequencies of coupled TE30,1 and TM29,1 for various bisphere with different size detuning. Dashed lines indicate the resonances of uncoupled modes. Large open circles show the result of wave optics calculation. Solid lines show the calculated normal modes of bisphere as a function of size detuning using the coupled harmonic oscillator model. Coupling constants are shown in Table 1, 2.

183

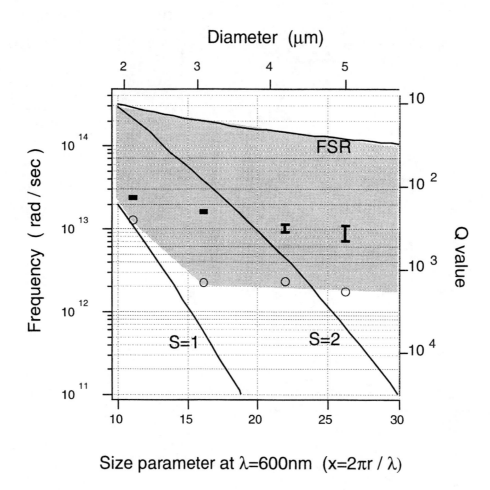

Figure 9. The size dependence of the coupling constants and Q-factors of the microspheres. Open circles show the measured Q-factors of the microspheres, while the inverse coupling constants are shown by solid squares and experimental bars. Dash line represents the Q-factor given by the Mie theory for $s = 1$ and $s = 2$ modes. The coherent coupling regime is possible within the shaded area, which lower boundary is given by the free spectral range (FSR).

Table 1. Best fit values of coupling parameters (TE modes) normalized by resonance frequency.

	TE26,s=2	TE30,s=1	TE25,s=2
TE26,s=2	3.4×10^{-3}	3.4×10^{-3}	-
TE30,s=1	3.4×10^{-3}	3.3×10^{-3}	3.4×10^{-3}
TE25,s=2	-	3.4×10^{-3}	3.4×10^{-3}

Table 2. Best fit values of coupling parameters (TM modes) normalized by resonance frequency.

	TM25,s=2	TM29,s=1	TM24,s=2
TM25,s=2	2.8×10^{-3}	3.5×10^{-3}	-
TM29,s=1	3.5×10^{-3}	3.3×10^{-3}	3.5×10^{-3}
TM24,s=2	-	3.5×10^{-3}	2.8×10^{-3}

Evanescent-wave sensor using microsphere whispering-gallery modes

A. T. Rosenberger[*] and J. P. Rezac

Department of Physics and Center for Laser and Photonics Research,
Oklahoma State University, 145 Physical Sciences, Stillwater, OK 74078-3072

ABSTRACT

The high Q of a microsphere whispering-gallery mode allows for sensitive resonant detection of atoms or molecules. The species being detected absorbs energy from the mode's evanescent field. It can be identified by knowing the resonant wavelength of the driving laser, and its concentration can be determined from the absorption signal on the light in reflection or transmission. High sensitivity results from the long effective absorption path length provided by the whispering-gallery mode's large Q. There are many possible implementations of and applications for such a sensor; several of each are described herein. In particular, for atmospheric trace-gas sensing, the microsphere has the potential to rival the performance of the multipass cell, but in a much more compact and rugged system. Our construction of a prototype system for detection of carbon monoxide, carbon dioxide, and ammonia is described.

Keywords: sensor, evanescent waves, microspheres, whispering-gallery modes

1. INTRODUCTION

The development of a novel optical sensor, using an individual dielectric microsphere as an optical evanescent-wave sensor, is currently underway in our laboratory. The whispering-gallery modes (WGMs) of these microspheres have high quality factors (Q), making possible microsensors orders of magnitude more sensitive than ordinary evanescent-wave sensors. Such microsensors will have many uses, and in particular will be able to detect environmentally important trace gases with a sensitivity equivalent to that of a typical multipass cell. In contrast to the multipass cell, the WGM microsensor is quite compact, using a sphere less than a millimeter in diameter. The proof-of-concept is being demonstrated by the detection of carbon monoxide, carbon dioxide, and ammonia using absorption on ro-vibrational overtone and combination-band transitions in the 1.51 - 1.58 μm wavelength region. The optics in this spectral region is well understood, so the microspheres can be made from standard optical fiber, and a commercial tunable diode laser is used as the light source. The molecular absorptions involved are relatively weak, yet the sensitivity is expected to be approximately a hundred parts per million for CO, CO_2, and NH_3. This corresponds to the level detected by a household CO sensor, and is about one-third of the atmospheric CO_2 concentration.

The sensitivities quoted above are for simple direct absorption measurements; using wavelength-modulation spectroscopy (WMS), which can easily be implemented with these microsensors, two or three orders of magnitude further improvement in sensitivity can be expected. WMS entails frequency modulation of the laser, at a modulation frequency small compared to the molecular absorption linewidth, and phase-sensitive detection. This technique will increase the sensitivity to such a degree that very compact commercial trace-gas sensors using inexpensive, rugged, and reliable semiconductor diode lasers may be developed using current technology. When the WGM microsensor technique has been established in the 1.51 - 1.58 μm spectral region, we will extend it to use the stronger transitions of these and many other molecules in the mid-infrared (2 - 8 μm), where sensitivities are expected to be about a part per billion for direct absorption and nearly a part per trillion using WMS.

---

[*] Correspondence: E-mail: atr@okstate.edu; Telephone: 405-744-6742; Fax: 405-744-6811

In *Laser Resonators III*, Alexis V. Kudryashov, Alan H. Paxton, Editors,
Proceedings of SPIE Vol. 3930 (2000) • 0277-786X/00/$15.00

A description of the usage of microspheres as WGM evanescent-wave microsensors is given in the next Section. In Section 3, particular aspects of trace-gas sensing in the 1.51 - 1.58 µm region are discussed. A summary and projection to future applications is provided in Section 4.

2. EVANESCENT-WAVE MICROSENSOR

2.1. Whispering-gallery modes

The so-called "whispering-gallery" modes of transparent dielectric spheres have extremely low losses, allowing such spheres to be used as microresonators with very high quality factor Q, as was pointed out several years ago by Braginsky, Gorodetsky, and Ilchenko.[1] Their report has motivated much recent work, in particular studies of the use of whispering-gallery microresonators for precision measurement of small displacements[2,3] and for single-molecule excitation and detection.[4-6] A whispering-gallery mode is essentially the limiting case of propagation, by total internal reflection, around a great circle of the microsphere, as the number of reflections becomes very large. Coupling of laser light into and out of such a mode can be accomplished through its evanescent component (that part of the mode outside the sphere's surface), by frustrating the total internal reflection at a point on the sphere. This input and output coupling can be done using internal reflection in a prism whose surface is brought close to the microsphere,[1,7-10] as shown in Fig. 1, or by overlapping the microsphere's evanescent field with that of an eroded or tapered single-mode optical fiber,[11-16] as shown in Fig. 2. For efficient coupling, the refractive index of the prism (or fiber) must be larger than that of the microsphere; light incident in the prism, at the critical angle of the prism-microsphere interface, will then propagate in the microsphere tangent to its surface, as desired. By matching the size of the incident beam's waist at the prism's reflecting surface, or by matching the cross-section of the fiber mode, to the cross-section of the WGM, the coupling can be made quite efficient, in the same sense as mode-matching to an ordinary optical resonator. The high Q of the microsphere means that a single atom or molecule interacting with a whispering-gallery mode can have a significant effect on the energy in that mode; that is, the microsphere's resonance makes it much more sensitive than a conventional integrated-optical evanescent-wave sensor.[17]

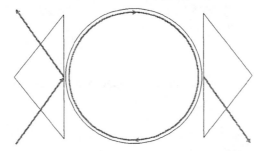

Fig. 1. Input and output coupling to a whispering-gallery-mode, using prisms.

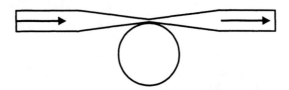

Fig. 2. Configuration for tapered-fiber coupling.

2.2. Estimate of sensitivity

The sensitivity of a whispering-gallery microsensor can be estimated by modeling the microsphere-couplers system as a unidirectional ring resonator, where each prism or fiber coupling is treated as a partially-transmitting mirror of reflectivity r^2. This reflectivity can be varied by changing the specifics of the coupling, such as the prism-sphere separation. In this simple model, r is assumed to be real and the incident light is assumed to have a frequency resonant with a lowest-order equatorial WGM. The intrinsic Q of the WGM (in the undercoupled limit where coupling losses are negligible), is determined by contributions from several types of loss,[7,8] but for the purposes of this model Q can be attributed to an intrinsic loss specified by the effective absorption coefficient α_i. Then Q is given by

$$Q = \frac{2\pi n_{eff}}{\alpha_i \lambda}, \tag{1}$$

where n_{eff} is the effective index of refraction of the sphere for the WGM being used and λ is the wavelength of the incident light. We further assume that the frequency of the laser coupled to the WGM is tuned to a homogeneously broadened resonant absorption of the molecule that we wish to detect. If molecules are present in the evanescent part of the mode, the absorption coefficient will increase to $\alpha = \alpha_i + \alpha_m$, where α_m is the intensity-weighted average of the molecular absorption coefficient over the mode volume:

$$\alpha_m = \frac{\lambda^2 N_r}{4\pi^2 \tau_{sp} \Delta\nu} f = \frac{2S}{\pi\Delta\nu} N f, \tag{2}$$

in which N_r is the density of resonant molecules, τ_{sp} is their spontaneous lifetime, $\Delta\nu$ is their absorption linewidth, f is the evanescent volume fraction of the WGM (typically about 10%), S is the spectral line intensity of the resonant transition, and N is the total molecular density. Tuning the laser through the molecular absorption resonance, while keeping it resonant with the WGM, then gives a detection signal in the form of an intensity change in the transmitted or reflected light.

If a single prism or fiber is used for coupling, and the reflected intensity is detected, the reflected intensity fraction is given by

$$R = \left(\frac{1 - x + s}{1 + x + s} \right)^2; \tag{3}$$

on the other hand, if two prisms or fibers are used and the transmitted intensity is detected, the transmitted intensity fraction is given by

$$T = \left(\frac{2x}{1 + 2x + s} \right)^2. \tag{4}$$

In Eqs. (3) and (4), x denotes the single-coupler relative coupling loss,

$$x = \frac{1 - r^2}{\alpha_i L}, \tag{5}$$

where L is the circumference of the sphere, and s denotes the relative molecular absorption loss,

$$s = \frac{\alpha_m}{\alpha_i}. \tag{6}$$

For low molecular density ($s \ll 1$), either T or $1-R$ can be written as proportional to the factor $\exp(-\alpha_m L_{eff})$. Here, L_{eff} is the effective absorption path length, which can be as large as $2/\alpha_i$ in the undercoupled case ($x \ll 1$), or tens of meters for Q on the order of 10^8. A typical direct absorption measurement such as described here has a sensitivity limit of about one part in 10^4 change in T or $1-R$, that is, $\alpha_m L_{eff} = 10^{-4}$; from this, the sensitivity can be calculated using the model above. Explicit examples will be given in Section 3.

2.3. Implementation

The microspheres that we use are 50-500 μm in diameter; we fabricate them from low-OH fused silica fibers by melting an end with a hydrogen-oxygen minitorch or a CO_2 laser. Each microsphere is left attached to its fiber stem, which can be used for manipulation. Light from a stabilized, tunable laser is coupled into whispering-gallery modes of the sphere using prisms (see Fig. 1). Use of prisms allows us to compare reflection and transmission responses, in contrast to using a bi-tapered fiber as in Fig. 2, which allows detection of reflection only. Because tapered-fiber coupling is envisioned for use in the ultimate version of the sensor, we are also developing a method for using two fibers tapered to points by an acid etch (as an alternative to a precision angle cut[16]), also allowing comparison of reflection and transmission. The light input and output (transmitted or reflected) are compared by a balanced receiver, and the resulting receiver signal is a measure of the strength of the molecular absorption and therefore of the concentration. The microsphere must be kept in resonance with the laser as the wavelength is scanned across the molecular absorption line; the ratio of signals on and off molecular resonance gives us the molecular concentration. To keep the microsphere resonant, it is mounted in a device (shown in Fig. 3) that squeezes the sphere at its poles, leaving the equatorial region free for optical coupling.[18] Squeezing the sphere allows for tuning through a free spectral range and means that the sphere can be kept in resonance as the laser is tuned, by superimposing a small-amplitude kilohertz modulation on the sphere's optical output using the PZT that drives the strain-tunable mounting, detecting this with a stabilization unit, and feeding back to the dc level of the PZT. (An alternative method is to frequency-modulate the laser with a modulation amplitude of a few MHz, scanning back and forth across the WGM resonance, and detecting the second harmonic of this to feed an error signal back to the PZT.) The mounting device also allows us to control the sphere's temperature, to keep it constant at a level somewhat above ambient, thus keeping the sphere's surface free of condensation. The sensor is characterized by using it in an airtight, variable-pressure enclosure in which the concentrations of CO, CO_2, and NH_3 can be varied.

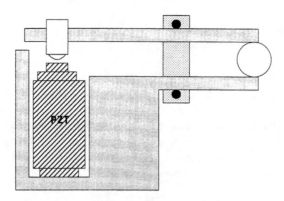

Fig. 3. Mounting device for tuning microsphere resonance frequency.

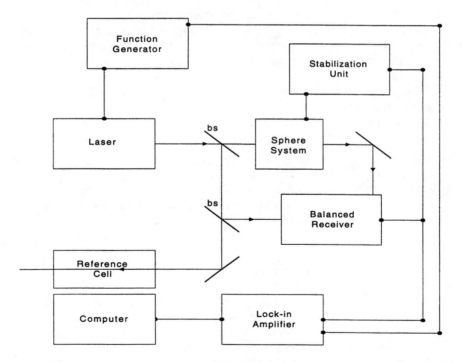

Fig. 4. Block diagram.

In a manner similar to that described above, a slower modulation of larger amplitude (a few GHZ, comparable to the pressure-broadened linewidth at atmospheric pressure) is imposed on the laser frequency, scanning back and forth across the molecular resonance, and then the derivative of the signal is detected. This constitutes wavelength modulation spectroscopy (WMS).[19-22] The operation of the system can be understood in terms of the block diagram in Fig. 4. The microsphere system includes the coupling optics and the gas enclosure; laser light is sent in, part of it being split off by a beamsplitter (bs) and sent to a reference cell containing pure gas, so we can tell when the laser is resonant with the desired molecule. Light coming out of the microsphere (we show only a single output, either reflection or transmission) is sent to a balanced receiver where it is compared with the input light; here, any input intensity variation is zeroed out, so that the variation remaining is a result of the microsphere-gas system interacting with the input laser light. The receiver signal goes to a stabilization unit, where the rapidly-varying component of the signal (representing the microsphere-light interaction) is used to feed back to the strain-tuner's PZT and keep the microsphere resonant with the slowly-wavelength-modulated light. The function generator causes the laser wavelength modulation, and the lock-in amplifier phase-sensitively detects this slower modulation, thereby extracting the gas-light interaction signal. This signal is then sent to an oscilloscope for observation, or to the computer for storage and analysis.

3. TRACE-GAS DETECTION

We are using a stable, tunable diode laser for the detection of trace amounts of CO, CO_2, and NH_3 in the 1.51 - 1.58 μm wavelength region, where we may use data from the HITRAN database to predict the sensitivity of the sensor in direct absorption meaurements.[23] Using WMS techniques, the sensitivity will be two or three orders of magnitude better. In this example we consider absorption on the 3ν overtone band of CO and the $2\nu_1+2\nu_2+\nu_3$ combination band of CO_2, which overlap near 1.57 μm,[19] as well as the overtone band of NH_3 near 1.53 μm.[20] For these transitions, the spectral line intensity per molecule is typically $S = 2.0\times10^{-23}$ cm^{-1}/(molecule·cm^{-2}), and the pressure-broadened linewidth is typically $\Delta\nu = 0.14$ cm^{-1}.[23] Assuming that $Q = 10^8$ and $n_{eff} = 1.40$,[16] the effective absorption length is

$$L_{eff} = \frac{2}{\alpha_i} = \frac{\lambda Q}{\pi n_{eff}} = 36 \text{ m}. \tag{7}$$

Note that this value depends on the size of the microsphere only through the dependence of n_{eff}.[16] The only other size-dependent factor that enters into the calculation is the evanescent volume fraction f. For the wavelength being used here, and for a sphere of about 200 μm diameter, a reasonable estimate for the volume fraction is $f = 0.10$; the value will be somewhat less than this for TE modes, and somewhat greater than this for TM modes.[1] It is desirable to use as small a sphere as practical, because L_{eff} and f both increase with decreasing sphere diameter (for a given wavelength).

Now, using the conservative direct-absorption sensitivity limit of $\alpha_m L_{eff} = 10^{-4}$ noted earlier, we estimate the minimum detectable molecular absorption coefficient to be $\alpha_m = 2.8\times10^{-8}$ cm^{-1}. For comparison, we can use Eq. (2) to write, assuming a concentration of one part per billion (ppb) in the atmosphere,

$$\alpha_m = 1.2\times10^{10} S \quad \text{(at 1 ppb)}, \tag{8}$$

where S in cm^{-1}/(molecule·cm^{-2}) gives α_m in cm^{-1}. The typical value of S then gives α_m(ppb) $= 2.4\times10^{-13}$ cm^{-1}. Thus our sensitivity is about 120 ppm (parts per million), which is approximately one-third the concentration of CO_2 in the atmosphere. Using WMS, we expect a sensitivity on the order of one ppm; this is nearly sufficient to detect typical atmospheric CO concentrations. Work is in progress in our laboratory to demonstrate the trace-gas evanescent-wave sensor application of microsphere WGMs using the three gases considered above.

4. SUMMARY AND OUTLOOK

In this work, we apply the techniques of wavelength modulation spectroscopy (WMS), in which a (relatively) slow modulation is imposed on the laser wavelength, scanning it back and forth across the molecular resonance; we then detect the derivative of the signal. WMS can improve the sensitivity by another two to three orders of magnitude,[19-22] to as much as one part in 10^7 (100 ppb) in the 1.5 to 1.6 μm wavelength range. This is very good sensitivity and will allow devices to be made, for commercial use, that incorporate standard semiconductor diode lasers. For research purposes, it will eventually be possible to do even better. The next step will be to extend the technique into the mid-infrared, making use of new tunable semiconductor lasers currently being perfected. In this spectral region, comprising wavelengths from 2 to 8 μm, another four orders of magnitude in sensitivity can be gained because of the much greater strength of fundamental (rather than overtone or combination-band) ro-vibrational transitions,[24-26] which have typical spectral line intensities of approximately $S = 4.0\times10^{-19}$ cm^{-1}/(molecule·cm^{-2}). The optics of this spectral region is not as well understood, so the microsphere fabrication and coupling will need to be developed, using new types of mid-IR-transmitting glasses.[27-29] Eventually, it is possible that WMS sensitivities better than a part per trillion will be achievable using these microsensors for certain strongly-absorbing atmospheric and biogenic trace gases (e.g., CO, CO_2, NH_3, CH_4, NO, N_2O, C_2H_6, H_2S, H_2CO, SO_2, HF, HCl, etc.). The microsensor has the potential to replace not only multipass cells but also other types of detector. Thus a much more compact sensor head can be engineered, at a substantially lower cost, and with less critical alignment tolerances than in present sensors.

Trace-gas detection is but one of many different types of potential commercial application of these sensors, but is itself a rather broad area with multiple possibilities, as all types of environmentally important trace gases lend themselves to detection by this method. For example, these sensors could be used in industry, in transportation, and in agriculture. In manufacturing and other industries, the sensors could detect leaks and/or be incorporated into process control systems; they could be used for monitoring of carbon monoxide levels at roadsides and in parking structures; and they could be used in measurements of emissions from soil, wetlands, crops, and livestock, including diagnostic veterinary applications. With subsequent development, and given recent progress in tunable laser sources, it is likely that within several years compact, rugged, and inexpensive sensors will be available for use in field instruments (the final configuration of the complete, self-contained sensor unit could be as compact as a typical pen-sized laser pointer). These novel sensors will permit precise monitoring of trace gases that affect the environment: CO_2 and other greenhouse gases, biogenic emissions from flora and fauna, industrial pollution, coastal and oceanic carbon and nitrogen compounds, and geothermal and volcanic emissions, for example. The data provided by networks of these sensors will aid in improving the modeling of ecosystems ranging from the very small to the very large, and will thus enhance the reliability of scientific assessments of the consequences of local and global environmental change.

ACKNOWLEDGEMENTS

We are indebted to Prof. Donna K. Bandy, of the Department of Physics, her graduate student Tigran V. Sarkisyan, and their collaborator, Prof. Anatoly N. Oraevsky of the Lebedev Physics Institute in Moscow, who have supplied encouragement and theoretical support. We have also benefited from consultation with Jeff Kimble, Leo Hollberg, Vladimir Velichansky, and Vladimir Ilchenko. This research is supported in part by grants from the Center for Sensors and Sensor Technologies and from the Energy Research Center of the Environmental Institute.

REFERENCES

1. V. B. Braginsky, M. L. Gorodetsky, and V. S. Ilchenko, "Quality factor and nonlinear properties of optical whispering-gallery modes," *Phys. Lett. A* **137**, pp. 393-397, 1989.
2. V. B. Braginsky, M. L. Gorodetsky, V. S. Ilchenko, and S. P. Vyatchanin, "On the ultimate sensitivity in coordinate measurements," Phys. Lett. A **179**, pp. 244-248, 1993.
3. V. S. Ilchenko, M. L. Gorodetsky, and S. P. Vyatchanin, "Coupling and tunability of optical whispering-gallery modes: a basis for coordinate meter," *Opt. Commun.* **107**, pp. 41-48, 1994.
4. D. J. Norris, M. Kuwata-Gonokami, and W. E. Moerner, "Excitation of a single molecule on the surface of a spherical microcavity," *Appl. Phys. Lett.* **71**, pp. 297-299, 1997.
5. M. D. Barnes, N. Lermer, C.-Y. Kung, W. B. Whitten, J. M. Ramsey, and S. C. Hill, "Real-time observation of single-molecule fluorescence in microdroplet streams," Opt. Lett. **22**, pp. 1265-1267, 1997.
6. S. Arnold, S. Holler, N. L. Goddard, and G. Griffel, "Cavity-mode selection in spontaneous emission from oriented molecules in a microparticle," Opt. Lett. **22**, pp. 1452-1454, 1997.
7. M. L. Gorodetsky and V. S. Ilchenko, "High-Q optical whispering-gallery microresonators: precession approach for spherical mode analysis and emission patterns with prism couplers," *Opt. Commun.* **113**, pp. 133-143, 1994.
8. S. P. Vyatchanin, M. L. Gorodetski, and V. S. Il'chenko, "Tunable narrow-band optical filters with modes of the whispering gallery type," *J. Appl. Spectrosc.* **56**, pp. 182-187, 1992.
9. S. Schiller and R. L. Byer, "High-resolution spectroscopy of whispering gallery modes in large dielectric spheres," *Opt. Lett.* **16**, pp. 1138-1140, 1991.
10. M. L. Gorodetsky and V. S. Ilchenko, "Optical microsphere resonators: optimal coupling to high-Q whispering-gallery modes," *J. Opt. Soc. Am. B* **16**, pp. 147-154, 1999.

11. N. Dubreuil, J. C. Knight, D. K. Leventhal, V. Sandoghdar, J. Hare, and V. Lefevre, "Eroded monomode optical fiber for whispering-gallery mode excitation in fused-silica microspheres," *Opt. Lett.* **20**, pp. 813-815, 1995.

12. G. Griffel, S. Arnold, D. Taskent, A. Serpengüzel, J. Connolly, and N. Morris, "Morphology-dependent resonances of a microsphere–optical fiber system," *Opt. Lett.* **21**, pp. 695-697, 1996.

13. A. Serpengüzel, S. Arnold, and G. Griffel, "Excitation of resonances of microspheres on an optical fiber," *Opt. Lett.* **20**, pp. 654-656, 1995.

14. A. Serpengüzel, S. Arnold, G. Griffel, and J. A. Lock, "Enhanced coupling to microsphere resonances with optical fibers," *J. Opt. Soc. Am. B* **14**, pp. 790-795, 1997.

15. J. C. Knight, G. Cheung, F. Jacques, and T. A. Birks, "Phase-matched excitation of whispering-gallery-mode resonances by a fiber taper," *Opt. Lett.* **22**, pp. 1129-1131, 1997.

16. V. S. Ilchenko, X. S. Yao, and L. Maleki, "Pigtailing the high-Q microsphere cavity: a simple fiber coupler for optical whispering-gallery modes," Opt. Lett. **24**, pp. 723-725, 1999.

17. E. F. Schipper, R. P. H. Kooyman, A. Borreman, and J. Greve, "The Critical Sensor: A New Type of Evanescent-Wave Immunosensor," Biosensors & Bioelectronics **11**, pp. 295-304, 1996, and references therein.

18. V. S. Ilchenko, P. S. Volikov, V. L. Velichansky, F. Treussart, V. Lefèvre-Seguin, J.-M. Raimond, and S. Haroche, "Strain-tunable high-Q optical microsphere resonator," *Opt. Commun.* **145**, pp. 86-90, 1998.

19. M. Gabrysch, C. Corsi, F. S. Pavone, and M. Inguscio, "Simultaneous detection of CO and CO_2 using a semiconductor DFB diode laser at 1.578 µm," Appl. Phys. B **65**, pp. 75-79, 1997.

20. M. Fehér, Y. Jiang, J. P. Maier, and A. Miklós, "Optoacoustic trace-gas monitoring with near-infrared diode lasers," Appl. Opt. **33**, pp. 1655-1658, 1994.

21. K. Uehara, "Dependence of harmonic signals on sample-gas parameters in wavelength-modulation spectroscopy for precise absorption measurements," Appl. Phys. B **67**, pp. 517-523, 1998.

22. D. S. Bomse, A. C. Stanton, and J. A. Silver, "Frequency modulation and wavelength modulation spectroscopies: comparison of experimental methods using a lead-salt diode laser," Appl. Opt. **31**, pp. 718-731, 1992.

23. L. S. Rothman, C. P. Rinsland, A. Goldman, et al., "The HITRAN molecular spectroscopic database and HAWKS (HITRAN Atmospheric Workstation): 1996 edition," J. Quant. Spect. and Rad. Transfer **60**, pp. 665-710, 1998.

24. T. Töpfer, K. P. Petrov, Y. Mine, D. Jundt, R. F. Curl, and F. K. Tittel, "Room-temperature mid-infrared laser sensor for trace gas detection," Appl. Opt. **36**, pp. 8042-8049, 1997.

25. M. Fehér and P. A. Martin, "Tunable diode laser monitoring of atmospheric trace gas constituents," Spectrochimica Acta A **51**, pp. 1579-1599, 1995.

26. P. Werle, "A review of recent advances in semiconductor laser based gas monitors," Spectrochimica Acta A **54**, pp. 197-236, 1998.

27. J. Wasylak, M. Laczka, and J. Kucharski, "Glass of high refractive index for optics and optical fiber," Opt. Eng. **36**, pp. 1648-1651, 1997.

28. J. Wasylak, "New glasses of shifted absorption edge in infrared as materials for optics and light fiber technique," Opt. Eng. **36**, pp. 1652-1656, 1997.

29. J. S. Sanghera and I. D. Aggarwal, "Development of chalcogenide glass fiber optics at NRL," J. Non-Cryst. Solids **213 & 214**, pp. 63-67, 1997.

The Application of Microresonators in Large Scale Optical Signal Processing Circuits

Brent E. Little

Massachusetts Institute of Technology, Cambridge MA 02139, and the Laboratory for Physical Sciences, University of Maryland, College Park MD 20742

Sai T. Chu

National Institute of Standards and Technology, Atomic Physics Division, Gaithersburg MD 20899

Abstract

Microring resonators are attractive for Very Large Scale Integrated Photonic Circuits (VLSI-PCs). They have been shown to be capable of many optical signal processing functions, and their small dimensions could lead to integration densities of 10,000 devices per square-cm. Here, the analytic theory of higher order mutually coupled resonators is derived. Experiments involving fabricated ring resonators and resonator arrays is described.

Keywords: microring, WDM, resonator

1. Introduction

Microring resonators have three key attributes which make them desirable for Very Large Scale Integrated Photonic Circuits (VLSI-PCs): (i) They provide numerous optical signal processing functions, including channel dropping filters[1-11], modulators, polarization rotators, phase/dispersion filters[12], true ON/OFF routing switches[13], notch filters[14], and for enhanced nonlinear effects, to name a few. (ii) The responses of the foregoing devices can always be improved by using multiple rings in coupled cavity arrangements[2,3,13]. (iii) With reported devices having radii of from 5 μm to 25 μm, device integration densities 10,000 devices per square-cm can be envisioned.

Microresonators require large refractive index contrasts in order to minimize curvature losses. To date, several high index contrast material systems have been investigated including Si/SiO_2[6], $GaAs/AlGaAs$[5,8,10], the compound glass Ta_2O_5/SiO_2[7,11], and polymer on Si_3N_4[9]. The bending loss in these high index contrast structures is negligible. The Q values are instead limited by sidewall roughness. Loaded Q values of approximately 500 to 15,000 have been measured. These correspond to linewidths of from 3 nm down to 0.1 nm, and are ideally suited to optical communications applications. Resonators can tolerate large losses. For example if ring losses could be reduced to below 10 dB/cm, Q values in excess of 50,000 could still be achieved. In the following, we describe the theory and fabrication of ring resonators and resonator arrays.

2. Resonator Response

A typical higher order ring resonator filter is depicted in Fig. 1. It is comprised of N rings each of radius R_n mutually coupled together. An input bus waveguide and a drop bus waveguide are coupled to the first and N^{th} ring respectively. At resonance, an input signal of amplitude s_i can be partially, or fully, transferred to the drop port (with amplitude s_d). Off resonance, most of the input signal will bypass the rings and appear at the through port (with amplitude s_t). A new signal (s_a) can be injected into the input stream by applying it to the add port. In the following analysis, the waveguide signals s_i, s_d, s_t, and s_a are assumed to be power normalized amplitudes. The optical bandwidth of the power transfer is largely determined by the device size and the interaction strength of the rings and bus waveguides. The response shape is determined by the mutual coupling between adjacent rings.

In *Laser Resonators III*, Alexis V. Kudryashov, Alan H. Paxton, Editors,
Proceedings of SPIE Vol. 3930 (2000) • 0277-786X/00/$15.00

193

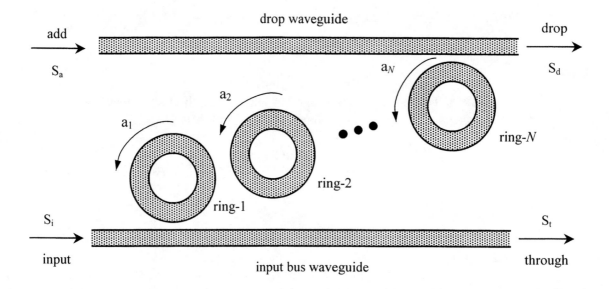

Fig. 1. Higher order microring resonator filter comprised of N mutually coupled rings.

Although rings and disks appear to be ideally suited as optical resonators due to the low intrinsic loss of the whispering gallery modes, we consider general resonator shapes by treating all resonators as lumped elements. This is accomplished by assigning to each resonator and amplitude a_n, where $|a_n|^2$ corresponds to the total energy stored in resonator-n. The response of resonator–1 (or ring-1) in a filter comprised of N resonators can be cast as the continued fraction[2]

$$a_1 = \cfrac{-j\mu s_i}{j\Delta\omega_1 + \cfrac{\mu_1^2}{j\Delta\omega_2 + \cfrac{\mu_2^2}{j\Delta\omega_3 + \cdots \cfrac{\mu_{N-1}^2}{j\Delta\omega_N}}}} \qquad (1a)$$

$$\Delta\omega_n = \begin{cases} \omega - \omega_n - j\dfrac{1}{\tau_n}, & n \neq 1, N \\[2ex] \omega - \omega_n - j\dfrac{1}{\tau_n} - j\dfrac{1}{\tau_{e,d}}, & n = 1, N \\[2ex] \omega - \omega_n - j\dfrac{1}{\tau_n} - j\dfrac{2}{\tau_e}, & N = 1 \end{cases} \qquad (1b)$$

The optical frequency of the input is ω, ω_n is the resonant frequency of resonator-n, and $\Delta\omega_n$ measures the complex frequency deviation. The terms $1/\tau_{e,d}$ and $1/\tau_n$ ($n = 1...N$) are energy decay rates, while the terms μ and μ_n ($n = 1...N$-1) are energy coupling coefficients. The mutual coupling of resonator-1 with the input waveguide leads to an external energy decay rate $2/\tau_e$ into the signal s_t. The mutual coupling of resonator-N with the drop waveguide leads to an external energy decay rate $2/\tau_d$ into the signal s_d. Absorption in resonator-n leads to an energy decay rate of $2/\tau_n$. The term μ describes the coupling between resonator-1 and the input bus. The term μ_n describes the coupling between resonator-$(n+1)$ and resonator-n. These coefficients are referred to an energy normalization, but for the case of waveguide ring resonators, they can be given in terms of the usual spatial coupling coefficients for mutually coupled waveguides[2]. The response of ring-N can be shown to assume a product of continued fractions[13]

$$a_N = s_i \prod_{n=1}^{N} T_N \qquad (2a)$$

$$T_N = \cfrac{-j\mu_{n-1}}{j\Delta\omega_n + \cfrac{\mu_{n-1}^2}{j\Delta\omega_{n-1} + \cfrac{\mu_{n-2}^2}{j\Delta\omega_{n-2}\cdots + \cfrac{\mu_1^2}{j\Delta\omega_1}}}}, \qquad n \neq 1 \qquad (2b)$$

$$T_1 = \frac{-j\mu}{j\Delta\omega_1} \qquad (2c)$$

The throughput signal is given by

$$s_t = s_i - j\mu a_1 \qquad (3a)$$

while the drop signal is given by

$$s_d = j\mu_d a_N \qquad (3b)$$

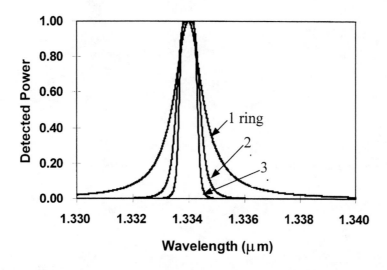

Fig. 2. Comparison of Maximally Flat filter response achieved by using 1, 2, and 3 rings.

Desirable filter shapes such as Maximally Flat or Chebyshev, are obtained by choosing the coupling coefficients μ and μ_n appropriately[2]. Figure 2 for instance, compares Maximally Flat response for filters comprised of 1, 2, and 3 rings. As the number of resonators increases, the filter shape becomes more box like.

3. Vertical Coupling

Most ring resonator configurations make use of lateral coupling, wherein the bus waveguides and ring are in the same lateral plane, (as depicted in Fig. 1). Typically, the coupling gaps between the rings and bus guides in this configuration are on the order of 100 nm to 300 nm. These coupling gaps determine the details of the filter response. Unfortunately, the realization of such small dimensions requires e-beam lithography, and the exact dimensions are difficult to control. Vertical coupling as depicted in Fig. 3a, offers several advantages over lateral coupling. Foremost among these is that the ring/bus interaction can be controlled to a finer degree, as the vertical separation is obtained by well-controlled deposition, rather than etching fine gaps. Further, buried guides suffer less scattering loss and offer better input/output coupling. Finally, if the guides are positioned directly below the ring, the interaction strength becomes insensitive to small alignment deviations, (the coupling strength is stationary).

Vertically coupled ring resonators have been fabricated by us at the Kanagawa Academy of Science and Technology in Japan[7,11,15-18]. A schematic cross section of the ring and buried waveguide, including dimensions, is depicted in Fig. 3b. The waveguides are fabricated from the compound glass Ta_2O_5/SiO_2, with the ratio of the two adjusted to achieve a refractive index of 1.653. Typical responses measured at the drop port are shown in Fig. 4. The Free Spectral Range (FSR) for a ring radius of 20 μm is about 11 nm. Channel dropping bandwidths (BWs) of from 2 nm to 0.1 nm have also been achieved.

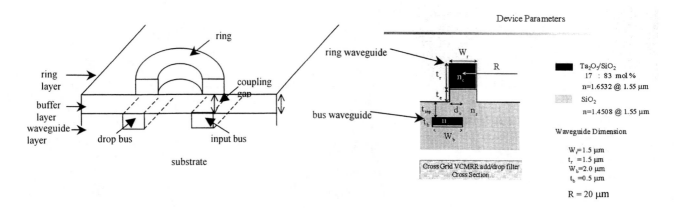

Fig 3. (a) A vertically coupled microring resonator (VCMRR) add/drop filter. (b) Cross sectional geometry of a typical fabricated device.

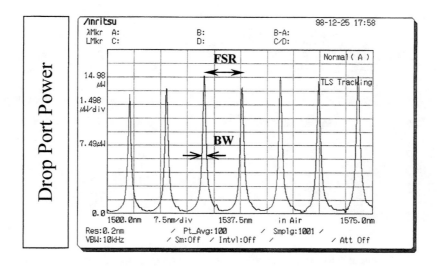

Fig. 4. Measured TM spectra at the drop port for the channel dropping filter with physical geometry depicted in Fig. 3b.

4. Cross-Grid Architecture

Owing to their small dimensions and versatility for optical signal processing applications, microring resonators are ideal candidates for Very Large Scale Integrated (VLSI) Photonics[17]. Vertical coupling is key to realizing general large scale architectures. In vertical coupling, the waveguide layer serves as the transport layer, interconneting the vertically integrated resonators. The two layers can be optimized independently, the bus layer for low loss propagation and low loss input/output fiber connections, while the ring resonator layer is optimized for tightly confined waveguides. Due to the lower index contrast of the buried bus waveguides, the guides can cross through one another with low scattering loss and low adjacent guide cross-talk. Fig. 5 shows a perspective schematic of a ring resonator vertically coupled above a pair of crossing waveguides. We call this a cross-grid node. The ring serves as a wavelength selective cross-connect, rerouting input signals at resonance to the drop port. Figure 6 shows implementations of second and third order filters based on the cross-grid node.

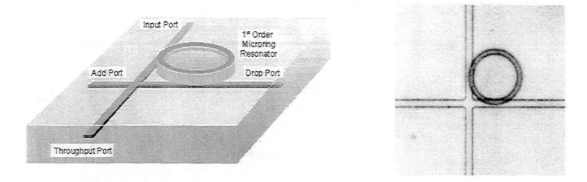

Fig. 5. Perspective schematic of a ring resonator vertically coupled above a pair of crossing waveguides. We call this a Cross Grid node. (b) Microscope image of a cross grid node incorporating a ring 10 μm in radius.

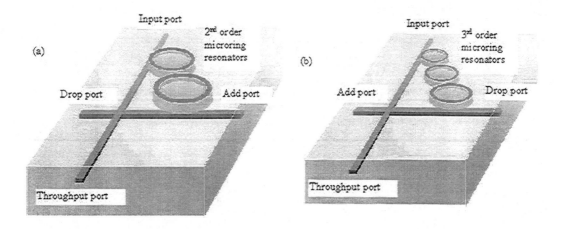

Fig. 6. Second and third order channel-drop filters based on the Cross-Grid node.

Larger scale architectures are built up from interconnections of cross-grid nodes. For instance, a linear array of cross-grid nodes, with one common bus waveguide and N crossing waveguides, leads to a compact 1xN WDM filter[11]. Figure 7a shows such a filter comprising 8 rings in this case. By adjusting the radius of each ring, different wavelength channels can be selected. Figure 7b shows the measured response in the TM polarization of all eight drop ports. Ring-1 has a radius of 10 μm, while the remaining rings increase in radius systematically by 50 nm. This 50 nm increment shifts the resonant wavelengths systematically by 5.7 nm.

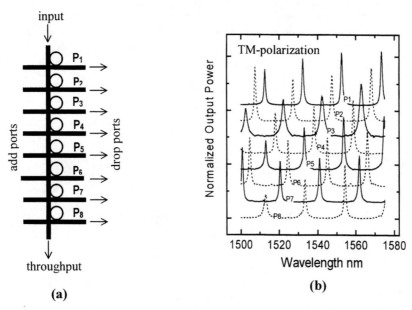

Fig 7 (a) A compact 1x8 add/drop filter. (b) Measured TM response at each port of the 1x8. The geometry of the individual rings is similar to that depicted in Fig 3.

The entire two-dimensional surface of a chip can be accessed by the cross-grid array depited schematically in Fig. 8. One may use crossing[18] or non-crossing[19] waveguides. An analytic formulation describing the complete scattering response of such two-dimensional arrays has been developed[18]. An 8x8 cross grid array of vertically coupled glass microring resonators has been fabricated for test purposes. A top-down schematic of the array is shown in Fig. 9a, while a photograph of part of the array appears in Fig. 9b. Fabrication details of similar devices can be found in Refs. [7,11]. The rings are nominally all 10 μm in radius, have a core refractive index of 1.6532, and a cladding index of 1.4508. The ring waveguide heights are 1.5 μm, and the widths vary from 1.0 μm to 1.7 μm in increments of 0.1 μm in such a way that along upward directed diagonals the rings are identical (see Fig. 9a). Spacing between bus waveguides is 250 μm. Two transmission measurements of this device are considered here, as shown by the labeled paths in Figs 9a. In one measurement, the cross-connect response of the upper left corner ring is measured. In the second, the signal is cross connected by the same ring, but travels the complete length of two sides of the array. In doing so, this second signal traverses sixteen junctions and interacts with fourteen additional rings. These two measurements are plotted in Figures 10a and 10b respectively. The signal that traverses the array periphery is attenuated due to the junction scattering loss and by partial extraction by other rings. The power scattering loss per junction is estimated experimentally at 15%, while the junction scattering into orthogonal waveguides is estimated at 1.5%. Scattering has been reduced to below -30dB experimentally in other similar devices[11]. The noise like ripple on the experimental data are due to the interference of multiple paths in the array.

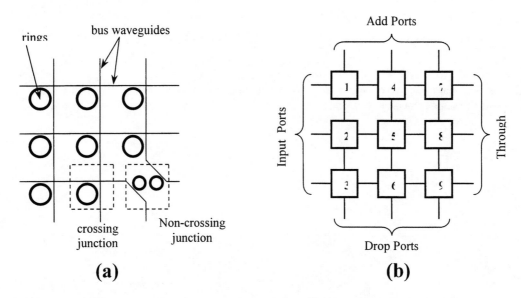

Fig. 8. (a) Cross-Grid array using crossing and non-crossing waveguides. (b) Functional description.

5. Conclusions

The feasibility of large scale arrays of microresonators has been investigated theoretically and experimentally. The resonator qualities (Q values from 500 to 15,000) are sufficient for practical applications. Applications calling for large scale integration are large channel count wavelength demultiplexers, and WDM cross connect switches.

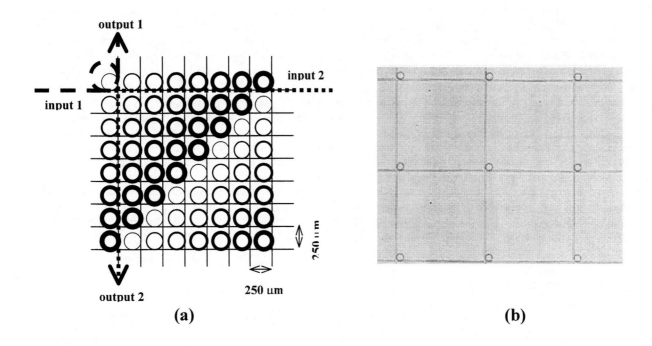

(a)

(b)

Fig. 9. (a) Schematic diagram of an 8x8 array of microring resonators. Rings with the same waveguide width are drawn with the same point size. (b) Micrograph of part of the 8x8 array.

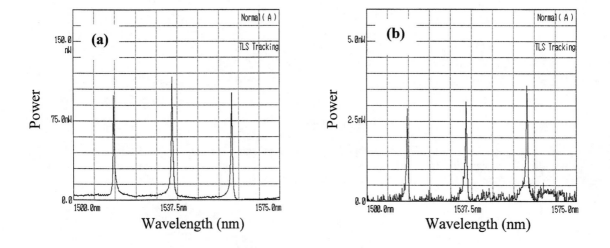

Fig. 10. Scattering response of the 8x8 microring resonator array in Fig. 9. (a) Measured response for path 1 in Fig. 9. (b) Measured response for path 2 in Fig. 9.

References

1. B. E. Little, S. T. Chu, and H. A. Haus, in LEOS 8th annual Meeting, Inst. Elec. Electron. Eng., NY, 1995, paper WDM 2.3.

2. B. E. Little, S. T. Chu, H. A. Haus, J. Foresi, and J.-P. Laine, "Microring resonator channel dropping filters," *IEEE J. Lightwave Tech.* **15**, pp. 998-1005, 1997.

3. R. Orta, P. Savi, R. Tascone, and D. Trinchero, "Synthesis of multiple-ring-resonator filters for optical systems," *IEEE Photonics Tech. Lett.* **7**, pp. 1447-1449, 1995.

4. M. K. Chin and S. T. Ho, "Design and modeling of waveguide-coupled single-mode microring resonators," *IEEE J. Lightwave Tech.* **16**, pp. 1433-1446, 1998.

5. D. Rafizadeh, J. P. Zhang, S. C. Hagness, A. Taflove, K. A. Stair, and S. T. Ho, "Waveguide-coupled AlGaAs/GaAs microcavity ring and disk resonators with high finesse and 21.6-nm free spectral range*," Opt. Lett.* **22**, pp. 1244-1246, 1997.

6. B. E. Little, J. Foresi, H. A. Haus, E. P. Ippen, and W. Greene, and S. T. Chu, "Ultra-compact Si/SiO$_2$ micro-ring resonator channel dropping filter," *Photonic. Tech. Lett.* **10**, pp. 549-551, 1998.

7. B. E. Little, S. T. Chu, W. Pan, D. Ripin, T. Kaneko, Y. Kokubun, E. Ippen, "Vertically coupled glass microring resonator channel dropping filters," *IEEE J. Photonic Tech. Lett.* **11**, pp. 215-217, 1999.

8. D. V. Tishinin, P. D. Dapkus, A. E. Bond, I. Kim, C. K. Lin, and J. O'Brien, "Vertical resonant couplers with precise coupling efficiency control fabricated by wafer bonding," *IEEE Photonics Tech. Lett.* **11**, pp. 1003-1005, 1999.

9. F. C. Bloom, H. Kelderman, H.J.W.M. Hoekstra, A. Driessen, Th. J. A. Popma, S. T. Chu, and B.E. Little, "A single channel dropping filter based on a cylindrical microresonator," Opt. Comm. **167**, pp. 77-82, 1999.

10. J. V. Hryniewicz, P. P. Absil, B. E. Little, R. A. Wilson, and P.-T. Ho, "Higher order filter response in coupled microring resonators," *IEEE Photonics Tech. Lett.* to be published March 2000.

11. S. T. Chu, B. E. Little, W. Pan, S. Sato, T. Kaneko, and Y. Kokubun, "An 8 channel add/drop filter using vertically coupled microring resonators over a cross-grid," *IEEE J. Photonic Tech. Lett.* **11**, pp. 691-693, 1999.

12. C. K. Madsen and G. Lenz, "Optical all-pass filters for phase response design with applications for dispersion compensations," *IEEE Photonics Tech. Lett.* **10**, pp. 994-996, 1998.

13. B. E. Little, H. A. Haus, J. S. Foresi, L. C. Kimerling, E. P. Ippen, and R. J. Ripin, "Wavelength switching and routing using absorption and resonance," *Photonic Tech. Lett.* **10**, pp. 816-818, 1998.

14. B. E. Little, S. T. Chu, H. A. Haus, "Second order filtering and sensing using partially coupled travelling waves in a single resonator," *Opt. Lett.* **23**, pp. 1570-1572, 1998.

15. S. T. Chu, W. Pan, S. Sato, T. Kaneko, Y. Kokubun, and B. E. Little, "Wavelength trimming of a microring resonator filter by means of a UV sensitive polymer overlay," *IEEE J. Photonic Tech. Lett.* **11**, pp. 688-690, 1999.

16. S. T. Chu, W. Pan, S. Suzuki, B. E. Little, S. Sato, and Y. Kokubun, "Temperature insensitive vertically coupled microring resonator add/drop filters by means of a polymer overlay*," IEEE J. Photonic Tech. Lett.* **11**, pp. 138-1140, 1999.

17. B. E. Little, S. T. Chu, W. Pan, and Y. Kokubun, "Microring resonator arrays for VLSI photonics," *IEEE Photonic Tech. Lett.* to be published March 2000.

18. S. T. Chu, B. E. Little, W. Pan, T. Kaneko, Y. Kokubun, "Cascaded microring resonators for cross-talk reduction and spectrum cleanup in add/drop filters," *IEEE J. Photonic Tech. Lett.* **11**, pp. 1423-1425, 1999.

19. R. Soref and B. E. Little, "Proposed N-wavelength M-fiber WDM crossconnect switch using active microring resonators," *Photonic Tech. Lett.* **10**, pp. 1121-1123, 1998.

SESSION 8

Fiber Lasers and Resonators

Experimental and numerical investigation of field propagation in multicore fibers

M. Wrage[1], P. Glas[1], M. Leitner[1], T. Sandrock[1], N. N. Elkin[2], A. P. Napartovich[2], A. G. Sukharev[2]

[1] Max-Born-Institut für Nichtlineare Optik und Ultrakurzzeitspektroskopie,Max-Born Straße 2a, D-12489 Berlin, Germany

[2] Troitsk Institute for Innovation and Fusion Research, Troitsk, 142092 Moscow region, Russia

ABSTRACT

Multicore fiber-lasers are designed to build high-power short length fiber-lasers. In our case the active single-mode cores (micro cores) are placed on a ring inside a big pump core. The micro cores are placed together very closely so that evanescent coupling between adjacent micro cores should be provided. To understand the coupling behavior in a multicore fiber in order to phase-lock all the micro cores we measured experimentally the coupling constant between the micro cores. Simultaneously we calculated the evolution of an injected field in a multicore fiber. In the experiment and in the simulations 38 micro cores with a diameter of approximately 6.9 µm are placed on a ring with a diameter of 115 µm. The distance between adjacent cores is about 2.6 µm. The measured coupling constant of $2c_{exp} = 0.82$ mm^{-1} is in good agreement with the value $2c_{th} = 0.83$ mm^{-1}. Further more the phase evolution in each micro core was evaluated.

Keywords : evanescent coupling, coupling constant, waveguides, field propagation, fiber laser

1. INTRODUCTION

Fiber lasers have opened a wide range of laser applications. The combination of easy handling in construction and use as well as the good beam quality make them a favorable candidate for many applications in laser physics. Further more high-power fiber-lasers have attracted recent interest. A standard way to obtain high output powers is the cladding pumped fiber laser. Due to absorption lengths of several meters this often results in more than 50 m active fiber length to obtain a high rate of pump absorption. In order to reduce the fiber length a new approach is to place many, active single-mode cores (micro cores) within a big pumpcore. This increases the absorption coefficient α of the pump wavelength by the number of micro cores compared to a single-core fiber. By placing the micro cores onto a centric ring the mode structure in the pump core is modified what again leads to an improvement in absorption. The cross-section of a multicore fiber is given in Fig. 1. The absorption of pump light in a multicore fiber has been calculated in [1]. Due to the incoherent overlap of radiation of each individual micro core the beam quality is reduced compared to the case of coherent emitting micro cores. The emission angle is no longer diffraction limited with respect to the ring structure like it is when coming from a one single-mode core. It is increased by a factor proportional to the ratio of the radius of the ring the micro cores are placed on and the radius of the micro cores themselves. In the employed geometries the beam quality factor M^2 is usually on the order of 20.

In order to increase the beam quality the individual micro cores have to be phase-coupled. Much work especially on coupling of CO_2 [2-4] and diode laser arrays [5-9] has been done. One possibility to achieve phase-locking is the evanescent coupling of adjacent lasers. Although this mechanism provides no spatial mode selection it yields large energy transfer. Spatial mode selection has to be introduced by some additional means. It is evident to find the preferred supermode of the system for obtaining effective mode selection. So, in order to design a phase-locked evanescently coupled multicore fiber-laser it is necessary to investigate the field propagation inside a multicore fiber.

In *Laser Resonators III*, Alexis V. Kudryashov, Alan H. Paxton, Editors,
Proceedings of SPIE Vol. 3930 (2000) ● 0277-786X/00/$15.00

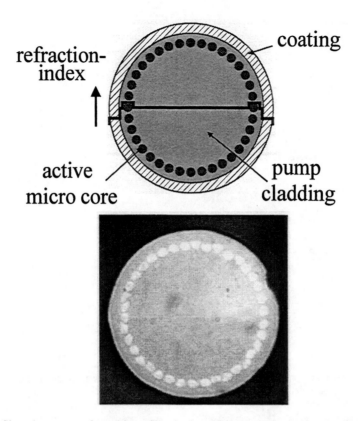

Fig. 1   Refractive index profile and geometry of a multicore fiber (top) and fabricated sample of such a fiber (bottom).

The spatial evolution of the fields in the micro cores of a multicore fiber is dominated by the coupling strength between adjacent waveguides. For linear interaction in the limit of weakly coupled waveguides, what is true in this case, the amplitude in the n-th core $A_n(z)$ is given by:

$$\frac{\partial A_n(z)}{\partial z} = ic(A_{n-1}(z) + A_{n+1}(z))$$

(1)

where c is the coupling constant. It is known that the phase constants of the supermodes depend on the coupling constant [10]. The coupling constant c can be estimated from approximations given e.g. by A.W. Snyder and J.D. Love [11]. But, as shown below, the approximated values do not fit well to the behavior of our system. So, to learn more about the evolution of the radiation field inside the laser system, it is important to measure the coupling coefficient.

2. EXPERIMENTAL SETUP

To determine the coupling length for neighboring cores experimentally, we launched radiation of a Nd:YAG laser, emitting at 1.064 µm, into one single core. Divergence and spot size of the beam were matched to diameter and numerical aperture of the micro core by a beam expanding telescope and an aperture in front of a focusing lens. A second lens at the distance of its focal length imaged the near-field pattern at the opposite end face of the fiber onto a CCD camera. The setup is given in Fig. 2. The wavelengths, at which each core would emit in the lasing state when being pumped with an appropriate source was between 1.056 µm and 1.066 µm. So, because the coupling constant is a function of λ for the investigations, we chose the wavelength $\lambda = 1.064$ µm. By that choice we made sure that we measured the coupling constant for the interesting wavelength, the fiber would emit when used as a laser.

We measured the coupling length of a multicore fiber with 38 Nd-doped micro cores. The diameters of the micro cores had a mean value of 6.9 µm. The ring had a diameter of about 115 µm and the distance of adjacent micro cores ran 2.6 µm. The index step between pump core and micro cores was 0.007 while the refractive index of the pump core was 1.456. We excited several cores for various sample lengths in the described way and detected the number of cores which emit at the opposite end of the fiber. Each additionally illuminated core to the left and to the right of the initially excited waveguide represented a coupling step. For each length we determined the

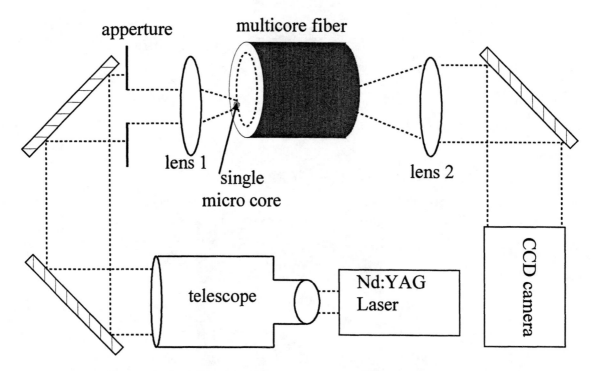

Fig. 2 Experimental setup for the determination of the coupling constant. The beam size of an Nd:YAG laser is expanded by a telescope. By choice of the size of the aperture and of the focal length of the focusing lens 1 the waist and divergence of the incident beam can be fitted to the NA and diameter of a micro core. The near-field pattern is collimated by lens 2 and recorded by a CCD camera.

average number of coupling steps that occur in the particular fiber. The lengths of the prepared fibers had to be in the order of only a few times of the expected coupling length. Rough analytical approximations gave a value of $2c = 0.67$ mm^{-1} in our case. Therefore, we prepared samples with a respective length between 3.5 mm and 23 mm.

3. NUMERICAL SIMULATIONS

The complex system „multicore fiber" was analyzed for a wide range of parameters such as core diameter, numerical aperture, etc. We developed a numerical code for the direct investigation of the scalar parabolic equation for a propagating wave field. This code allows the determination of the coupling coefficients of weakly coupled waveguides and check the limits for its applicability with high precision. The 3-D diffraction code is based on splitting into diffraction and refraction /

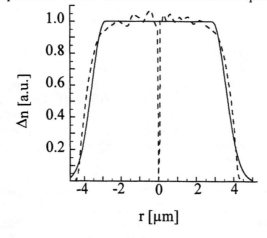

Fig. 3 Numerically employed (solid line) and measured (dashed line) index profile of the micro cores.

gain operators after each propagation step. We employed the 2- D Fast Fourier Transformation technique (FFT) in calculations of diffraction effects. A nonuniform refraction index was accounted for by direct integration over the light-propagation direction (z-axis) along the fiber. A restriction on the resulting phase variations of the wave field after each calculation step in order to reduce the numerical error sets an upper limit to the step size. Therefore, in our calculations we chose a step size along z-direction of 2.5 µm. The calculations were done on a rectangular mesh with a square cross section and a size of 512 * 512 grid points. The index profile of the multicore fiber (Fig. 1 top) was embedded into this grid. To avoid any influences by periodic boundary conditions, we introduced absorption on the grid points outside the pump core as recommended in [14]. Special tests proved the numerical accuracy of the chosen formalism. With respect to the experiment, the index profile of each micro core was taken to be a flat top distribution with smooth boundaries. The index profile of the micro cores used in the simulations and the profile measured experimentally are shown in Fig. 3.

4. EXPERIMENTAL AND NUMERICAL RESULTS

Fig. 4 shows two near-field patterns for multicore fiber lengths of 4.35 mm and 7.5 mm (Fig. 4a, 4b). An arrow marks the initially excited core. One can recognize the light guiding cores. Additional intensity peaks result from scattering and the formation of speckle patterns due to the spatial coherence of the Nd:YAG laser. These speckle patterns fill the whole pump core. To gain contrast all speckles are removed in the second row of Fig. 4.

a) b)

c) d)

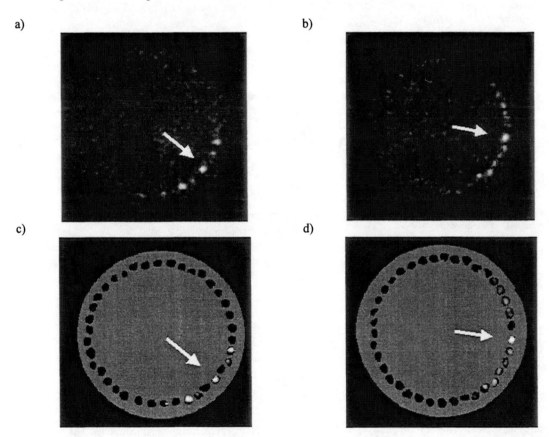

Fig. 4   Different near-field patterns of multicore fibers with lengths of 4.35 mm (left) and 7.5 mm (right) respectively. The excited cores are highlighted by arrows. The first row shows the original near-field distribution. In the second row the all speckles are removed.

The code, described above, was applied to the experimental situation with a micro core diameter of 6.9 µm and an index step of 0.007. As result the calculated near-field distribution the two multicore fiber lengths are shown in Fig. 5. A good agreement between the experimental and numerical results is obtained. It proves the reliability of the numerical simulations.

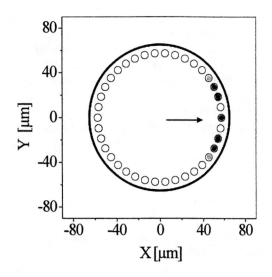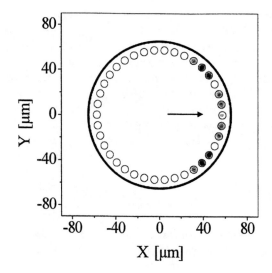

Fig. 5 Calculated intensity distribution at the output endface for fiber lengths of 4.35 mm (left) and 7.3 mm (right) corresponding to the situations shown in Fig. 4.

In the experiment we illuminated different single micro cores for each length of multicore fibers and determined the number of coupling steps. The determined numbers of coupling steps for different fiber lengths are plotted in Fig. 6. The graph was fitted by a linear function. The coupling constant 2c corresponds to the slope of this function. For the value of the coupling constant we derived $2c = 0.82$ mm^{-1}. The experimental error was about 14%. We calculated the near-field patterns at the output plane of a multicore fiber for various lengths when only one core is excited. Similar to the evaluation of the experimental data the number of coupling steps for each length calculated was determined. The resulting graph is also shown in Fig. 7. It agrees with the experimental one.

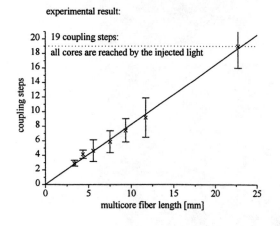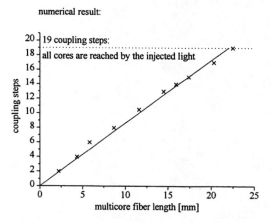

Fig. 6 Measured coupling steps determined by the number of illuminated cores at the endface of the multicore fiber. The experimental data are shown in the right panel while the numerical results are shown in the left. For fiber length more than 23 mm the light has crossed to the core opposite to the firstly excited one so that all cores have been reached by the injected light.

In the experiment the near-field pattern for a multicore fiber length of 4.35 mm shows a very symmetric pattern. This is what one would expect from the ideal theory of coupled waveguides with a homogeneous structure and what was measured e.g. in the case of linear structures [12]. The near-field distribution after a length of 7.5 mm is not as symmetric as for a fiber length of 4.35 mm. Every core is more or less illuminated (cf. Fig. 4) and the near-field structure deviates from the calculated near-field pattern. This can be explained as follows. The cores are not exactly circular but show variations in shape and size (Fig. 1 bottom). The variation in geometry lead to slightly different wave propagation constants in the micro cores. As a result we get a mismatch between propagation constants of neighboring micro cores. It is known [13] that, as a result of this mismatch, the efficiency of energy exchange is smaller than unity. The energy is only partly transferred to the adjacent waveguides. The rest remains in the micro core so that every core is more or less illuminated. Therefore, we achieve an almost ideal symmetric near-field pattern only for the first coupling steps. With increasing length, more cores take part in the coupling procedure and the mismatch in wave-propagation constants increases. The symmetric near-field pattern which one would receive from an ideal structure is washed out in the experiment by the imhomogeneous structure of the multicore fiber. The variations in the shape of micro cores have a large influence on the exchange of energy between them while the coupling constant is hardly affected by such imperfections. This can be deduced from the linear dependence of the number of coupling steps from the fiber length.

The calculated intensity inside the micro cores for different fiber lengths is shown in Fig 8. Core number 0 is the initially excited one. Core number 1 is its first neighbor, core number 2 its second and so on. The problem is symmetrically. Therefore it does not matter whether one regards the second neighbor to the left or to the right. It can be seen from the calculations, that a front of energy propagates across the micro cores. This front is extended over about 4 micro cores. Within this front the phase relation between adjacent cores is well defined. The phase step is about $\pi/2$. The corresponding phase relation between the micro cores is also shown in Fig. 8. As soon as the front has crossed through the micro cores the fixed phase relation between adjacent waveguides vanishes. The phase distribution becomes random. As described below (cf. equation 2) the field amplitude A_n inside the nth micro core is described by the Bessel function of nth order $J_n(2c \cdot z)$. Every time the Bessel function changes its sign the corresponding amplitude A_n gets a phase shift. Because adjacent waveguides are described by different Bessel functions with different periodicity the established phase relation between neighboring cores breaks up after a short distance. This can be deduced from Fig. 8.

To extract the value of the coupling constant from the results of the simulations we compared also the calculated intensity distribution along z-axis inside a micro core with the behavior described by equation (1). The solution of equation (1) gives analytically the evolution of the amplitude A_n in the nth core in z-direction:

$$A_n(z) = J_n(2c \cdot z). \tag{2}$$

Here, J_n is the Bessel function of nth order. We fitted the function $J_0^2(2c_{th} \cdot z)$ to the calculated intensity distribution along z-axis inside the 0th micro core which corresponds to the initially excited one and to its first neighbor. The fit is shown in Fig. 7a. A good agreement between the numerical simulations and the analytical expressions is obtained. The resulting value is $2c_{th} = 0.83$ mm^{-1}. The same evaluation has been done for the intensity distribution within the first core. The result is

a)   b)

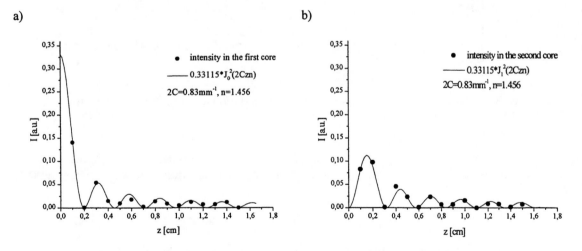

Fig. 7   The calculated intensity over the distance z inside the initial micro core (a) and its first neighbor (b). The fits show the corresponding Bessel functions with resulting coupling constants.

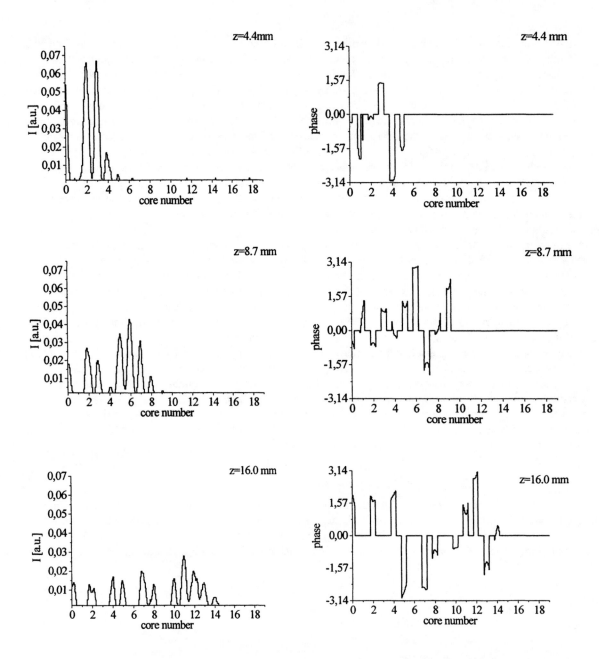

Fig. 8   Intensity and phase in each core after certain propagation distances. Core number 0 represents the initially excited core. The fixed phase relationship of $\pi/2$ between adjacent waveguides breaks up after a short time as described in the text.

shown in Fig. 7b. We receive the same coupling constant. Estimations of the coupling constant following the expression given in [16] gave a value of $2c = 0.67$ mm^{-1}. The coupling constant found from our numerical code is remarkably closer to the experiment.

5. CONCLUSIONS

In conclusion, precise values of the coupling constants for a novel fiber design (multicore fiber) comprising a number of active micro cores inside a large pump core were obtained from the passive case. Numerical simulations of the system were done. The experimental value of the coupling coefficient $2c_{exp} = 0.82$ mm^{-1} is in good agreement with the numerical value of $2c_{th} = 0.83$ mm^{-1}. This proves the reliability of the chosen numerical model. The calculated field evolutions were compared to the analytical solutions and have shown a good agreement with the theoretically predicted dynamics. The phase evolution in the micro cores was obtained. Based on the presented results, further numerical simulations of the multicore fiber used as high power short length fiber laser can be done in order to achieve stable phase-locking conditions and to optimize output power and beam quality.

ACKNOWLEDGMENT

We thank Th. Pertsch and D.Vysotsky for discussions. We acknowledge the company Fiber-Tech in Berlin for pulling the fiber with high precision and quality.

This work was supported by the German ministry of education and research BMBF under contract number 13 N 7146 and by the Russian basic research foundation RBRF under project No. 99-02-17469.

REFERENCES

1. P. Glas, M. Naumann. A. Schirrmacher, and Th. Pertsch, Opt. Com. **151**, 187 (1998)
2. V. V. Antykhov, A. F. Glova, O. R. Kachurin, F. V. Lebedev,V. V. Likhanskii, A. P. Napartovich, V. D. Pis'mennyi, JETP Letters **44**, 78 (1986)
3. A. F. Glova, S. Y. Kurchatov, V. V. Likhanskii, A. Y. Lysikov, A. P. Napartovich, S. B. Shchetnikov, V. P. Yartsev, and U. Habich, Quantum Electron. **27**, 309 (1997)
4. K. M. Abramski, A. D. Colley, H. J. Baker, and D. R. Hall, Appl. Phys. Lett. **60**, 530 (1992)
5. M. Cronin-Golomb, A. Yariv, and I. Ury, Appl. Phys. Lett. **48**, 1240 (1986)
6. M. P. Nesnidal, T. Earles, L. J. Mawst, D. Botez, and J. Buus, Appl. Phy. Lett. **73**, 587 (1998)
7. M. Løbel, P. M. Petersen, and P. M. Johansen, Opt. Lett. **23**, 825 (1998)
8. J. R. Leger, G. Mowry, and Xu Li, Appl. Optics **34**, 4302 (1995)
9. D. Botez, P. Hayashida, L. J. Mawst, and T. J. Roth, Appl. Phys. Lett. **53**, 1366 (1988)
10. E. Kapon, J. Katz, A. Yariv, Opt. Lett. **10**, 125 (1984)
11. A. W. Snyder and J. D. Love, *Optical Waveguide Theory*, chapter 18, Chapman and Hall, London, New York, 1983
12. S. Somekh, E. Garmire, A. Yariv, H. L. Garvin and R. G. Hunsberger, Appl. Phys. Lett. **22,** 46 (1973)
13. A. Yariv, *Quantum Electronics*, chapter 22, John Wiley & Sons, Inc., New York, Chichester, Brisbane, Toronto, Singapore, 1989
14. R. Kosloff, D. Kosloff, J. Comput. Phys. Rev. **63**, 363 (1986)

Reconstruction of field distributions of an active multicore fiber in multimode fibers

M. Wrage[1], P. Glas[1], M. Leitner[1], T. Sandrock[1], N. N. Elkin[2], A. P. Napartovich[2], D. V. Vysotsky[2]

[1] Max-Born-Institut für Nichtlineare Optik und Ultrakurzzeitspektroskopie,
Max-Born Straße 2a, D-12489 Berlin, Germany

[2] Troitsk Institute for Innovation and Fusion Research, Troitsk,
142092 Moscow region, Russia

ABSTRACT

Multimode interference (MMI) is fairly known from the one-dimensional case of slab waveguides. We present for the first time to our knowledge the reconstruction of the two-dimensional radial symmetric structure of a multicore fiber laser in a multimode fiber. In the concept of multicore fiber, rare earth-doped single mode waveguides (micro cores) are placed on a ring inside a big pump core. The situation of injecting radiation from N incoherent emitting sources into a multimode waveguide is described analytically. Experimental and numerical results for various multimode diameters and fiber lengths dealing with the reconstruction of the injected near-field pattern and the corresponding far-field patterns are presented. We propose that the reconstructed field could be re-injected into the multicore fiber-laser in order to introduce parallel coupling of all emitters. Additionally, using the multimode fiber as a passive element, without re-injection, the on-axis intensity of the multicore laser radiation is significantly increased by a single pass through a multimode fiber with a certain length. This effect takes place without any loss of energy.

Keywords: fiber laser, mode propagation, multimode interference, field reproduction

1. INTRODUCTION

In recent years, the development of high power table-top laser systems has attracted a lot of interest. These compact laser systems like laser array, especially diode lasers but also fiber laser arrays, suffer from bad beam quality. The on axis intensity can be increased by phase-locking of all laser emitters within the array. Many work has been done to achieve stable phase-locking conditions for different types of laser arrays [1-4]. Often a lot of optical equipment is needed to introduce coupling between the individual emitters and therefore the necessary adjustment becomes complicated [5-10].

Diode pumped fiber lasers are known to be easy to adjust. Advantages as compact setup, wavelength tunability and availability of many different wavelengths open up a wide area of applications for fiber lasers. In order to decrease the fiber length while increasing the output power the concept of multicore fiber lasers has been developed. In this concept not only one core but many active single-mode cores (micro cores) are placed on a ring inside a big pump core. This geometry leads to an improved pump absorption and is analyzed in [11]. As mentioned above, the increase of the number of emitters is followed by growth of the beam quality factor M^2. In our configurations M^2 is on the order of 20 - 30.

The reconstruction of field distribution by multimode interference (MMI) can possibly be used to change the phase distribution within the wave field. This would lead to an increase of spatial coherence. The effect of MMI and the resulting reconstruction of initial fields were studied in integrated optics [12-14]. In the following we investigate numerically and experimentally the situation when the wave field distribution coming from a multicore fiber laser propagates through a multimode fiber.

212

In *Laser Resonators III*, Alexis V. Kudryashov, Alan H. Paxton, Editors,
Proceedings of SPIE Vol. 3930 (2000) ● 0277-786X/00/$15.00

2. THEORY

The injection of the near-field pattern of a multicore fiber laser into a multimode fiber (MMF) can be described as incoherent summing of N beams in a multimode waveguide. The incoherent summing results from the fact that no fix phase relationship between the N emitting micro cores within the multicore fiber laser is established. The intensity $I_k(r, \varphi - \varphi_k)$ in the k-th micro core, which has the symmetry $I_k(r, -\varphi) = I_k(r, \varphi)$, can be expanded in a Fourier series. It yields

$$I_k(r, \varphi - \varphi_k) = \sum_{m}^{\infty} a_{km}(r)\cos(m(\varphi - \varphi_k)) \qquad \text{with} \qquad \varphi_k = \frac{2\pi}{N}k, \qquad (1)$$

where $a_{kn}(r)$ is the Fourier coefficient. If all micro cores have the same intensity, the coefficients $a_{mk}(r)$ do not depend on k. Then the summing over all micro cores gives the intensity distribution

$$I(r, \varphi) = \sum_{k}^{N} I_k(r, \varphi) = \sum_{m}^{\infty} a_m(r) \sum_{k}^{N} \cos(m(\varphi - \varphi_k)) \qquad (2)$$

Carrying out the summation leads to the following expression for the intensity

$$I(r, \varphi) = N \sum_{j=0}^{\infty} a_{jN}(r)\cos(jN\varphi) \qquad (3)$$

The equation (3) yields, that the intensity distribution resulting from this incoherent summing has a form of rings with modulation with angular frequencies N, $2N$,... It is to emphasize that despite the absence of interference the summed intensity is modulated periodically with the same symmetry as the injected field.

Expressing the injection of a given field distribution in term of multimode fiber eigenfunctions it results the following analytical formalism.

Generally, evolution of the wave field in the MMF can be described in terms of expansion of the injected field over modes of MMF. The field distribution $u_0(r, \varphi)$ of one single micro core can be approximated as a truncated Bessel function of the 0-th order. Eigenfunctions normalized to unity of beam power can be written as [15]

$$\Psi_{mk}(r, \varphi) = \sqrt{\frac{2}{\pi}} \frac{J_m\left(\mu_m^{(k)} \dfrac{r}{R_0}\right)}{\left|J_{m+1}\left(\mu_m^{(k)}\right)\right| R_0} \cos(m\varphi) \qquad (4)$$

where R_0 is the radius of the MMF. Propagation of the eigenfunctions in z - direction through the fiber is described by

$$\Psi_{km}(r, \varphi, z) = \Psi_{km}(r, \varphi, 0)\exp(i\beta_{mk}z), \qquad (5)$$

where

$$\beta_{mk} = \frac{\left(\mu_m^{(k)}\right)^2 \lambda}{4 n_0 R_0^2 \pi} \qquad (6)$$

is the index depending part of the propagation constant of the eigenfunction and $\mu_m^{(k)}$ is the k-th root of the Bessel function of m-th order. Regarding now the propagation of the injected wave field, which is expressed in terms of $\Psi_{mk}(r, \varphi)$, through the multimode fiber it can be written as result of the propagation of different multimode waveguide modes. Thus after a propagation distance of z the injected wave field is of the following form

$$u(r, \varphi, z) = \sum_{m,k} C_{mk} \Psi_{mk}(r, \varphi)\exp(i\beta_{mk}z), \qquad (7)$$

In equation (7) the coefficients C_{mk} are given by

$$C_{mk} = \frac{2\sqrt{2}\, \mu_0^{(1)} \dfrac{a}{R_0} J_m\left(\mu_m^{(k)} \dfrac{a}{R_0}\right) J_m\left(\mu_m^{(k)} \dfrac{R_C}{R_0}\right)}{\left(\mu_0^{(1)}\right)^2 - \left(\mu_m^{(k)} \dfrac{a}{R_0}\right)^2} \frac{}{\left|J_{m+1}\left(\mu_m^{(k)}\right)\right|}, \qquad (8)$$

where a is the radius of the micro core, R_c is the radius of the ring on which the micro cores are placed, and R_0 is the radius of the multimode waveguide. The dependence of C_{mk} on k for two different values of m is plotted in Fig. 1. The number of modes essentially involved in the series is about 100 (for $m=0$). The dependence on k is a combination of quasi-periodical and diminishing ones. For high-order angular harmonics ($m=61$) components with $k<20$ can be neglected. The far-field distribution for the mode $\Psi_{mk}(r,\varphi)$ is given by the Fourier transformation of equation (4). It is of the form

$$\Psi_{mk}(\kappa,\varphi) = \frac{2\sqrt{2\pi}\, R_0 i^m \mu_m^{(k)} J_m\left(R_0\kappa\right)}{\left(\mu_m^{(k)}\right)^2 - \left(R_0\kappa\right)^2} \frac{J_{m-1}\left(\mu_m^{(k)}\right)}{\left|J_{m+1}\left(\mu_m^{(k)}\right)\right|} \cos(m\varphi) \qquad (9)$$

with $\quad \kappa = \dfrac{2\pi}{\lambda}\theta$. $\qquad\qquad\qquad\qquad\qquad\qquad\qquad\qquad (10)$

Therefore the far-field distribution resulting from a field distribution which is given by equation (7) can be written as

$$u(\kappa,\varphi) = \sum C_{mk}\Psi_{mk}(\kappa,\varphi)\exp(i\beta_{mk}z) \qquad (11)$$

Experimentally observed near- and far-field intensity patterns can be calculated as an incoherent sum of wave fields given by expression (7) and (11), respectively.

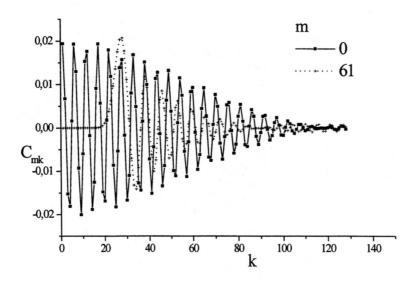

Fig. 1   The dependence of the coefficients C_{mk} on k for $m=0$, and $m=61$ is shown.

Regarding a multicore fiber it is evident that the radial symmetry is advantageous for reconstruction in a radial symmetric MMF. In the following we present near- and far-field distributions after a certain distance of propagation in different MMF. The on axis intensity can be increased just by propagation through the fiber. Additionally it is proposed that this scheme can be employed to introduce phase-locking between the emitting micro cores of the multicore fiber laser.

3. EXPERIMENTAL SETUP

For investigations of field reconstruction the near-field distributions of different multicore fiber-lasers are mapped onto the endface of a piece of the MMF by two lenses. The near-field patterns at the opposite endface of the MMF are observed by a lens and a CCD-camera. For the sake of reducing losses and diffraction and best image transfer the distance between the lenses is equal to the sum of their focal lengths. The endface of the multicore fiber-laser is in the left focal plane of the first lens and the end face of the MMF is in the right focal plane of the second lens. The setup is shown in Fig. 2. The use of two lenses is preferable to direct injection of the laser radiation into the MMF for the sake of investigation the multimode interference properties. Principally, it shouldn't make any difference whether the near-field is mapped onto the multimode fiber by two lenses or directly injected into the multimode fiber. For confirmation pieces of multimode fibers have been placed right in front of the multicore fiber-laser. The resulting near- and far-field distributions have been observed. They correspond to those recorded when using two lenses with equal focal lengths. So the use of lenses means no restriction to the general validity of the presented results. On the other hand the focal lengths can be varied and additional mechanisms such a spatial filters can be introduced in order to gain more information about the investigated mechanisms. In all experiments it is guaranteed that the numerical aperture (NA) is always equal or greater than the divergence of the injected wave field.

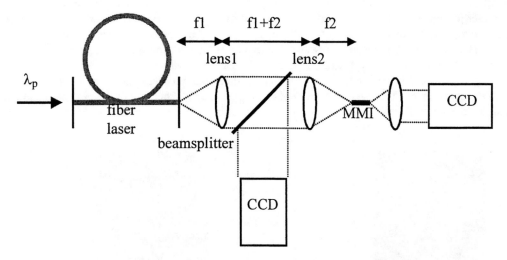

Fig. 2 : Experimental setup. The fiber laser is pumped by a laser diode at the wavelength λ_p. The near-field distribution of the multicore fiber is imaged by two lenses onto the end face of the MMF. Position and shape of the image is controlled by a CCD camera. The resulting near- and far-field patterns at the opposite end face of the multimode fiber are recorded by a second CCD camera.

Different kinds of multicore fiber-lasers were employed in the experiments. All of them were Nd-doped emitting at wavelengths between 1.056 µm and 1.066 µm. They are characterized as follows:

Fiber	number of cores	micro core diameter	micro core NA	ring diameter
MC1	42	3.8 µm	0.21	95 µm
MC2	61	6.7 µm	0.16	150 µm
MC3	61	11 µm	0.16	290 µm
MC4	47	7.7 µm	0.16	140 µm

Fig. 3 shows two near-field patterns of the multicore fiber-lasers MC2 and MC4. MMFs with diameters of 150 µm, 200 µm, and 400 µm are cut to various lengths. The experiments are very sensitive to adjustment. Therefore the MMF is placed on a translation stack which can be tilted and shifted in all three directions with micrometer precision.

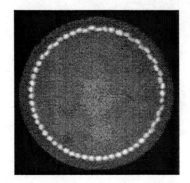 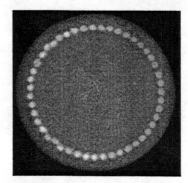

Fig. 3   Near-field patterns of the multicore fibers MC2 (left) and MC4 (right)

4. RESULTS

Evolution of the wave field along z described by expressions (7), (8) is non-monotonous. In the experiments for every length of MMF a ring, that corresponds in width and radius to that injected from the multicore fiber, is formed at the end face of the MMF. Furthermore, at special distances the ring became similar in symmetry to the near-field pattern of the multicore fiber laser (cf. Fig 4, 7, 8). The reconstruction of field distribution is a result of the symmetry as described above. An indispensable condition for this reconstruction is the coincidence of the centers of the radial symmetric injected field and of the MMF. This corresponds to the necessity that all micro cores have the same distance from the center of the MMF. The more one shifts the center of the injected field against the center of the MMF, the more the resulting near-field pattern is blurred. Shifting the injected beam just a few micrometers in the plane transverse to the propagation direction leads to destruction of the reconstruction. This again is a result of the corresponding symmetry of the multicore and the multimode fibers. Shifting of the centers leads to destruction of the symmetry. As a result, the reconstruction of the intensity distribution vanishes.

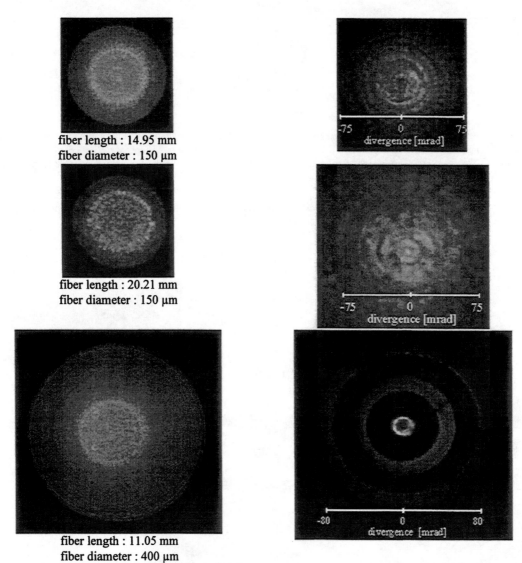

fiber length : 14.95 mm
fiber diameter : 150 μm

fiber length : 20.21 mm
fiber diameter : 150 μm

fiber length : 11.05 mm
fiber diameter : 400 μm

Fig. 4   Near- (left) and corresponding far-field patterns (right) after different lengths of propagation in MMFs. Lengths and sizes of the MMFs are written below the near-field patterns.

In Fig. 4 the near-field patterns and corresponding far-field distributions are shown. The far-field distributions depend on the lengths of the MMF. They always exhibit ring structures which indicate an increase of spatial coherence. Specially for a MMF length of 20 mm the far-field distribution shows a concentric ring structure which is similar to the far-field distribution of a phase-locked multicore fiber laser with zero phase step between adjacent micro cores. The central lobe and the rings of the first order in this far-field pattern have the same size and width as in the case of a phase coupled multicore fiber-laser. The bad contrast results from the large part of incoherent radiation. Nevertheless, this far-field distribution coming from a reconstructed near-field pattern of the multicore fiber-laser indicates a transformation of energy from an uncoupled system into the image of a partly coupled system. From the point of view of phase coupling it seams to be promising to re-inject the near-field distribution, reconstructed at a multimode fiber length of 20 mm, into the multicore fiber-laser in order to introduce phase coupling between adjacent waveguides. After a propagation distance of 20 mm a near-field distribution is reconstructed that is equivalent to the in-phase supermode of the multicore fiber laser. This can be deduced from the far-field pattern. The change of phase distribution within the wave field can be explained by the propagation in the MMF. The eigenmodes of MMF, in which the energy of the injected field distribution is deposited propagate with different phase velocities (cf. eqns. 5, 6). Therefore, the phase distribution within the propagating wave field is changed. Reinjection of this near-field pattern with a changed phase relation between the intensity peaks could lead to a parallel coupling of all micro cores. This proposal has to be tested in order to confirm the ability of introducing parallel phase-locking.

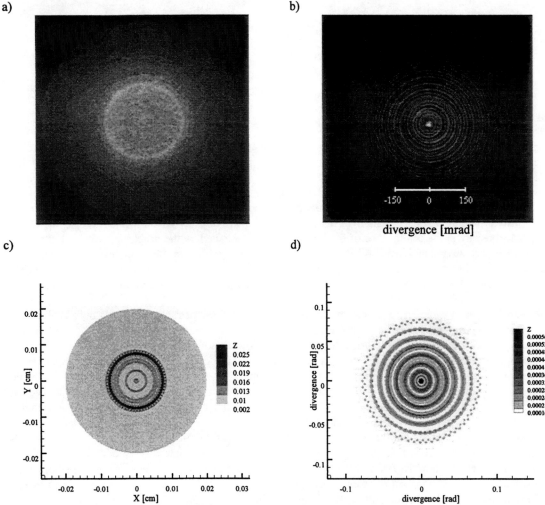

Fig. 5   Near- and corresponding far-field patterns after a 49.84 mm-long multimode fibers. A good reconstruction is obtained in the output plane of the multimode fiber (a) while the far-field distribution (b) shows a well defined structure and high intensity in the central lobe. The experimental intensity distributions equal the numerical results for the near- (c) and far- (d) field patterns.

Also the far-field patterns coming from other lengths of multimode fibers exhibit ring structures. After different distances of propagation in MMF different concentric systems of rings with a high contrast are observed. After special lengths of MMF a central lobe with an increased intensity compared to the far-field pattern coming from a multicore fiber laser can also be observed. In contrast to that one mentioned above these far-field distributions can not be identified as certain supermodes of

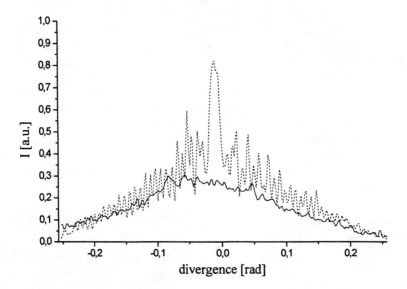

Fig. 6   Cross section of the intensity distribution in the far-field pattern coming from a MMF of a length of 50 mm (dashed line) and coming from the multicore fiber (solid line).

a multicore fiber laser. Nevertheless, these structures also indicate an increase in spatial coherence. For example for all MMFs with a length of about 50 mm a well defined far field structure is obtained, when the wave field coming from any multicore fiber laser (MC1, MC2, MC3, or MC4) is injected. Fig. 5 shows the experimental and numerical field distributions for the near- and far-field patterns resulting from propagation through a 49.84 mm-long MMF. In Fig. 6 cross-sections of the intensity profile of the experimentally measured far-field distributions with and without propagation through MMF are shown. The intensity in the central lobe of the far-field pattern obtained by the propagation through the MMF is three times larger than the maximum intensity coming directly from the multicore fiber. In this context it is worth mentioning that no energy is lost while propagating through MMF. We determined the power in front of and behind MMF. We measured exactly equal values. This confirms the assumption that the wave field is not lost but transformed into a state of higher spatial coherence.

When the wave field is injected symmetrically into the MMF the quality of reconstruction depends only slightly on the length of MMF. It rather depends on the distance between injected image of the multicore fiber and the face of the MMF. Injecting the image of multicore fiber MC2 (150 µm diameter) into a MMF with a diameter of 150 µm no reconstruction takes place. Injecting the same image into a MMF with a diameter of 200 µm only poor reconstruction can be observed. Full reconstruction takes place when injecting this near-field distribution into a MMF with 400 µm diameter. This behavior does principally correspond to the numerical simulations (cf. Fig. 7). The vanishing of reconstruction with decreasing multimode diameter can be explained by increasing influence of the radial boundaries of the MMFs. A radial field distribution, which is injected symmetrically into the inner part of the MMF, can be adapted by modes which have most part of the wave field in the inner area of the multimode fiber. These modes are hardly affected by border effects. In contrast to that a radial field distribution with a size equal to the multimode waveguide is adapted by modes which guide a large amount of energy in the outer region of the fiber. These modes are strongly influenced by the existence of the border. In the experiments imperfections and inhomogenities at the transition from core to cladding lead to losses and perturbations of theses modes. Thus the reconstruction additionally suffers from these losses.

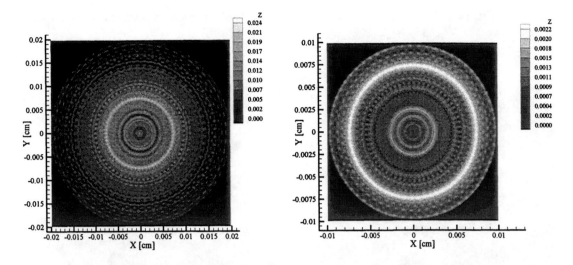

Fig. 7   Numerically simulated near-field distributions for fiber laser MC2 injected into MMF. The left MMF has a diameter of 400 μm while the right one has a diameter of 200 μm. Z corresponds to the intensity. The reconstruction suffers from the reduced size of the MMF.

The imaging lenses are varied in order to change the size of the injected field distribution. As a result, we always obtained a good reconstruction of the initial field distribution at the opposite end of the MMF. The reconstruction of an image of the MMF for different degrees of reduction in size is shown in Fig. 8. By the reduction of size down to an appropriate value it was also possible to obtain reconstruction of near-field patterns of the fibers MC2, MC3, and MC4 inside the MMF with a diameter of 150 μm. This result the assumption that the energy when propagation through the MMF is mainly restricted to a ring which equals in size and shape the injected near-field pattern of the multicore fiber. Additionally it can be deduced from the fact, that the boundary of thewaveguide structure leads to perturbations of the propagating modes. These perturbations can be neglected when the distance between boundary and injected ring is large enough. With decreasing distance the perturbations grow and reconstruction suffers until it totally vanishes.

a)   b)   c)

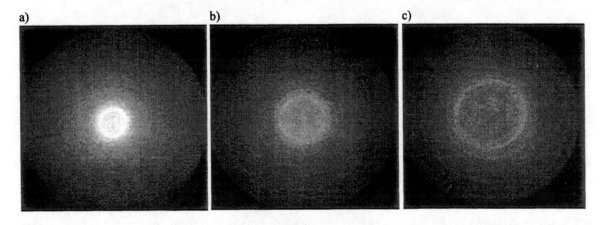

Fig. 8   Reconstructions of field distributions in a 400 μm-wide, 47.43 mm-long MMF. The size of the injected image was scaled by a factor of 8/18 (a), 12.5/18 (b), and 18/18 (c) corresponding to the two chosen imaging lenses.

As mentioned above, for the given injected field geometry of a multicore fiber a large part of energy is restricted to a ring when propagating through the MMF. This restriction of energy to a ring is demonstrated by the injection of the radiation of one single micro core into a multimode fiber. The distance between the point of injection and the center of the MMF was 100 μm. The situation was also simulated numerically. The results are shown in Fig. 9. In the experiments, as well as in the numerical calculations, a ring of 100 μm radius can be identified. Due to the coherence additionally the formation of speckle patterns take place.

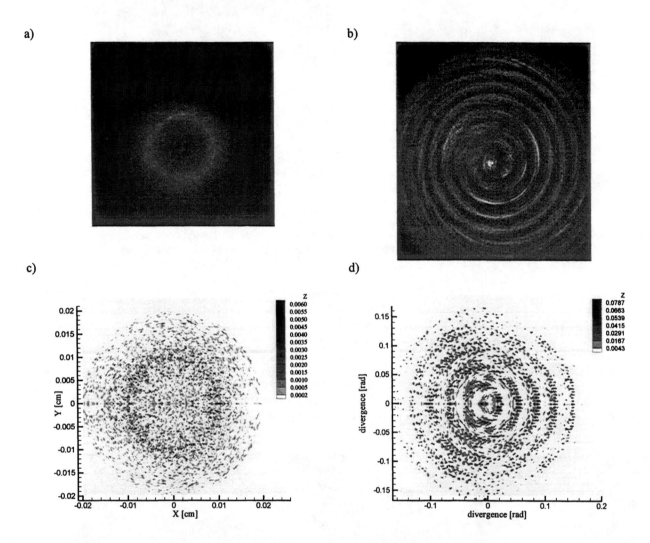

Fig. 9   Demonstration of the restriction of intensity onto a fixed ring resulting from launch conditions. The radiation of one single micro core is injected at a distance of 100 μm from the center of the MMF. Both, the experimental (a) and numerical (c) near-field patterns show a ring of 100 μm radius with some speckles. The corresponding far-field distributions (experimental (b) and numerical (d)) are shown.

5. CONCLUSION

To conclude, we have demonstrated for the first time to our knowledge the reconstruction of a radial symmetric field distribution (multicore fiber) in a two dimensional structure (multimode fiber). As a matter of symmetry, the main part of injected energy is always restricted to a ring. Based on the received far-field patterns one can conclude that by propagation through the MMF the spatial coherence is increased. The presented numerical simulations confirm these statements. The increase of spatial coherence does not affect the performance of the laser system. In contrast to spatial filters, no energy is lost. Some far-field patterns could be identified as far-field distribution coming from a reconstructed field that is close to a supermode of the multicore fiber-laser. It is proposed to re-inject these reconstructed near-field patterns back into the multicore fiber laser in order to introduce parallel phase coupling between all micro cores.

The presented results open a wide range of beam shaping by this new concept. Further investigation have to be done in order to optimize the presented effects.

ACKNOWLEDGMENT

This work is supported by the German ministry of education and research BMBF under No. 13 N 7146 and Russian Basic Research Foundation under project No. 99-02-17469.

REFERENCES

1. K. M. Abramski, A. D. Colley, H. J. Baker, and D. R. Hall, Appl. Phys. Lett. **60**, 530 (1992)
2. M. P. Nesnidal, T. Earles, L. J. Mawst, D. Botez, and J. Buus, Appl. Phy. Lett. **73**, 587 (1998)
3. J. R. Leger, G. Mowry, and Xu Li, Appl. Optics **34**, 4302 (1995)
4. D. Botez, P. Hayashida, L. J. Mawst, and T. J. Roth, Appl. Phys. Lett. **53**, 1366 (1988)
5. J. Morel, A. Woodtli, and R. Dändliker, Opt. Lett. **18**, 1520 (1993)
6. J. R. Leger and G. Mowry, Appl. Phys. Lett **63**, 2884 (1993)
7. A. F. Glova, N. N. ELkin, A. Y. Lysikov, and A. P. Napartovich, Quantum Electron. **26**, 614 (1996)
8. M. Løbel, P. M. Petersen, and P. M. Johansen, Opt. Lett. **23**, 825 (1998)
9. A. F. Glova, S. Y. Kurchatov, V. V. Likhanskii, A. Y. Lysikov, A. P. Napartovich, S. B. Shchetnikov, V. P. Yartsev, and U. Habich, Quantum Electron. **27**, 309 (1997)
10. M. Cronin-Golomb, A. Yariv, and I. Uri, Appl. Phys. Lett. **48**, 1240 (1986)
11. P. Glas, M. Naumann. A. Schirrmacher, and Th. Pertsch, Opt. Com. **151**, 187 (1998)
12. O. Bryngdahl, J. Opt. Soc. Am. **63**, 416 (1973)
13. L. B. Soldano and E. C. M. Pennings, J. Lightwave Technol. **13**, 615 (1995)
14. J. Leuthold, R. Hess, J. Eckner, P. A. Besse, and H. Melchior, Opt. Lett **21**, 836 (1996)
15. A. W. Snyder and J. D. Love, *Optical Waveguide Theory*, chapter 18, Chapman and Hall, London, New York, 1983

Invited Paper

Analytical and Numerical Resonator Analysis of Spectral Beam Combining of Yb Fiber Lasers

E. J. Bochove[*a] and C.A. Denman[b]

[a] Air Force Research Laboratory/DELO, KAFB, NM 87117 and
Center for High Technology Materials, University of New Mexico, Albuquerque, NM 87106

[b] Air Force Research Laboratory/DELO, KAFB, NM 87117

ABSTRACT

An analytical/numerical approach, based on the Van Vleck Green function, was developed in order to analyze spectrally multiplexed beam combining of a linear array of fiber lasers, in which a blazed diffraction grating located in an external cavity plays the role of a spatially dispersive element. The focus is on the laser-external cavity coupling, which determines excess cavity loss, as affected by primary aberrations of the transform lens. Other issues are also touched upon, however, such as beam quality and bandwidth dependence on element location, and the optimization of the latter. Analytical results were supported by a more general numerical implementation. Typical values for bandwidth were found optimally as low as a few GHz, which, being substantially narrower than the free-running fiber laser line-width, maximum power limits are foreseeably determined by stimulated light scattering. A rough but encouraging degree of agreement of the resonator theory with independent lens aberration calculations and preliminary experiments performed at MIT/LL using triplet and quadruplet transform lenses was encountered so far.

1. INTRODUCTION

Using spectral beam combining, multiple fiber lasers of good beam quality and relatively low power can be scaled to high power while maintaining the beam quality of single lasers. Cavity loss and bandwidth are primary concerns.

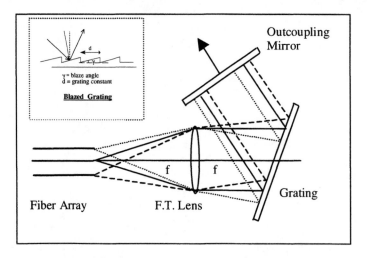

Fig.1. Schematic illustration of incoherent beam combining method

---

* Correspondence Email: bochovee@plk.af.mil; Telephone: (505) 846-5937.

222

In *Laser Resonators III*, Alexis V. Kudryashov, Alan H. Paxton, Editors,
Proceedings of SPIE Vol. 3930 (2000) ● 0277-786X/00/$15.00

1.1. Grating properties

In their normal operation a diffraction grating or prism behave like "beam-fanning" devices in which a multi-frequency laser beam is split into a number of monochromatic beams in a pitchfork-like geometry. Following an innovation of T.Y. Fan, under development at MIT-Lincoln Laboratories,[1] we consider the reverse operation in which a grating combines multiple monochromatic beams into one multi-frequency beam. If a gain medium (e.g. a Yb-doped fiber) is placed in each of the fan-out branches (i.e., the "prongs") of the resonator shown in Fig.1, and in addition mirrors placed at each "prong tip" and at the end of the fan-in portion (the "handle"), then independent lasing may be achieved in each sub-cavity consisting of the "handle" and a "prong". The output coupling takes place at the partially transmitting mirror at the end of the "handle". The one-to-one correspondence between position and k-vector direction, required for the satisfaction of the resonator requirement using a plane out-coupling mirror, is accomplished with a fourier-transform lens placed between the fiber array and the grating.

While a diffraction grating has greater dispersion power than a glass prism, it is subject to additional loss from multi-order diffraction and <100% reflection or transmission. As suggested by Lord Rayleigh (1888), a diffraction efficiency (DE) approaching the theoretical limit of unity is in principle achievable with a blazed grating (Fig.1, insert) that couples into a single order[2]. In practice, however, DE values of available metal-coated reflection gratings are typically below 90% at optimum wavelengths, and since that value depends strongly upon polarization, the fabrication of single-polarization state fiber lasers is a desirable end for this application.

In this paper the focus is on the fiber-external cavity coupling and the optical properties of the grating and transform lens that affect the resonator's performance. To this end we assume the validity of the grating equation. In our notation, it reads

$$\sin \beta - \sin \alpha = \frac{m\lambda}{d}, \quad |m| = 1,2,\ldots , \tag{1}$$

where α is the angle of the incident plane wave, and β is the angle in the direction of the mth diffraction order. For given blaze angle, γ, and fixed m, there is an optimum incident angle, α, defined by the condition that the diffraction envelope function attains a maximum value. Accordingly, Fraunhofer-Fresnel diffraction theory yields

$$2\gamma = \alpha - \beta . \tag{2}$$

Fig.2 plots the optimum α, satisfying Eqs.(1,2), as function of $m\lambda / d$. This figure is included only for illustrative purposes. Since the validity of results based on F-F theory is dubious when $\lambda / d \approx 1$, only putative quantitative validity should be assumed for Eq.(2).

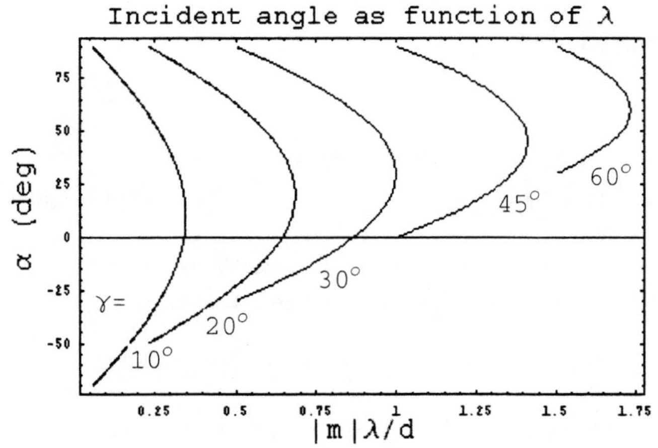

Two useful relations follow immediately from Eq.(1). Thus the transverse displacement, X_n, of the nth element is given by $X_n \approx (\alpha_n - \alpha_0)f$, where f is the lens' focal length,

Fig. 2. Optimum incident angle in Fresnel-Fraunhofer model

is accompanied by a wavelength shift given by

$$\lambda_n - \lambda_0 = \frac{d\lambda}{dX} X_n = -\frac{d \cos \alpha_0}{mf} X_n . \tag{3}$$

If $d \approx \lambda = 1 \mu m$, $f = 5 cm$, $m = 1$, $X_n = 0.25 cm$, and $\cos \alpha_0 \approx 1$, then $\Delta\lambda$ approximates the likely estimated gain bandwidth of, say, $50 nm$. In addition, using $\lambda / \Delta\lambda = |m| W_g / d$ for the resolving power, where W_g is the grating width, one has min $W_g = f\lambda / \cos \alpha_0 \upsilon$, where υ is the required spatial resolution. Taking $\upsilon \approx \lambda / 4$ yields min $W_g \approx 4f / \cos \alpha_0$. Hence, the grating width sets an upper limit for the focal length of the transform lens.

Eq. (1) shows that a nearly collimated beam expands in the spatial frequency domain when M, defined by

$$M \equiv \frac{d\beta}{d\alpha} = \frac{\cos\alpha}{\cos\beta} = \frac{\cos\alpha}{\sqrt{1 - (m\lambda/d + \sin\alpha)^2}}$$

(4)

exceeds unity, and contracts when $M < 1$. $M = 1$ at the Littrow condition, $\sin\beta = -\sin\alpha = m\lambda/2d$.

After reflection from the out-coupling mirror and a second grating pass the distortion expressed in Eq.(4) is replaced by near-specular reflection. The incident direction, $\alpha = \alpha_0 + \Delta\alpha$, goes over into $\alpha' = \alpha_0 + \Delta\alpha'$, where α_0 is the direction of incidence for which the return ray is perfectly reversed. To second-order,

$$\Delta\alpha' \approx -\Delta\alpha + A\Delta\alpha^2,$$

(5)

where $A = -m\lambda(1 + \sin\alpha_0 \sin\beta_0)/d \cos\alpha_0 \cos^2\beta_0$. The angular beam divergence at the grating, given by $\Delta\alpha \approx w_o/f \approx 10^{-5} \, rad$, in terms of mode radius w_0, is small enough to be neglected except possibly near normal or grazing incidence. Hence, to a very good approximation, the grating and out-coupling mirror assembly acts jointly like a mirror with wave length-dependent orientation.

1.2. Discussion of the analytical method

The "transmission function"[3] of the external cavity is based on Van Vleck's[4] asymptotic Green function for wave mechanics, modified for adaptation to optics. This procedure is permitted by motive of the property possessed by Hamilton's point characteristic as the generating function for a canonical transformation[5] linking the ray coordinates in the object and image planes. The resulting Green function is then

$$G(r', r) = (2\pi)^{-1} \sqrt{\det \frac{-ik\partial^2 S(r', r)}{\partial x_i \partial x'_j}} e^{ikS(r', r)},$$

(6)

where $k = 2\pi/\lambda$. $S(r', r)$ (the point characteristic) is the optical path measured along a light ray connecting point $r' = (x', y')$ in the plane $z = z_1$ to $r = (x, y)$ in the plane $z = z_2$. Hence, $G(r', r)$ is obtained entirely from geometrical optics.

A significant simplification results from expansion of the coordinates in $S(r', r)$ in a second-order Taylor series about a pair of fixed points, denoted (X', X). Writing only the non-vanishing terms for a system having bilateral symmetry about the $y = 0$ plane, the expansion is of the form: [6]

$$S(r', r) = u_x x + w_x x' + \frac{1}{2} U_x x^2 - V_x x x' + \frac{1}{2} W_x x'^2 + \frac{1}{2} U_y y^2 - V_y y y' + \frac{1}{2} W_y y'^2.$$

(7)

In order that this is justified it is necessary that the points (X', X) be located on a ray, while, moreover, this ray, which we call the *"central ray"*, must at least nearly coincide with the axis of the physical light beam, being subject to the Rayleigh criterion[7]. In addition to G being of convenient gaussian form, the expansion, Eq.(7), has the advantage that most of the resonator aberrations are included in the coefficients' ($u_x, ..., W_y$) dependence on the fixed vectors (X', X). Our procedure was to compute first the characteristic for a single pass through the transform lens, next to perform the above expansion, and then to derive the expression for the round trip characteristic, including the reflection from the effective tilted mirror. The error resulting from the neglect of higher-order terms will be investigated in the future.

The approximate validity of the above expansion was assumed in both the analytical and numerical approaches of this work, in which the above mentioned precautions were carefully observed. *Experimentally*, the approximation can be tested, since it would be confirmed if it can be shown that the maximum on-axis coupling, defined below, is only slightly less than unity.

2. RESONATOR PROPERTIES

In this section expressions for the bandwidth and beam profile are derived using simplified arguments that will be reinforced in Sect.3 by a more rigorous calculation.

2.1. Bandwidth

Taking α and β determined by laser element and mirror positions, say at values α_0, β_0, where β_0 corresponds to the direction normal to the mirror, then Eq.(1) defines a central wave length, λ_0, which satisfies $\sin \beta_0 - \sin \alpha_0 = m\lambda_0 / d$. A small difference in wavelength, $\Delta\lambda = \lambda - \lambda_0$, causes the field to be slightly shifted after a round trip in the external cavity. A ray leaving the element at the angle $\alpha = \alpha_0$, but having wave length $\lambda_0 + \Delta\lambda$ will reflect from the grating at the angle $\beta = \beta_0 + \Delta\beta$, satisfying $\sin(\beta_0 + \Delta\beta) - \sin \alpha_0 = m(\lambda_0 + \Delta\lambda) / d$, so that $\Delta\beta = m\Delta\lambda / d \cos \beta_0$ to first order in $\Delta\lambda$. Upon its return the incident angle will be $\beta' = \beta_0 - \Delta\beta$, and the reflected ray's angle, $\alpha_0 + \Delta\alpha$ is obtained from $\sin(\alpha_0 + \Delta\alpha) - \sin(\beta_0 - \Delta\beta) = -m(\lambda_0 + \Delta\lambda) / d$. This yields $\Delta\alpha = -2m\Delta\lambda / d \cos \alpha_0$. The displacement of the field at the fiber is therefore given by $\delta \approx f\Delta\alpha = -2mf(\lambda - \lambda_0) / d \cos \alpha_0$. Describing the fiber near-field by a gaussian function of width w_0, $E(r) = E_0 Exp(-r^2 / 2w_0^2)$, for a fiber located at an axial distance $f + \varepsilon$ from the lens, the overlap integral $J(\lambda; \varepsilon)$ of the shifted return field with the original field can be calculated in the normal paraxial optics approximation. Its absolute value squared yields the intensity coupling, given by

$$J(\lambda; \varepsilon)J(\lambda; \varepsilon)^* = \frac{\eta^2 \, e^{-\frac{\delta^2 k\eta}{2(\varepsilon^2 + \eta^2)}}}{\varepsilon^2 + \eta^2} \quad , \tag{8}$$

where $\eta = k_0 w_0^2 = 2\pi w_0^2 / \lambda_0$ is the diffraction distance of the near field. Plots of Eq.(8) are shown in Fig.(3). The values used were $d = \lambda = 1\mu m$, $w_0 = 2\mu m$, $m = 1$, $f = 5$ cm, $\cos \alpha_0 \approx 1$. The above yields the band width as given by:

$$\Delta\lambda = \left| \frac{d\lambda}{dX} \right| \frac{\sqrt{\varepsilon^2 + \eta^2}}{\sqrt{2k\eta}}, \tag{9}$$

where the "dispersion parameter" $d\lambda / dX$ is defined in Eq.(3). Taking $\varepsilon = 0$, this equals about 0.01 nm (3 GHz).

2.2. Far-field beam profile

The beam profile at arbitrary wavelength, λ, in a given exterior plane is determined by the evolution of the field in the reference plane by its passage through the lens in Fig.(1), the grating, and any external optics. The grating's effect on a plane wave is given by the shift Δq_n of the transverse wave vector, q, in which M is defined in Eq.(4):

$$\Delta q = \underbrace{(M - 1)q}_{\text{geometric distortion}} - \underbrace{\frac{2\pi m(\lambda - \lambda_0)}{\lambda_0 \, d \cos \beta_0}}_{\text{chromatic distortion}} , \tag{10}$$

where λ_0 is the central wavelength. After performing the mentioned transformations the absolute-squared field amplitude is multiplied by the spectral window function of Eq.(5), and integrated over wavelength. The result is a sum of mutually centered Gaussian functions, one corresponding to each of the elements in the array.

$$I(x, y) = \frac{1}{\pi\sigma_y} \sum_n \frac{P_n}{\sigma_{xn}} e^{-x^2/\sigma_{xn}^2 - y^2/\sigma_y^2}. \tag{11}$$

The transverse width σ_{xn} is

$$\sigma_{xn} = \sqrt{\sigma_{\lambda n}^2 + \sigma_{x0}^2}, \quad \text{where} \quad \sigma_{x0}^2 = \frac{1}{k\eta}\left[\left(\frac{B}{Mf}\right)^2(\eta^2 + \varepsilon^2) + A(M^2 f^2 A - 2\varepsilon B)\right], \quad \text{and} \quad \sigma_{\lambda n} = B\Delta\lambda_{Nn}/d, \tag{12}$$

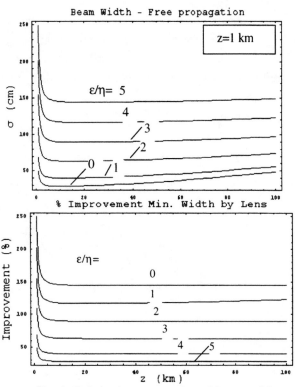

Beam Width – Free propagation

% Improvement Min. Width by Lens

Fig. 4. Relative beam narrowing with external lens

where $\Delta\lambda_{Nn}$ is the normal incidence ($\alpha_0 = 0$) bandwidth, given in Eq.(9) for on-axis-, and Eq.(22) for off-axis elements. σ_{x0} is the narrow-band beamwidth, and $\sigma_{\lambda n}$ broadening due to grating dispersion. A and B are ray-matrix elements describing external propagation from the out-coupling mirror to the observation plane. The expression for the lateral width, σ_y, is obtained from Eq.(12) by substituting $M \to 1$, $\sigma_\lambda \to 0$, leading to a minimum when $f \to f_{opt} = \sqrt{B\eta/A}$. A mini-mum transverse width is achieved for $M = M_{opt}$, where

$$M_{opt} = \sqrt{B\sqrt{\varepsilon^2 + \eta^2}/Af^2} \tag{13}$$

If both directions are optimized, $\sigma_{x0} = \sigma_{y0} = \sqrt{2AB/k}$.

The transverse beam-width, σ_x, for free propagation over distance z ($A = 1$, $B = z$), is plotted in Fig.(3) as function of M. Note the sharp rise as M decreases below M_{opt}, but the curves are almost level when $M > M_{opt}$. The width for this case, with M subject to Eq.(13), is given by

$$\sigma_{x,opt}^{free} = \frac{1}{f\sqrt{k\eta}}\sqrt{\frac{1}{2}\left((\varepsilon^2 + \eta^2)z^2 - 4\varepsilon f^2 z\right) + 2zf^2\sqrt{\varepsilon^2 + \eta^2}}. \tag{14}$$

.At large distance the beam width in the x-direction is controlled by the grating term:

$$\sigma_{x,opt}^{free} \to \sigma_{\lambda n} = \frac{w_0}{f\eta}\sqrt{\frac{1}{2}(\varepsilon^2 + \eta^2)}\; z \xrightarrow[\varepsilon=0]{} \frac{w_0 z}{f\sqrt{2}}, \tag{15}$$

On the other hand, if the observation plane is in the focal plane of an external lens ($A = 0, M_{opt} \to \infty$) some beam-narrowing is possible in the transverse direction. This is plotted in Fig.(4). It is seen that the lens does little to decrease the spot size beyond about 1 km. The reason for this is that the beam width approaches then again just the long-distance limit of the free-propagation value, given in Eq.(15).

3. EFFECTS OF LENS ABERRATIONS

3.1. Solution for the cavity transmission function

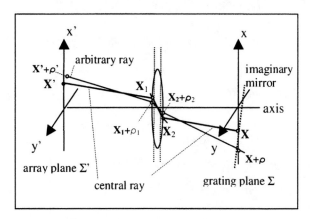

Fig. 5. Ray paths illustrating single pass through lens

A result of paraxial optics (in relation to the axis in Fig.5) is that the coupling, J, is independent of the transverse distance, X. In reality, lens aberrations cause J to decrease rapidly when X exceeds one or two millimeters. The primary aberrations connected with the transverse offset should be considered first for fiber lasers with good beam quality, but a contribution by spherical aberration (i.e., the contribution of non-quadratic terms to $S(x',x)$) should not necessarily be ruled out. We approached the former problem in the manner discussed in section 1.2. Consider an arbitrary wavelength λ, and let a vector X' define a point in the array plane Σ' (the reference plane), and X a vector in a plane Σ parallel to Σ' defining a point on the virtual "mirror" associated with λ (as discussed in sect.1.1) which is orthogonal to the ray connecting X' to X. This ray is by definition the "central ray" for $S(x',x)$. X' gives the position of the central ray in Σ' after its second pass through the lens. Next, let X_1 and X_2 define the intersection of this ray with the lens surfaces (see Fig.5). We use the notation of Eq.(7) for the single-pass characteristic from plane Σ' to plane Σ, and use the superscript R to identify the round trip $\Sigma' \rightarrow \Sigma \rightarrow \Sigma'$. In order to make the algebra tractable we assumed $X = -X'$ in the analytical model that follows, which unfortunately restricts the geometry of the array to one in which the off-axis elements point (approximately) to the center of the lens. This condition is hardly more than formal, since it yields results that differ only slightly from those predicted by a parallel arrangement. Moreover, the restriction is not necessary in the numerical approach described in Sect 4.

Writing the single-pass characteristic as $S = S(x',y',x_1,y_1,x_2,y_2,x,y)$, the intermediate coordinates are determined by "Fermat's principle", defined by $\partial S / \partial q_i = 0, q_i = x_i, y_i ; i = 1,2$. The transverse (x) and lateral (y) coordinates are expanded about the central ray coordinates as follows, taking $Y'=Y=0$:

$$x' = \delta X + \varpi \Delta x', \ y' = \varpi \Delta y', \quad x_1 = \delta X_1 + \varpi \Delta x_1 + \varpi^2 \Delta x_{12}, \ y_1 = \varpi \Delta y_1 + \varpi^2 \Delta y_{12},$$
$$x_2 = -\delta X_1 + \varpi \Delta x_2 + \varpi^2 \Delta x_{22}, \ y_2 = \varpi \Delta y_2 + \varpi^2 \Delta x_{22}, y = -\delta X + \varpi \Delta x, \ y = \varpi \Delta y, \tag{16}$$

in terms of expansion parameters δ and ϖ, to be equated to unity at the end. Eq.(16) contains nine unknowns, X_1, and $\Delta x_i, \Delta y_i, \Delta x_{ij}, \Delta y_{ij} \ (i,j=1,2)$. The central ray is defined by $\varpi = 0$. We found the solution to orders δ^3 and ϖ^2 to be sufficiently accurate (Sect.4).

In the next step, the solutions obtained for the intermediate coordinates and X_1 are substituted into the expression for S, which is next expanded in δ and ϖ. This leads to an expression of the form of Eq.(7). The set of single-pass coefficients is listed as follows for general availability:

$$u_x = \frac{(2nR+t-nt)X}{np(2R-t)+(p+R)t} + X^3\Big[4n^3p(2R-t)^3t+(p+R)t^4+2n^2(3p+R)(2R-t)^2t^2$$
$$+n^4p(2R-t)^4+nt(2R-t)(4R^3-2R^2t+4pt^2+3Rt^2)\Big]\Big/\Big[2\big(np(2R-t)+(p+R)t\big)^4\Big],$$

$$U_x = \frac{-n(2p+R)(R-t)+n^2p(2R-t)-(p+R)t}{((n-1)p-R)(np(2R-t)+(p+R)t)} -$$
$$X^2\Big[3(p+R)^3t^4+3n^6p^3(2R-t)^4-6n^5p^2(2R-t)^3(2pR+R^2-3pt-Rt)+$$
$$n(p+R)^2t\big(12R^4-14R^3t+18R^2t^2+3R(8p-3t)t^2-18pt^3\big)+$$
$$n^4p(t-2R)^2\big(3R^2(2R^2-2Rt+t^2)+3p^2(4R^2-20Rt+15t^2)+2pR(7R^2-27Rt+18t^2)\big)+$$
$$n^2(p+R)t\big(2R^2t(7R^2-9Rt+3t^2)+3p^2t(24R^2-40Rt+15t^2)-2pR(16R^3-54R^2t+63Rt^2-21t^3)\big)+$$
$$n^3p(2R-t)\big(12p^2t(4R^2-10Rt+5t^2)+2R^2t(14R^2-27Rt+12t^2)+pR(-4R^3+84R^2t-174Rt^2+81t^3)\big)\Big]\Big/$$
$$\Big[2(p-np+R)^2(np(2R-t)+(p+R)t)^4\Big],$$

$$V_x = \frac{-nR^2}{((n-1)p-R)(np(2R-t)+(p+R)t)} +$$
$$X^2nR^2\Big[3n^4p^2(t-2R)^2-n^3p(p(5R-9t)+3R(R-t))(2R-t)^2-(p+R)^2t(6R^2-5Rt+3t^2)-$$
$$n^2p(2R-t)(R(10R-9t)t+2p(R^2+6Rt-6t^2))+n(p+R)t(R(-5R+3t)+p(8R^2-18Rt+9t^2))\Big]\Big/$$
$$\Big[(p-np+R)^2(np(2R-t)+(p+R)t)^4\Big], \qquad w_x = -u_x, \qquad W_x = U_x,$$

$$U_y = \frac{-n(2p+R)(R-t)+n^2p(2R-t)-(p+R)t}{((n-1)p-R)(np(2R-t)+(p+R)t)} -$$
$$X^2\Big[(p+R)^3t^4+n^6p^3(2R-t)^4-2n^5p^2(2R-t)^3(2pR+R^2-3pt-Rt)+$$
$$n(p+R)^2t\big(4R^4-6R^3t+6R^2t^2+R(8p-3t)t^2-6pt^3\big)+$$
$$n^4p(t-2R)^2\big(R^2(2R^2-2Rt+t^2)+p^2(4R^2-20Rt+15t^2)+6pR(R^2-3Rt+2t^2)\big)+$$
$$n^2(p+R)t\big(2R^2t(3R^2-3Rt+t^2)+p^2t(24R^2-40Rt+15t^2)-2pR(8R^3-20R^2t+21Rt^2-7t^3)\big)+$$
$$n^3p(2R-t)\big(2R^2t(6R^2-9Rt+4t^2)+4p^2t(4R^2-10Rt+5t^2)+pR(-4R^3+32R^2t-58Rt^2+27t^3)\big)\Big]\Big/$$
$$\Big[2(p-np+R)^2(np(2R-t)+(p+R)t)^4\Big],$$

$$V_y = \frac{-nR^2}{((n-1)p-R)(np(2R-t)+(p+R)t)} +$$
$$X^2nR^2\Big[n^4p^2(t-2R)^2-n^3p(p(R-3t)+R(R-t))(2R-t)^2-(p+R)^2t(2R^2-Rt+t^2)-$$
$$n(p+R)t^2(4pR+R^2-3pt-Rt)-n^2p(2R-t)(R(2R-3t)t+2p(R^2+Rt-2t^2))\Big]\Big/$$
$$\Big[(p-np+R)^2(np(2R-t)+(p+R)t)^4\Big], \quad W_y = U_y. \tag{17}$$

where R is the radius of curvature of the symmetrical lens, t the lens thickness, n is the refractive index of the lens, $p \equiv f+h+\varepsilon$, where $1/f = (n-1)\big(2/R-(n-1)t/nR^2\big)$ is the inverse focal length, $h \equiv (n-1)ft/nR$, and $f+\varepsilon$ is the distance from the plane Σ' to the lens vertex, which also formally equals the distance from the second vertex to the plane Σ. Much precision of the latter value is actually not important when the inequality $\varepsilon \ll f$ is satisfied.

The "mirror"'s action is described using the Gaussian ray matrix formalism. Its ray matrix is

$$M_{m} = \begin{bmatrix} 1 & -D_{eff} & -D_{eff}\sin\phi_0 \\ 0 & 1 & 2\sin\phi_0 \\ 0 & 0 & 1 \end{bmatrix}, \tag{18}$$

where ϕ_0 is the tilt angle, and D_{eff} is the efective distance of the grating from the outcoupling mirror, accounting for a small shift of the reflected ray. Terms in D_{eff} are negligible subject to the condition $D_{eff}\eta << f^2$. The round trip ray matrix is given by $M^R = M_L'M_mM_L$, in terms of the single-pass ray matrix, M_L, and its partner for reverse propagation, M_L'. M_L is obtained directly from $S(x',x)$, and M_L' from $S'(x',x) \equiv S(x,x')$. The result yields the round-trip ray matrix from which the coefficients for the round-trip point characteristic $S^R(x',x)$ are calculated:

$$u_x^R = w_x^R = \frac{u_x V_x + w_x U_x + V_x \sin\gamma_0}{U_x} \approx \frac{(n-1)(2nR+t-nt)X(\lambda)}{nR^2 + (n-1)\varepsilon(2nR+t-nt)},$$

$$U_x^R = W_x^R = W_x - \frac{V_x^2}{2U_x}$$

$$\approx -V_x^R \approx nR^3(2nR+t-nt)^2 \Big/ \Big[2\varepsilon nR^3(2nR+t-nt)^2 + X(\lambda)^2 \Big(3n^4(R-t)(2R-t)^3 -$$
$$n^3(2R-t)^2(4R^2-18Rt+9t^2) + Rt(12R^2-14Rt+3t^2) + nt(-32R^3+18R^2t+6Rt^2-3t^3) +$$
$$n^2(-8R^4-8R^3t+54R^2t^2-42Rt^3+9t^4)\Big)\Big]$$

$$u_y^R = w_y^R = 0, \quad U_y^R = W_y^R = W_y - \frac{V_y^2}{2U_y}$$

$$\approx -V_y^R \approx nR^3(2nR+t-nt)^2 \Big/ \Big[2\varepsilon nR^3(2nR+t-nt)^2 + X(\lambda)^2 \Big(n^4(R-t)(2R-t)^3 +$$
$$3n^3(2R-t)^3t + Rt(4R^2-6Rt+t^2) - nt(16R^3-10R^2t-2Rt^2+t^3) +$$
$$n^2(-8R^4+8R^3t+14R^2t^2-14Rt^3+3t^4)\Big)\Big] \tag{19}$$

where γ_0 is the angle of the central ray with the axis, i.e., $\tan\gamma_0 = (X(\lambda)-X_1)/Z_1$. The wavelength dependence of X, being important to the spectral determination (below), is expressed explicitly.

3.2. Coupling of fiber to cavity

We consider two applications: the coupling efficiency, which determines the cavity loss, and the off-axis laser spectrum.

For a Gaussian mode profile at wavelength λ, having offset X_0, spot radius w_0, and tilt angle χ, the field is

$$E(x,y;\lambda) = E_0 e^{-(x-X_0)^2/2w_0^2 - y^2/2w_0^2 + ikx\sin\chi}. \tag{20}$$

We define $X_0 \equiv X(\lambda_0)$, which gives the laser element's location, and λ_0 is the central wavelength. $X(\lambda)$ is the central ray coordinate at wavelength λ in the Σ' plane. The tilt angle satisfies $\sin\chi = w_x^R$ for optimum alignment of the beam with the central ray direction. For simplicity, below are presented only results in the limit $t \to 0$. From Eq.(1), $X(\lambda) \approx X(\lambda_0) + mf(\lambda-\lambda_0)/d\cos\alpha_0$. Using the transmission function discussed in Sect. 1.1 to calculate the field modified by a round trip in the external cavity, the absolute square of the round-trip overlap integral is found as

$$|J(\lambda)|^2 = \frac{\exp\left[-\dfrac{4k\eta}{(1+\varepsilon/f)^2}\dfrac{X_0{}^2}{f^2}\right]\exp\left[-(\lambda-\lambda_0)^2/\Delta\lambda^2\right]}{\sqrt{\left[1+\left(\dfrac{2\varepsilon f+(3+1/n)X_0{}^2}{2\eta f}\right)^2\right]\left[1+\left(\dfrac{2\varepsilon f+(1+1/n)X_0{}^2}{2\eta f}\right)^2\right]}} \ . \tag{21}$$

where $\Delta\lambda$ is the bandwidth in units of length, given by

$$\Delta\lambda = \left|\frac{d\lambda}{dX}\right|\frac{\sqrt{n^2(\varepsilon^2+\eta^2)+n(3n+1)\varepsilon X_0{}^2/f+(3n+1)^2 X_0{}^4/4f^2}}{\sqrt{2k\eta}\left(n-3(3n+1)X_0{}^2/2f^2\right)} \ . \tag{22}$$

The above coupling function has a peak on-axis value of unity (*viz* Sect.1.2). Off-axis, $|J|$ peaks at values of X_0 at which $\varepsilon < 0$ which correspond to the minima of the denominator of the expression in Eq.(21). Their location is determined by the tangential and sagittal foci. For $\varepsilon > 0$ the peak is on-axis, for $0 > \varepsilon > -(2+1/n)\eta$ the peak is located at

$$\varepsilon = -\frac{(2n+1)X_0{}^2}{2nf} \ . \tag{23}$$

If $\varepsilon < -(2+1/n)\eta$ there are two solutions, given by

$$\varepsilon = \frac{-(2n+1)X_0{}^2}{2nf} \pm \sqrt{X_0{}^4/4f^2-\eta^2} \ . \tag{24}$$

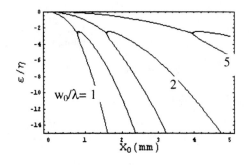

Fig. 6. Laser-cavity coupling peaks location as function of element position.

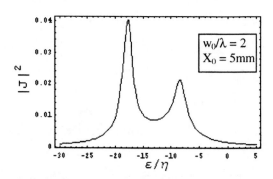

Fig.7. Coupling strength as function of longitudinal offset

Fig. 6 shows the locations of the maxima in $|J|$ in the $\varepsilon - X_0$ plane. As the laser element is moved away from the focal plane the point of peak coupling moves farther away from the axis, until it bifurcates into two peaks. These are effects of field curvature. For positive ε (array behind focal plane) the (single) peak is always on the axis.

Fig. 7 shows a plot of the coupling efficiency as function of normalized longitudinal displacement for the values $w_0 = 2\lambda, X_0 = 5\,mm$. The presence of the double peak confirms the bifurcation effect given in Eq.(24) that is also illustrated in Fig. 6.

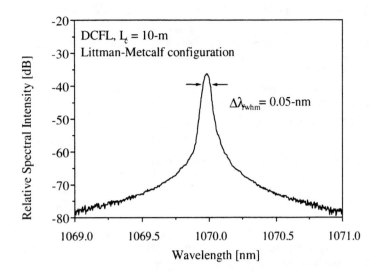

Fig. 8. Experimental spectrum[1] of grating external cavity-locked Yb fiber laser, narrowed to $\Delta\lambda_{fwhm} = 0.05\,nm$.

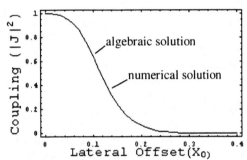

Fig. 9. Numerical and analytical solutions compared

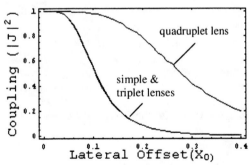

Fig. 10. Performance of three lenses compared

Eq.(22) shows that the bandwidth is an increasing function of the transverse distance X_o, as may be expected from the opposite property of the coupling function (Eq.(21)). When $X_o = 0$ Eq.(22) reduces to Eq.(9), which supports the elementary calculation given in Sect.2.1, based on the simple displacement argument of overlapping fields.

In Fig. 8 is shown a strongly narrowed spectrum, which was recorded at MIT/LL[1] of an external cavity grating-locked Yb-doped Polaroid-fiber laser. Assuming that the measured line-width of 0.05 nm is not of predominantly instrumental origin, with $f \approx 5\,cm, d \approx 1\,\mu m$, we obtain from it the mode radius $w_0 \approx 5\,\mu m$, with a smaller value being more likely if the measured line width is significantly instrument broadened, or if significant longitudinal or transverse offset was present. Using the stated value for w_0, we predict a drop in the coupling, $|J|^2$, of only about 2.3% for the 4-element lens described below, at a transverse offset of 2 mm, which agrees well with independent wave-front error calculations on the same quadruplet[1]. Next, using the manufacturer's value for the numerical aperture, $NA = 0.12$, and the above estimate for w_0, we obtain from conventional circular fiber mode theory[8] a normalized frequency, $V \approx 1$. The definition $V \equiv ka\,N.A.$, yields then $a \approx 1.5\,\mu m$ for the core radius. The low V-value corresponds to a mode that is approximately 80% cladding-guided[8], which, by employing the argument of Sect. 2.1, could provide an explanation of the wide "wings" in the spectrum of Fig.8, provided these also are not an artifact of the apparatus, which remains to be checked[1]. If such is not the case, their shape would be indicative of an exponential cladding field at large radius (which fits the asymptotic behavior of the zero-order Hankel function of imaginary argument).

4. NUMERICAL APPLICATION TO SIMPLE- AND MULTI-ELEMENT TRANSFORM LENSES

These analytical studies using a one-element transform lens indicate ways in which the method may be extended to more complex systems. They proved to be computationally prohibitive, however, when applied to the triplet and quadruplet lenses used in MIT/LL experiments. We resorted therefore to a numerical means of computing the coefficients in the expansion of Eq.(7), which was based on ray bundles of 1-10 μm radius. This technique was first applied to the simple lens resonator, so that it serves as test for the algebraic expressions. A sample result is shown in Fig. 9, for which $\lambda = 1.1 \mu m$, $a = 4 \mu m$ ($w_0 = 2.87 \mu m$), NA=0.12, and $f = 5.48 cm$, and $\varepsilon = 0$ were assumed, in addition to a small tilt angle of off-axis fiber laser elements, so that the optical beam axis passes through the center of the lens, as discussed above. X_0 is measured in centimeters in this figure and those that follow. The difference between numerical and analytical results, shown in Fig.9, is imperceptible.

A single-pass ray-trace for an N-element lens leads us to having to solve $4N$ coupled equations in $4N$ unknown ray-intercept coordinates, thus being 16 for the quadruple*t* lens. For this reason it was deemed more practical to determine the round-trip transmission function in terms of intermediate functions. The above equations were solved using a generalized Newton-Ralphson iterative approach, in which the analytically determined paraxial ray solution was used as the initial value set. This method required no more than three or four iterations in all cases to obtain far better than required accuracy.

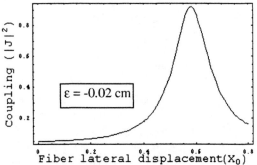

Fig. 11. Effect of field curvature in quadruplet lens

A comparison of the performance of a simple lens with the triplet and quadruplet lenses of comparable focal lengths, is shown in Fig.10. Almost identical coupling values are predicted for the singlet and triplet lenses, which is apparently due to the triplet lens not being corrected for the chief primary aberration responsible for its performance limits (see below). On the other hand, the drop in coupling efficiency as function of off-set from the axis is significantly reduced in the quadruplet, which agrees with independent spot-diagram and wave-front error plots made by using commercial software.[1]

Fig. 11 illustrates the coupling dependence as function of transverse displacement of the fiber for a cavity containing the quadruplet lens, given a fiber that is displaced 0.2 mm behind the focal plane. The calculations were performed using the numerical technique described above. Note the significant effect of field curvature that is predicted here, since by moving the laser element 0.2 mm in the direction of the lens only a few percentage points drop in coupling is predicted even when X_0 is as large as 8 mm. This is compared to $X_0 \approx 1$ mm for a comparable coupling drop for an element located in the focal plane (see Fig. 10). The conclusion follows that by using a *curved* fiber-array the maximum number of fiber elements contained in it may be increased as much as by a factor of eight, or likely even more, as compared to that by using a linear array.

The above comparisons were independently checked with a computer program for the primary aberrations corresponding to a double passage through the lens, in which a mirror was placed in the focal plane (corresponding closely to the actual set up). The distortion and field curvature for the singlet and triplet lenses where virtually identical, while they were also found to be the largest aberrations. Consequently, these findings agree well with the described behavior of the coupling parameter.

5. SUMMARY

We constructed a mathematical model to study spectral beam combining that was developed from Van Vleck's asymptotic Green function, with the specific objective to determine the effect of aberrations of the fourier transform lens on the coupling efficiency of the individual lasers to their respective external cavities. An analytical version applied to a system using a simple transform lens proved to apply also to a cavity containing an inadequately corrected triplet lens. A more general numerical implementation of the method was applied to a cavity containing a quadruplet lens. Qualitative agreement with preliminary measurements was found in the case of the triplet and better quantitative agreement with independent calculations on the improved quadruplet lens. Measurements of the external cavity-locked laser spectrum, using the quadruplet, were found to be consistent with our theory, where the measured line width was only a little larger than its theoretical value predicted for optimum design conditions. To be conclusive, however, further theoretical investigations and comparisons with experiment should be made when more experimental data becomes available.

Independent computer calculations of the primary aberrations of the three transform lenses modeled yielded results in qualitative agreement with those of the resonator theory. Among them was the finding that distortion and field curvature are the dominant primary aberrations affecting the performance of uncorrected transform lenses. In addition, based on the resonator model, it was determined that the use of a properly corrected transform lens will significantly reduce coupling loss of off-axis elements, while a curved array is potentially able to yield an additional order-of-magnitude increase in the maximum number of beam-combined elements. We hope that these effects will eventually be confirmed by experiments.

We have shown that beam quality is affected by lens quality. The asymptotic dependence at long distance of the width in the transverse direction (i.e., normal to the grating lines) of the term contributed by a single element is proportional to the external cavity-locked line-width. The optimally focused beam shape is elliptical in a given reference plane exterior to the laser system, with excess broadening in the transverse direction over that in the lateral direction that is equal to the above asymptotic limit.

Provisionary obstacles to attaining power-scalability in this system include the overcoming of internal cavity losses as contingent on grating and transform lens performance. In the long range, the achievement of this goal is likely decided by the ability to increase SBS/SRS thresholds.

ACKNOWLEDGMENTS

The authors are grateful to T. Y. Fan for useful conversations, critical commentary, and permission to use unpublished experimental data.

REFERENCES

1. T.Y Fan, MIT Lincoln Laboratory (current developmental research project).
2. R. W. Ditchburn, *Light* (Interscience Publishers, New York, 1953), § 6.34.
3. M. Born and E. Wolf, *Principles of Optics*, 5th ed. (Pergamon Press, Oxford, 1975), § 9.5.
4. J. H. Van Vleck, Proc. Nat. Acad. Sci. U.S. 14, 178-188 (1928).
5. R. K. Luneburg, *Mathemathical Theory of Optics* (University of California Press, Berkeley, 1966), §18 *et seq*,
6. J. A. Arnaud, *Beam and Fiber Optics* (Academic Press, New York, 1976), Ch.4.
7. Ref. 2, § 8.12; Ref. 3, § 8.6.3.
8. D. Gloge, Weakly guiding fibers, Appl. Opt. 10, 2252-2258 (1971).

Addendum

The following papers were announced for publication in this proceedings but have been withdrawn or are unavailable.

[3930-01]   **Enhancing the performance of unstable resonators with diffractive optics**
J. R. Leger, S. Makki, Univ. of Minnesota/Twin Cities

[3930-02]   **Characterization of excimer laser beams**
R. D. Jones, National Institute of Standards and Technology

[3930-05]   **Analysis and correction of aberrations in high-power diode-pumped Nd:YAG rod lasers**
N. Kugler, A. Vazquez, Laser- und Medizin-Technologie gGmbH (Germany); H. Laabs, Technische Univ. Berlin (Germany); H. Weber, Laser- und Medizin-Technologie gGmbH (Germany) and Technische Univ. Berlin (Germany)

[3930-06]   **Nonstandard optical elements with nonuniform surface performance for laser resonators**
A. M. Piegari, ENEA Thin-Film Optics Lab. (Italy)

[3930-12]   **Do we correctly and completely measure all laser beams by using the procedure of the ISO 11146 document?**
G. Nemes, Stanford Univ.

[3930-14]   **Design of high-brightness all solid state 1-μm MOPA systems with phase conjugation**
H. J. Eichler, O. Mehl, Technische Univ. Berlin (Germany)

[3930-17]   **High-brightness Yb:YAG lasers**
E. C. Honea, R. J. Beach, S. C. Mitchell, Lawrence Livermore National Lab.; P. V. Avizonis, Boeing Co.; R. S. Monroe, J. A. Skidmore, M. A. Emanuel, S. B. Sutton, Lawrence Livermore National Lab.; D. G. Harris, Boeing Co.; S. A. Payne, Lawrence Livermore National Lab.

[3930-19]   **Edge-pumped thin-disk laser for high-average power**
L. E. Zapata, R. J. Beach, E. C. Honea, S. A. Payne, Lawrence Livermore National Lab.

[3930-27]   **Novel microcavity resonators based on quantum chaos theory**
A. D. Stone, Yale Univ.; G. Hackenbroich, Univ. of Essen (Germany); P. Jacquod, Yale Univ.; E. E. Narimanov, Lucent Technologies/Bell Labs.; J. U. Noeckel, Max-Planck-Institut für Physik Komplexer Systeme (Germany)

[3930-28]   **Quantum optics in silica microspheres: from classical lasing to single photon-single atom devices**
V. Lefevre-Seguin, Ecole Normale Supérieure (France)

[3930-29] **Coupling semiconductor nanostructures to a dielectric microsphere: semiconductor microcavity with extremely high Q-factors**
H. Wang, X. Fan, S. Lacy, M. C. Lonergan, Univ. of Oregon

[3930-33] **Optimization of high-power cascade dye laser amplifier system**
S. A. Vasiliev, M. A. Kuzmina, V. A. Mishin, General Physics Institute (Russia)

[3930-34] **Master oscillator of narrowband tunable dye laser system**
S. A. Vasiliev, M. A. Kuzmina, V. A. Mishin, General Physics Institute (Russia)

[3930-35] **Tuning technique of multiple-electrode-pairs TEA CO_2 laser**
L. Ye, X. Li, A. He, Nanjing Univ. of Science and Technology (China)

[3930-36] **Mode of resonators with spatial-temporal phase modulation**
Q. Lin, L. Wang, G. Zhou, X. Jiang, S. Wang, Zhejiang Univ. (China)

[3930-39] **Influence of guiding structures created at the electrode surface on the slab laser resonator quality**
V. A. Saetchnikov, S. Leshkevich, Belarusian State Univ.

[3930-42] **Highly efficient coupling to silica microsphere whispering-gallery modes using fiber tapers**
K. J. Vahala, M. Cai, California Institute of Technology

[3930-43] **6-kW CO_2 laser using the novel stable-unstable resonator with external filtration of output beam**
A. M. Zabelin, E. V. Zelenov, A. V. Korotchenko, V. N. Tchernous, Technolaser Ltd. (Russia)

[3930-44] **Large laser interferometric TAMA 300 and ultra-high quality and large aperture optics**
K. Ueda, Univ. of Electro-Communications (Japan)

[3930-45] **300-W Nd:YAG laser with active resonator automatic control**
Y. I. Malashko, A. S. Rumyantsev, V. V. Valuev, ALMAZ (Russia);
G. V. Smirnov, TRINITY (Russia)

Author Index